Growing American Rubber

Studies in Modern Science, Technology, and the Environment

Mark A. Largent, Series Editor

The increasing importance of science and technology over the past 150 years—and with it the increasing social, political, and economic authority vested in scientists and engineers—established both scientific research and technological innovations as vital components of modern culture. The Studies in Modern Science, Technology, and the Environment series focuses on humanistic and social science inquiries into the social and political implications of science and technology and their impacts on communities, environments, and cultural movements worldwide.

Mark R. Finlay, *Growing American Rubber: Strategic Plants and the Politics of National Security*

Gordon Patterson, *The Mosquito Crusades: A History of the American Anti-Mosquito Movement from the Reed Commission to the First Earth Day*

Growing
American Rubber

Strategic Plants and the
Politics of National Security

MARK R. FINLAY

RUTGERS UNIVERSITY PRESS
New Brunswick, New Jersey, and London

First paperback printing, 2013
Library of Congress Cataloging-in-Publication Data

Finlay, Mark R.
    Growing American rubber : strategic plants and the politics of national security /
Mark R. Finlay.
        p. cm. — (Studies in modern science, technology, and the environment)
    Includes bibliographical references and index.
    ISBN 978-0-8135-6157-8 (pbk : alk. paper)
    ISBN 978-0-8135-4483-0 (hardcover : alk. paper)
    1. Rubber industry and trade—United States—History.   2. Rubber plants—
Research—United States.   I. Title.   II. Series.
    HD9161.U52F56 2009
    633.8'95—dc22   2008029190

A British Cataloging-in-Publication record for this book is available
from the British Library

Visit our Web site: http://rutgerspress.rutgers.edu

Manufactured in the United States of America

Contents

Illustrations

Tables

Acknowledgments

I would like to acknowledge the many scholars, colleagues, archivists, and friends who have helped make this book possible. Its origins may lie with Alan Marcus, then at Iowa State University, who first introduced me to the rather obscure ideology of "chemurgy," whose proponents proclaimed that agricultural products were essential both as raw materials for industrial production and to the politics of national security. Thus the relationships between plants and power, and the history of human manipulation of plant organisms, have interested me long before biotechnology and renewable resources emerged as buzzwords of the early twenty-first century.

Yet I came to the issue of domestic rubber somewhat suddenly, when my colleague at Armstrong Atlantic State University (AASU), Anne Yentsch, proposed that I participate in a project with the Edison & Ford Winter Estates in Fort Myers, Florida. I soon became immersed in Thomas Edison's research on domestic rubber crops, working in fruitful and collaborative relationships with historic preservation consultants Suzanna Barucco and Marty Rosenblum. The product of this research—mainly evident in chapter 3 of this book—appeared as a substantial section in their report: Martin Jay Rosenblum, R. A., and Associates, *Historic Structures Report* (Edison & Ford Winter Estates, 2002).

A desire to put Edison's work into a broader context impelled me to broaden the project. Along the way, countless scholars have provided further guidance. The list begins with Peter Neushul of the University of California, Santa Barbara, whose 1992 dissertation and personal expertise with domestic rubber enthusiasts provided essential source material as well as valuable suggestions on how the argument might be framed. Others who have provided important guidance include Doreen Valentine of Rutgers University Press; the late Philip Pauly of Rutgers University; David Wright of Michigan State University; Paul Israel of the Thomas Edison Papers; Deborah Fitzgerald of the Massachusetts Institute of Technology; Jonathan Harwood of the University of Manchester; Gabriela Petrick of New York University; Audra Wolfe, now with the Chemical Heritage Foundation; Anne B. W. Effland of the U.S. Department of Agriculture; Bill Kovarik of Radford University; Jeffrey Jones of Purdue University; Shane Hamilton and his colleagues at the University of Georgia; Uwe Luebke and Frank Uekötter of the German Historical Institute; copyeditor Nicholas Taylor; among others. Nearly as important as the intellectual support, I also benefited from financial support and travel assistance from the Edison & Ford Winter Estates, the William C. Clements Center for Southwest Studies and DeGolyer Library at Southern Methodist University, the Baylor University Collections of Political Materials, and a Research and Scholarship Grant as well as other support from AASU.

I am further grateful for the help of Frank Akira Kageyama, who participated in the search for a domestic rubber crop while interned at the Manzanar War Relocation Center during World War II. Katrina Cornish, Jeff Martin, Mike Finley, Betsy Brottlund, and others, helped bring the story into the present century with information on rubber crop research currently under way. Several descendents of important actors in this history also provided vital documents, memories, and genuine enthusiasm for my project. These include Thomas Dudley Fennell, Glenn Kageyama, Richard Anderson, Wolf Dieter Hauser, and the late Margaret Crowl MacArthur, each of whom understood that the search for a domestic rubber crop is an important yet largely untold story in twentieth-century American history.

In addition, I would like to thank just a few of the many archivists who have helped guide me through important manuscript sources during my visits to their facilities. These include Doug Tarr and Leonard DeGraaf of the Edison National Historic Site; Tom Smoot, Pam Miner, Lisa Sbutonni, MaryAnna Carroll and especially Chris Tenne Pendleton of the Edison & Ford Winter Estates; Andrea Boardman of the Clements Center for Southwest Studies and Cynthia Franco of the DeGolyer Library at Southern Methodist University; Ben Rogers of the Collections of Political Materials, Baylor University; John V. Miller of the Department of Special Collections, University of Akron; Mark Hatchmann and Richard Potashin of the Manzanar National Historic Site; Tom Rogero of the Fairchild Tropical Garden; David Pfeiffer of the National Archives; Dale Mayer of the Herbert Hoover Presidential Library; David Farrell of the Bancroft Library at the University of California; John Weeren of the Seely G. Mudd Manuscript Library, Princeton University; Amy Cooper of the

Special Collections Department, University of Iowa Libraries; Lian Partlow of the Pasadena Historical Museum, Pasadena, California; and Carl Van Ness of the University of Florida Department of Special Collections.

The following scholars and archivists had the difficult task of helping me secure manuscript materials from afar: Becky Jordan of the Special Collections Department, Iowa State University; Beth Porter of the Eastern California Museum; Katherine Morrissey and graduate student Marcus Burtner of the University of Arizona; Angela Todd of the Hunt Institute for Botanical Documentation, Carnegie Mellon University; Art Hansen of Special Collections at California State University, Fullerton; Ruthanne D. Vogel of the University of Miami Department of Special Collections; Tamara Kennelly of the Virginia Polytechnic University and State University Archives; Darlene Wiltsie of the University of Guelph; Susan Lemke of the National Defense University; Susan Searcy of the Nevada State Library and Archives; Erika Castaño of Oregon State University; Carolyn G. Hanneman of the Carl Albert Center of Congressional Research and Studies, Norman, Oklahoma; Jennifer L. Martinez of Special Collections and University Archives, San Diego State University; Jennifer Albin of the Arizona State Library; John Konzal, Western Historical Manuscript Collection; and George Gilbert Lynch of Bakersfield, California.

Many past and present colleagues and students at Armstrong Atlantic State University have helped with research, proofreading, and careful criticism of my thinking. Foremost among them are my former student Angelina Long and my colleague Michael Price, both of whom have read every word of every draft and brought countless improvements to the manuscript. I also would like to acknowledge the following AASU and Savannah-area colleagues for their support: Ed Wheeler, June Hopkins, Allison Scardino Belzer, Richard Wallace, Will Lynch, John Brewer, Melissa Jackson, Caroline Hopkinson, Judy Dubus, Brittany Holman, Chris Shirley, Juan Aragón, Katya Isayev, Tania Sammons, Lori Andrew Weaver, Laura Kaliakperova, Whitney Palefsky, Frank Linton, Chica Arndt, Sherry Cortes, and George Rogers.

Above all, I wish to acknowledge the three generations of my family who, perhaps unknowingly, have provided profound support for my interest in and passion for—of all things—domestic rubber crops: my parents, Roger and Dee Ann Finlay; my wife, Kelly Applegate; and my fine young sons, Greyson and Ellis. The book is dedicated to them.

Growing American Rubber

Introduction

O<small>N THE SAME DAY</small> that the *New York Times* reported on page 6 that the Nazis had slain seven hundred thousand Polish Jews, a different headline appeared above the fold of the front page: "Lehman Ends Tennis; Shoes to Rubber Pile." As part of his commitment to respond to the U.S. rubber crisis, New York governor Herbert Lehman and his family had donated tennis shoes and other household items to the nation's scrap-rubber drive.[1] As the juxtaposition of these stories demonstrates, no domestic issue generated more attention or caused more public anxiety in early 1942 than looming shortages of rubber. For decades, many Americans had been as concerned about foreign control of rubber as people are today about petroleum. The situation proved especially grave in 1942. Government officials had no real plan to obtain and fairly distribute rubber. Vital for war production, military operations, and sustaining the civilian standard of living, rubber was the nation's most valuable agricultural import. The United States consumed about 60 percent of the world's rubber, some six hundred thousand tons each year, yet it produced virtually none. About 97 percent of the nation's supply came from lands in Southeast Asia that had fallen to Japan after Pearl Harbor. Government officials had prepared for this possible calamity in only haphazard and halfhearted ways, stockpiling barely enough rubber to get through the year. Prospects for 1943 were far worse. In response, officials launched scrap-rubber drives and unveiled several desperate programs to find alternate rubber supplies, including a rushed development of Latin American sources, an ambitious increase in the production of synthetic rubber, and a renewed effort to develop domestically grown rubber crops.

This was not the first time that some Americans had argued that domestic rubber crops such as guayule, cryptostegia, goldenrod, and kok-sagyz (also known as the Russian dandelion) could serve vital national security needs. Thirty years of hard lessons had taught that an industrial nation was only as strong as the weakest link in its chain of strategic materials. In particular, it became clear that the nation's reliance on imported rubber carried a fundamental economic and military vulnerability. The 1911 Mexican Revolution had shown that local political upheavals could quickly disrupt the flow of imports. World War I had proved rubber a vital and strategic commodity in wartime, and shortages had contributed to the Central Powers' defeat. During the 1920s, American anger over the rubber policies of Winston Churchill and other British leaders had demonstrated that tensions over raw materials could disrupt the consumer economy even in peacetime.

These crises had sparked efforts to grow and collect rubber from plants in California, Arizona, Florida, and elsewhere—schemes that gained national attention when Thomas Edison, Henry Ford, Harvey Firestone, and others touted domestic rubber crops as vital for the nation's economic development and military preparedness. With synthetic rubber still an unproved technology, the search for rubber focused on plant species that could develop into a natural and reliable domestic source of rubber. Although many diplomats and military-preparedness experts of the 1930s had issued explicit calls for Americans to develop a domestic rubber crop, few listened to their ominous warnings.

The rubber issue resurfaced in the national consciousness immediately after Pearl Harbor. As much of the Pacific fell under Japanese control, wartime shortages exposed all the penuriousness and shortsightedness of the past. Even in wartime, many Americans citizens rose to criticize their leaders for the "rubber mess" and their failure to prepare for the emergency.[2] Support for rubber crop research galvanized rapidly, and in March 1942 the federal government funded the Emergency Rubber Project (ERP). An effort that resembled the Manhattan Project in its scale, scope, and urgency, the ERP enlisted over one thousand scientists, engineers, and technicians in rushed yet thorough research on guayule and other alternative domestic rubber crops. Thousands of politicians, scientists, and ordinary citizens also poured their energy into the effort. Over time some of the ERP's drawbacks became apparent, and domestic rubber crops became embroiled in increasingly intense political debates and scientific controversies. In the end, none of the alternative crops, nor synthetic rubber derived from grain, nor even the apparent panacea—synthetic rubber derived from petroleum—proved to be a viable long-term solution for this quandary. To this day, the United States still relies on a potentially risky combination of imported natural rubber and synthetic rubber derived largely from imported petroleum products. In face of these risks, a small but persistent group continues the effort to develop American rubber crops.

A Brief History of Rubber

Nowadays, rubber seems almost invisible despite its everyday importance. But what is rubber? Chemically speaking, it is a hydrocarbon polymer found naturally in the milky latex of thousands of plant species. Latex may serve a variety of evolutionary purposes in the plant, such as stored food, as protection in case of wounds or disease, or as means to discourage animals from eating it. Natural latex comes in hundreds of different forms, but most are soft, tacky, and impractical for commercial use. It lacked real industrial application before Charles Goodyear (and others) devised and improved the vulcanization process in the 1830s and afterward. Through the addition of various combinations of heat, sulfur, and other compounds, vulcanized rubber gains tensile strength, shelf life, chemical tolerance, and other desirable characteristics. Vulcanized rubber can be shipped, stored, and manipulated into countless industrial products that range from hospital tubing to tires to bowling balls.

Since the early twentieth century, scientists have developed numerous means to produce rubber synthetically. Synthetic rubber can be made from many different feedstocks and by employing many different catalysts. In general, the goal is to induce a relatively inexpensive monomer—such as butadeine derived from petroleum—to polymerize and gain the elasticity (and other properties) of natural rubber. Several synthetic rubbers have become less expensive to produce than natural rubber, and rubber derived from petroleum now serves about 70 percent of the market. But only natural rubber has properties vital for airplane tires, the sidewalls of automobile tires, latex gloves, contraceptives, and many other applications. As a result, both synthetic rubber derived from petroleum and natural rubber produced in the tropics have remained complementary sources of supply since World War II. As industrialization and automobility spreads around the globe, world demand for both natural and synthetic rubber has never been greater.

The main source of natural rubber is *Hevea brasiliensis*, a tree native to the upper Amazon basin. Found widely scattered in its native jungle habitat, the wild plants require tremendous effort and expense to bring the product to market; as such, the species initially seemed ill suited for the modern industrial economy. By the late nineteenth century, botanists and industrialists in Britain saw that *Hevea* could be far more lucrative if grown in tightly managed plantation environments. Replacing expensive and dispersed Amazonian labor with a cheap and centralized workforce became a priority. Establishing a plantation system presented hurdles, however, primarily because the trees thrive only in the tropical climates that surround the equator. Attempts to grow the plant more than a few degrees latitude north or south proved quite unsuccessful. Moreover, trees grown in close proximity are at risk of a natural blight that does not occur among the scattered wild plants of the jungle habitat. Further, *Hevea*'s harvest remains wedded to preindustrial methods; to this day, rubber tappers in Southeast Asia manually incise the bark to induce the flow of its white latex. Workers repeatedly collect raw rubber from the dozens of trees under their watch, pouring the milky substance from small cups into larger containers. The bark heals and each tree can be tapped again just a few days later, perhaps 120 times each year. Meanwhile, field workers quickly add ammonia and other chemicals to the raw rubber as it begins its travel through a chain of processing plants, warehouses, and factories that eventually bring the product to the industrial world.[3]

As demand grew in the mid-nineteenth century, Brazil opened the Amazon River basin to foreign traders who brought disease, abuse, and warfare to many indigenous populations. Yet rubber also brought immense wealth. Rubber made the Amazon port at Manaus one of the most opulent cities in the world; its rubber barons were purportedly rich enough to water their horses with French champagne and send their clothes to Europe for laundering. Yet Brazil's control of about 90 percent of the rubber industry hung from a fragile thread. As rubber capitalists pushed deeper into the Amazon jungle, difficulties with locating new trees, building transportation infrastructure, and controlling a labor force made Brazilian rubber ever more expensive to harvest. Brazilian rubber barons' ability to exploit their monopoly to manipulate market prices impelled Western industrialists and statesmen to search desperately for new sources of supply.[4]

Meanwhile, the booming automobile, bicycle, and electric industries caused global demand for rubber to increase sixfold between 1890 and 1910. American investors hastily formed several corporations that promised to develop plantations of an alternative species, *Castilloa elastica*, in southern Mexico, Central America, and Venezuela. Leopold II, king of the Belgians, engaged in an especially ruthless search for rubber from the *Landolphilia* vine native to his possessions in the Congo. British promoters touted *Funtumia elastica* as an easy-to-grow alternative that would bring prosperity to investors in Gold Coast, southern Nigeria, and other colonies, while Ceara (*Manihot glaziovii*) was the favored species in British East Africa. Other colonial governments and business syndicates also sent teams of plant explorers to Madagascar, Angola, Río Muni, India, South Africa, and other exotic locales in search of species that could serve as alternatives.

All of these schemes failed. Most projects in southern Mexico were riddled with corruption, banditry, disease, and labor shortages; they were abandoned when it became clear that *Castilloa* could not compete with *Hevea*. In Madagascar, India, and elsewhere, the ruthless exploitation of wild rubber species led to the rapid denuding of landscapes that rendered the desired plants virtually extinct. In Angola, the rubber trade was implicated in the transfer of slaves to the cocoa plantations operated by the Cadbury brothers, English businessmen who built a chocolate empire. Journalists revealed that rubber barons in the Belgian Congo commonly flogged, imprisoned, and even cut off a hand of rubber workers who resisted or failed to meet their production quotas. Similarly, in the Putumayo region of eastern Peru, scandal surrounded the harsh methods that rubber traders used to force indentured laborers and criminals to toil in the remote jungle environment, as well as the rape, murder, starvation, and other abuses that rubber barons wrought on the native populations.[5]

In all, a combination of economic incentives, moral pressures, and botanic discovery converged in ways that shifted the rubber industry from the search for wild rubber to the establishment of plantation rubber in Southeast Asia. In 1876, in a classic case of "botanical imperialism," the British planter Henry Wickham deceived Brazilian customs officials and secured enough *Hevea* seeds to eventually wrest the rubber trade from Brazilian control. Twenty-two of these seeds reached Britain's Malay States colony in 1877. There, botanists launched a long and systematic effort to turn the wild plant into an agricultural commodity. After developing successful methods for propagating new trees, tapping mature ones, and clear-cutting local jungles, British colonists had significant tracts of *Hevea* rubber in place by 1897. By 1907, clones of those original twenty-two seeds—which fortunately proved to be immune from the devastating blight—had spawned some ten million rubber trees on plantations in the Malay States, Ceylon, and other British possessions. Dutch colonial officials followed suit and launched plantations on Sumatra by 1906.

These changes came at just the right moment for the industrial world, for the automotive age brought a sudden and lasting increase in rubber demand. As one eyewitness to the rubber boom put it, "Probably no single discovery in the whole course of human history has changed the world so rapidly as the extensive use of rubber."[6] Vast plantation acreages, efficient agricultural methods, regimented control of an inexpensive labor force (largely made up of indentured servants imported from India, Java, and China), and a blight-free environment allowed these plantations to efficiently and profitably produce *para* rubber, the highest grade. East Indian plantation rubber offered such advantages in terms of labor costs and economies of scale that it quickly flooded

global markets and drove down rubber prices. Within just a few years, virtually all the industrial world's rubber came from the genetically identical and
neatly arranged trees of East Indian plantations. Wild rubber hunted and gathered from the jungles of Latin America could not compete and Brazil's role as
an important rubber producer ended forever. As a whole, the rubber industry
emerged as a model of a highly centralized, intensely systematized, and heav-
ily capitalized business, just at the moment in history when industrialists like
Henry Ford were creating an insatiable demand for its product.[7]

Yet one flaw remained in this triumph of industrial capitalism. The new
British and Dutch monopoly left the countries that consumed but did not
produce rubber, like the United States, hanging from a still fragile thread.
Industrial economies remained dependent on local circumstances in the colonies, favorable trade and diplomatic relations among the metropole nations
and their peripheries, and their experts' ability to continually enhance their
mastery of tropical agriculture. Further, the supply lines that separated tropical
producers and Western consumers were thousands of miles long and vulnerable to disruption during times of war. Cases of local rebellion, diplomatic
crises, poor weather, and deliberate manipulation of markets could bring further volatility to this vital commodity. Although many Americans ignored
the implications of this vulnerability, others became committed to the effort
to find a reliable domestic source of natural rubber.

Plants and Power in American History

Rubber offers an opportunity to examine the interplay of war and military preparedness through the lenses of science, technology, agriculture, and the environment. Wars involve more than ideological struggles, military campaigns,
political maneuvers, and mobilization of the home front. Behind the ideological or political justifications that frame most military conflicts, the core issues
in war often are materialist concerns like industrial capacity, mineral wealth,
and agricultural productivity. Concomitantly, success on the battlefield often
depends on unfettered access to strategic resources. Modern wars also have
had tremendous impacts on environments around the world, not only through
battlefield destruction, but also through their demands for intensified exploitation of vital commodities. In all, war-preparedness experts and military planners came to understand that wars cannot be won without natural resources
and the massive mobilization of the bureaucratic, scientific, technological,
and agricultural tools required to exploit them.[8]

The emergence of the United States as an industrial power early in the
twentieth century brought the issue of strategic resources to the fore. Illusions

of an endless frontier and limitless access to natural resources began to give way as scientific and technical experts brought attention to new notions of limits and vulnerability. Likewise, many assumed that impartial experts would develop new strategies to exploit native resources in new and efficient ways. As one of the few crucial commodities that the United States was unable to produce, rubber's significance became evident during World War I, when demand increased, supplies tightened, and enemy forces patrolled the shipping lanes. Once British officials put further pressure on rubber prices and supplies in the early 1920s, prominent Americans such as Thomas Edison, Henry Ford, and Harvey Firestone responded by mobilizing a systematic effort to find alternative domestic rubber crops, at least as security in the case of war. The question of domestic rubber sources loomed unanswered on the eve of the next war, and many American leaders had yet to come to terms with the fragile reliance of the United States on other nations for strategic raw materials.

The search for a domestic rubber crop also underscores the links between war and broader discussions about the relationship of scientific enterprise to geopolitics.[9] Their scientific, technical, and agricultural expertise enabled Western nations to exploit global natural resources for industrial and military applications. Increasingly, scientists began to rationalize their work through the rhetoric of national security.[10] During World War II, scientists probably reached the apogee of their connections with national policy. While tens of thousands worked on national defense projects, others became directly involved with political issues; scientists as a whole received considerable credit for their contribution to the military victory. The case of physicists and their involvement in the Manhattan Project is perhaps best known, but hundreds of botanists, chemists, and agricultural scientists also mobilized to search for an American rubber crop.[11]

The work to develop a domestic rubber crop also illustrates the increasing level of state intervention into scientific, technical, and agricultural research. Before Pearl Harbor, most studies of domestic rubber crops were relatively isolated and poorly funded projects that fell far short of success. Much of Thomas Edison's work, for instance, took place in small plots and laboratories across the street from his winter home. Despite financial support from his millionaire friends, he lacked the resources necessary to systematically introduce a new crop into American agriculture. Similarly, government officials continually rebuffed, for nearly thirty years, efforts by the Intercontinental Rubber Company (IRC) to develop guayule as a new American crop and strategic resource. Despite the potential for rubber shortages, most leaders concluded that the rubber supply should follow the ups and downs of the free market.

This situation changed during World War II, when the scale and urgency of the rubber project transcended traditional geographic, economic, disciplinary, and national boundaries. The government's creation of the ERP in 1942 signaled a dramatic shift in rubber research from the private to the public sphere—officials agreed to an unprecedented bailout of the IRC, an unprofitable American company. But when an important consumer commodity had suddenly become a war necessity, it seemed that the nation had to obtain the IRC's prized genetic resources and know-how. Through the ERP, the government mobilized land, labor, and capital on an unprecedented scale in order to develop agricultural crops for the American war effort. The forced relocations of Japanese Americans from the California coast, and the rapid mobilization of women, children, Mexican braceros, and prisoners of war to work with domestic rubber crops, represented a dramatic reshaping of California's demography and economy. The ERP also reveals how government pressure, sometimes coercive and invasive, pushed landowners to experiment with untried rubber crops even when demand for proven agricultural commodities remained high. In a branch of the project that extended into Haiti, officials condemned choice farm plots and lured formerly independent farmers to work on American-funded rubber projects. In all, the story of domestic rubber underscores the expansive interconnections between scientific legitimacy and government power in the context of war.

The case of rubber also reveals the significance of agricultural and botanical experts as actors in geopolitical issues. Although unappreciated in most accounts, efforts to control strategically important plants have been at the core of imperial power since the very origins of Western dominance.[12] Plants have been a focal point in the relationship between science and power, but there is no inevitable trajectory indicating which plants will be grown in which regions. These decisions are the results of many complex scientific experiments and controversial political negotiations. Thus botanists and agriculturists with expertise on strategic and valuable species have been indispensable actors in issues of political intrigue, economic rivalries, and geopolitical power.[13] Western institutions that served the imperial mission, such as Britain's Kew Gardens, sought to "nationalize nature" and systematically remove valuable plants like rubber from local environments, thereby substituting scientific understanding of botanic properties for indigenous knowledge. Thus scientific agriculturists rarely stood by as neutral participants in the transfer of valuable knowledge; rather, national governments used such experts as tools of empire and power.[14]

Plants and agricultural products also stood at the core of Western industrial power and wealth. Cotton, sugar, timber, flax, hemp, indigo, madder, wheat, and other plants proved as vital to Western industrialization as coal and iron. Because synthetic substitutes for many of these natural products did not exist until the twentieth century, agricultural shortages could be devastating to the industrial economy or military operations.[15] Investment in agricultural science often went hand in hand with industrialization and war preparedness.[16]

The desire to obtain valuable crops from tropical climates provided Western governments an even greater incentive to shift from the plunder of wild plants to the scientific control of cultivated crops.[17] As World War II broke out, synthetic nitrogen fertilizers, chemical pesticides, hybrid seed, internal combustion engines, and increasingly mechanized methods brought most American farmers in close connection with developments in science and engineering. Indeed, American farmers' special allocation of tires and other rubber resources during World War II offers an appropriate illustration of agriculture's prominence in the nexus of industry, science, and geopolitics.[18]

Agricultural concerns long have been linked with American expansion, power, and a desire for what the late Philip Pauly labeled "horticultural independence."[19] In 1898, the same year the nation acquired new tropical territories following the "splendid little war" with Spain, the U.S. Department of Agriculture organized a Section of Seed and Plant Introduction to help bring economically and strategically valuable plant resources from around the world under national control. In 1899, President William McKinley—who as a congressman had represented Akron, Ohio, center of the American rubber industry—called for rubber cultivation in the new U.S. territories. In a departure from earlier methods that relied on amateurs stationed at consular offices, the government employed professional botanists and agriculturists ("bioprospectors") who systematically identified, extracted, exploited, and transferred foreign botanic resources to the United States. In an effort to secure American ecological independence through new crops, botanists were actively participating in this imperial mission by the turn of the century.[20]

Yet rubber presented an unusual problem. The plant explorers who successfully introduced soybeans, citrus fruits, avocados, dates, hearty wheat varieties, and countless other commercially valuable plants into the United States made little headway with *Hevea* and the other rubber plants. Indeed, during these same decades, rubber became increasingly concentrated in the politically vulnerable and very distant territories in Southeast Asia.[21] Leaders of other nations, most notably Nazi Germany and the Soviet Union, also recognized the tremendous role that plant scientists could play in controlling resources for the construction of new national identities and as preparation for war. Rubber became a central concern for those nations as well, and they demonstrated relatively more success in their own campaigns for alternative rubber crops.[22] In contrast, the U.S. government did not enlist its botanic and agricultural experts in the search for a domestic rubber crop until March 1942. With a confidence built on past successes, many officials were betting that agricultural scientists and technologists would solve the crisis through domestic and renewable rubber plants. These efforts did not fully succeed.

The struggle for a domestic rubber crop marks a transitional phase in the history of American technology and science, a new era in which artificial and imported solutions to economic and political problems suddenly offered suitable alternatives to the traditional dependence on domestic natural resources. Before the crisis of World War II emerged, those who sought alternatives to imported rubber, for all practical purposes, could only frame the issue in agricultural terms. With synthetic rubber an unproved possibility, debates revolved around which crops, which growing regions, and which geopolitical circumstances could best be altered to develop new agricultural sources of rubber. In this same context, a small but vocal group of activists emerged in the 1930s, the "chemurgists," a fundamentally nationalist group that pushed for the industrial utilization of farm products. Like today's promoters of bio-based and renewable raw materials such as fuel ethanol, chemurgists argued that the development of new crops and the industrial utilization of surplus farm products would reduce the nation's dependence on imports. Yet the chemurgists failed to win widespread support for their notion that natural, domestic, and reliable sources of agricultural raw materials should become a crucial element of American strategic policy.[23]

Patent and intellectual property law also thwarted efforts to develop valuable resources from plants. Because nature dictates that unusually valuable specimens change their genetic identity with each generation, controlling improvements in the organic world proved far different from manipulating the inorganic. Plant breeders could engineer varieties through hybridization, grafting, and seed collection, and hope to sell their improvements to farmers and gardeners. But because living things can reproduce themselves, buyers would have little incentive to make a repeat purchase; indeed the customer could become the innovator's competitor just one generation later.[24] Courts had ruled that patent law could not extend into the natural world, so it is not surprising that those who sought to manipulate American rubber crops found themselves at the center of intellectual property debates that remain pertinent today.

Many hurdles stood in the way of developing natural and domestic rubber sources. In some cases, resistance came from the plants themselves, for some of the domestic rubber plants presented more complex agricultural problems than their promoters had predicted. In other cases, agricultural experts achieved considerable success with the botanic and scientific questions, but political circumstances thwarted the successful adoption of domestically grown rubber. Although officials invested millions of dollars and committed thousands of workers to the development of a domestic rubber crop, they devoted much more funding and labor to synthetic-rubber research and development. Domestic rubber crops also faltered because synthetic rubber better fit the emerging cultural and political milieu of the United States in the postwar world. In contrast to the chemurgists, many Americans began to embrace the idea that synthetic and nonrenewable products offered important advantages over those derived from plants.

Rubber derived from petroleum products emerged during World War II as a solution to the crisis, but not simply because it offered certain economic or technological advantages over alternatives. Synthetics seemed to free consumers from nature's variability and instability, while applied organic chemistry seemed to promise universal access to nature's resources. Some argued that traditional constraints of nature and geography would evaporate in a new international order, for power would no longer be linked to the accident of which lands were endowed with natural resources.[25] A new rhetoric emerged that highlighted the role of American laboratories in general, and synthetic rubber in particular, as keys to victory in World War II. Chemists, engineers, and government planners all claimed that their dedicated investment in synthetic rubber and petroleum chemistry would help solve the wartime crisis. As nylon replaced silk, DDT replaced natural insecticides, and plastics became ubiquitous, the notion that agriculture could provide solutions to the nation's resource needs faded from view.

The struggle for domestic rubber crops also lost resonance in the context of the nation's new position in international affairs. In contrast to the nationalist considerations that had driven earlier promoters of domestic rubber crops, World War II resulted in broader acceptance of the nation's commitment to scientific internationalism in the postwar world.[26] A new consensus emerged that envisioned an international order in which some plant resources could be imported while others could be replaced with synthetic substitutes. For these "agricultural internationalists," the solution lay in expanding reciprocal trade agreements for goods that foreign nations could purchase with revenue from their sales of rubber, petroleum, and other resources valuable to the United States.

The case of rubber reinforced both. As the war drew to a close, federal authorities ordered the destruction of millions of healthy domestic rubber plants that no longer had a place in their vision of a postwar world. With complete command of the world's synthetic-rubber production in 1945, the United States might have elected to exercise monopolistic control of that technology.

Instead, the government quickly closed or sold many synthetic-rubber factories and rejected calls for long-term price supports to protect the domestic synthetic-rubber industry. The government also allowed Southeast Asian *Hevea* rubber producers to regain their preeminence by offering various forms of support to the British and Dutch colonial officials who depended on revenue from natural-rubber exports.[27] Today, about 70 percent of world rubber production is a synthetic product derived from petroleum; about 30 percent remains the natural product of *Hevea brasiliensis*, now concentrated in Indonesia, Thailand, and Malaysia.[28] Both sources reflect cultural assumptions common in the postwar world. Synthetic rubber derived from petroleum offers a stable source of supply, unaffected by seasonality, the unreliability of weather, and uncertain crop yields of the natural product. The renewed dependence on imported natural rubber also accords well with the nation's leadership role in a global economy. As long as petroleum supplies remain accessible and political and environmental circumstances in Indonesia, Thailand, and Malaysia remain stable, American dependence on these twin solutions will likely continue.

Yet such circumstances cannot be assured, which explains why this issue remains pertinent. Interest in natural-rubber plants, including guayule, tends to arise whenever rising petroleum prices threaten the economic viability of synthetic rubber. And should the rubber blight ever emerge on the natural-rubber plantations of Southeast Asia—a distinct possibility under the current threat of bioterrorism—the search for alternative rubber sources may again become paramount. If and when a real rubber crisis materializes, alternative rubber crops may again reemerge because scientists have already completed much basic research in the plants' genetics, physiology, and biochemistry.

In the end, none of the proposals to develop a rubber industry based on domestically grown rubber plants achieved commercial viability. Neither Edison's favored crop, goldenrod, nor guayule, nor the other alternative crops, nor synthetic rubber derived from grain, proved a long-term solution. In trying to understand reasons for this outcome, it is important to recognize the political, cultural, and other issues that contributed to the defeat of this technology. These also had long-term implications in the development of the U.S. postwar hydrocarbon-based society. Unlike today, when most Americans routinely expect chemical and "high-tech" solutions to social and economic problems, this case illustrates the alternative trajectories that might have prevailed.

The Plants and the People

This story focuses on several plants and the people who believed in them. At one time or another, each plant emerged as a panacea in the American struggle for a domestic rubber crop. Each possessed unique botanical, ecological, and agricultural characteristics that seemed to promise an easy answer to the American demand for rubber. Each also possessed significant impediments—political, cultural, botanical, and agricultural—that stood in the way of success. Rubber from domestic plants was not an ideal solution, and none of these plants offered a reasonable hope of defeating the alternatives of imported rubber or synthetic rubber derived from petroleum in the commercial marketplace. That each failed to emerge as a legitimate rubber substitute offers some lessons in the context of current efforts to find to bio-based alternatives to synthetic imports.

Although scientists studied hundreds of potential rubber plants, four species deserve special introduction.[29] The most significant, guayule (*Parthenium argentatum* Gray), is a woody shrub, about two feet tall, with silvery-olive leaves, small white flowers, and deep and extensive root systems. A drought-resistant native of the elevated deserts of northern Mexico and a small corner of southwestern Texas, guayule is a long-lived plant that reproduces quite slowly. During the guayule boom of the early twentieth century, it took little time for harvesters to render wild guayule virtually extinct. To keep the industry afloat, the preeminent guayule firm, the IRC, hired botanists and agricultural experts who tried to find new propagation techniques and cultivation methods. After decades of trial and error, IRC experts developed tedious and expensive strategies to germinate selected seed in nursery beds, transplant seedlings to the field, and tend the plants for years in semiarid and isolated environments. Because the plant thrived in lands that competed with no other agricultural crops, it seemed an ideal addition to the agricultural development of the American Southwest.

Guayule grows and photosynthesizes normally in warm or humid conditions, but it forms its rubber during cooler or drought conditions. Most of the plant's latex is found in its woody stems and stalks, but because a third of the rubber is in the roots, the only efficient means to extract it is to uproot the entire plant at harvest. Manufacturers then subject the shrub to a complex process of mechanical crushing and chemical extraction. Guayule requires at least four to five years to reach its maximum rubber yield, and most growing regimens call for the crop to be planted on large acreages that can be harvested on a rotating basis over this four- or five-year cycle. This created a fundamental hurdle because few farmers—accustomed to annual cycles of markets, loan payments, and price supports—were prepared to wait such a long time without some kind of guaranteed return. Moreover, guayule rubber typically contains an unsatisfactory amount of resin, and the final product generally lacks the tensile strength and elasticity of *Hevea* rubber. Despite

Table 1.1
Comparison of Alternative Domestic Rubber Crops

Plant Name	Rubber % (avg ; max)	Yield: # rubber/acre	Advantages	Disadvantages
Guayule	7–16; 20	900–1,500[a]	• commercial potential demonstrated • mechanical harvesting and extraction demonstrated; reduced labor costs • grows in marginal lands not suitable for most other crops • could be planted and left as a living rubber reserve until needed	• 4–5 years needed before harvest • commercial efforts rarely profitable • rubber quality poor (before WWII improvements)
Goldenrod	2–6; 12	200–300[b]	• ready for harvest within one year • could grow readily throughout the USA • could potentially be mechanically harvested • alternative to cotton in the South • by-products potentially useful	• poor germination; seedlings better than seeds; thus emergency start-up would be difficult • no mechanical harvesting or extraction methods developed • could take food and fiber lands out of production
Cryptostegia	2–4; 9	175–300[c]	• ready for harvest within 1–2 years • very easy to grow • many seeds, good germination rates	• labor intensive harvest • no mechanical harvesting or extraction methods • difficult extraction process • invasive, nuisance species • sensitive to frost and soil and moisture conditions
Kok-sagyz	4–10; 23	60–200	• ready for harvest within one year • rubber quality high • wide range of habitat • could potentially be mechanically harvested	• requires fertile land needed for other crops • high cost of harvest • susceptible to disease

Plant			Advantages	Disadvantages
Milkweed	2–4; 8	125	• ready for harvest within one year • by-products useful for other industries • wide range of habitat	• best variety requires three years before harvest • no mechanical harvesting or extraction methods developed
Rabbitbrush	1–3; 7	Low	• grows in marginal lands not suitable for most other crops • wider range of habitat than guayule • could be maintained as natural reserve until emergency	• rubber yield poor; quality moderate • harvest difficult
Pinguay	1–4	Low	• grows in marginal lands not suitable for most other crops	• rubber quality very poor • harvest difficult

Sources:

Blythe, Samuel G. "Taming the Wild Guayule." *Saturday Evening Post*, 2 May 1931, 28, 30, 106, 108–109.

"Desert Milkweed Producer of Rubber." *New York Times*, 22 May 1932.

Fuetsel, Irvin C., and Frederick E. Clark. "Opportunities to Grow Our Own Rubber." *Crops in Peace and War: The Yearbook of Agriculture, 1950–1951*. Washington, DC: Government Printing Office. 1951, 367–374.

Hall, Harvey Monroe, and Thomas Harper Goodspeed. *A Rubber Plant Survey of Western North America*. Berkeley and Los Angeles: University of California Press, 1942.

Hall, Harvey Monroe, and Frances L. Long. *Rubber-Content of North American Plants*. Washington, DC: Carnegie Institution of Washington, 1921.

Jenkins, Dale W. "Cryptostegia as an Emergency Source of Rubber." Board of Economic Warfare, Technical Bulletin No. 3. n.p., 1943.

"U.S. Grows Own Latex." *New York Times*, 7 December 1941.

a. Some experts suggested guayule yielded only 300 pounds per acre in actuality.

b. Predictions based on highest yielding varieties. Actual results in Fort Myers indicated 60 to 210 pounds per acre.

c. Estimated yields for mature plants. For younger plants, which would be more pertinent during a war emergency, experts estimated yields to be 100 to 150 pounds per acre.

Figure I.1. Guayule (*Parthenium argentatum*), a rubber-bearing shrub native to arid climates in northern Mexico and southwest Texas.

these complications, the net yield of rubber from guayule can be high and has improved after decades of scientific manipulation. It remains the most feasible of the domestic rubber crops. Whenever *Hevea* rubber was unusually expensive, guayule's production costs approached those of imported natural rubber. Guayule's enthusiasts remain active today and the plant may fit a niche in the market for nonallergenic latex gloves. To this day, guayule is perhaps the nation's best-understood plant in technical and scientific terms that still is not in widespread commercial production.[30]

The vine *Cryptostegia grandiflora* emerged repeatedly as a potential rubber source. Native to the India Ocean basin, cryptostegia has been prized as an ornamental variety by gardeners for its large purplish flowers and bright green leaves since the nineteenth century. Prominent Americans, including Ford, Edison, David Fairchild, Bernard Baruch, and Henry A. Wallace, as well as many journalists, rubber-industry officials, and lesser-known botanists, all became convinced, at least momentarily, that cryptostegia could become an ideal emergency rubber source. Indeed, cryptostegia had several advantages: it flourished in almost every climate; it could be propagated easily from cuttings; its seeds were "exceptionally virile"; it had few insect enemies; and it required no fertilizer. Its white latex exuded from the cambium beneath the bark when broken and yielded a pure, high-quality rubber. In terms of yield per acre, promoters claimed that cryptostegia could produce more high-quality rubber than any other annual plant. Yet the vine presented numerous obstacles, none more problematic than engineers' failure to develop mechanical means to harvest and extract the rubber. The only method of harvest, it seemed, was to collect the latex drop by drop from each stem of the vine. Today, this "exceptionally virile" plant is better known as a nuisance, one that costs governments, especially in Australia and Hawaii, substantial funds and effort to control. In view of climate-change trends, cryptostegia's invasive nature is only likely to intensify.[31]

With the personal endorsement of Thomas Edison, a variety of goldenrod also emerged as a promising rubber alternative in case of a war emergency. Edison claimed to have investigated some seventeen thousand plant species from around the world in his search. He eventually focused on just one, *Solidago leavenworthii*, a plant relatively common in the pinelands and hammocks of southern Florida near his winter home in Fort Myers. A wispy wildflower, typically three to five feet tall, the plant featured shiny elongated leaves and showy, fragrant yellow blossoms. Its latex was found in the leaves. Processors could obtain a substantial amount of a decent-quality rubber by drying the plant, stripping and crushing the leaves, and extracting the rubber with benzene, acetone, or other solvents. Edison declared goldenrod the solution for

Figure 1.2. Haitian President Élie Lescot in front of a stand of *Cryptostegia grandiflora,* a rubber-bearing vine native to Madagascar, 1943. Photograph by Thomas A. Fennell. Courtesy of Thomas Dudley Fennell.

several reasons: it was relatively easy to grow; it produced a large number of leaves per plant and many plants per acre; its rubber yield could be increased through selective propagation; and it seemed possible to spread the plant into a large portion of the Southeast. Further, he was confident that engineers could develop mechanical means to sow the seeds and harvest the plant, and that it could be harvested within one year of a national emergency. Once again, actual results proved disappointing. Engineers had trouble designing a machine that could efficiently remove leaves from the stem, fertilization proved more expensive and more complex than anticipated, and the crop did not acclimate well north of Florida. Despite modest efforts that improved goldenrod's rubber yields after Edison's death, the inventor's plant fell from favor rather quickly during World War II.[32]

Kok-sagyz (*Taraxacum kok-saghyz*) emerged as another highly touted rubber plant.[33] A native of higher elevations in Central Asia, kok-sagyz is commonly known as the Russian dandelion and certainly resembles the backyard weed in its appearance. Like other dandelions, the leaves of kok-sagyz are arranged at the top of a hard root, from which slender stalks with yellow flowers shoot upward eight to ten inches. The rubber is found in the root, particularly the outer phloem tissues that are sloughed off at the end of the year. By the late 1930s, Soviet scientists had developed an effective extraction process that used a pebble mill to grind the root, combined with filtration devices that separated rubber from the plant material. Quick to boast of the young nation's successes, Soviet officials announced that kok-sagyz could help end their dependence on imported rubber.

In the United States, kok-sagyz fever reached a peak in the spring of 1942. In a dramatic episode that helped solidify the American and Soviet alliance of World War II, seeds of kok-sagyz arrived in the United States at the height of the rubber crisis. Because it could endure much cooler temperatures than all other rubber-producing plants, kok-sagyz seemed especially attractive as a tool for the long-range economic development across a swath of states from Vermont to Oregon. Kok-sagyz had other advantages as well: it could tolerate a broad range of soil types; it could be sown and harvested with slight modifications to the machinery already developed for the sugar beet or potato

Figure I.3. USDA assistant geneticist Bayard L. Hammond with a stand of goldenrod (*Solidago leavenworthii*) grown at the USDA plant introduction station near Savannah, Georgia, 1940. National Archives.

Figure 1.4. Taraxacum kok-saghyz, commonly known as the Russian dandelion or kok-sagyz, the rubber plant that was an important symbol of U.S.-Soviet collaboration during World War II. Photograph by J. H. Stoecker, August 1942. National Archives.

industries; and it yielded a high-quality rubber. Perhaps its greatest attraction was that it could be harvested in the same year that it was planted, unlike the years required for guayule and *Hevea* rubber to mature. As with the other plants, however, several complications emerged. Kok-sagyz was highly susceptible to fungal diseases, its heavy roots meant that transportation costs were high, and evidence emerged that Soviet scientists had simply exaggerated the plant's rubber yield. Hopes for kok-sagyz dissipated even before the war was over.[34]

If these plants provide the raw material for this story, the people involved provide its drama. Prominent politicians, unknown agricultural scientists, state governors, crackpot inventors, greedy capitalists, real estate promoters, aggressive journalists, war planners, pacifist Quakers, and ordinary citizens all became convinced that few issues were more pressing than the development of an American rubber crop. And the struggle attracted a long list of fascinating personalities: William McCallum, a Canadian-born scientist who began his thirty-year research on guayule by smuggling seeds into the United States to keep them from Mexican revolutionaries; Harvey Monroe Hall, a Berkeley botanist who trekked the High Sierra for weeks in search of plants that could help defeat the German kaiser; the aged inventor Thomas Edison, who made domestic rubber his last project with the assistance of funds from Henry Ford and Harvey Firestone; Barukh Jonas, an eccentric Turkish-American-Jewish philosopher who sought new rubber resources while railing against automobiles and other symbols of modernity; Dwight Eisenhower, a then-obscure army major who concluded in 1930 that the vast acreages of guayule could help the nation survive a wartime rubber emergency; Chaim Weizmann, the Zionist biochemist who linked the idea of renewable rubber sources with his political cause; Shimpe Nishimura, one of several Japanese Americans who developed important new guayule rubber production techniques while confined to the Manzanar internment camp; Hugh Anderson, a polio-stricken humanitarian who hoped that guayule could form the basis for cooperative agricultural communities in the deserts of the American Southwest; and Phil and Pete Ohanneson, California landowners who, despite the wartime rubber shortages, openly defied the government's rubber crop program and eventually helped kill the project altogether. For these and many others, domestic rubber crops represented more than just a few interesting botanic species; they demonstrated the importance of strategic plants at the center of national security.

Chapter 1

The American Dependence on Imported Rubber

The Lessons of Revolution and War, 1911–1922

O~N~ 30 MARCH 1911, a band of revolutionaries loyal to Francisco Madero attacked two sites that belonged to the Intercontinental Rubber Company (IRC), the American firm that dominated the guayule rubber industry of northern Mexico.[1] That night, rebels stole and destroyed merchandise, corn, and hay worth over $2,100; over the next several weeks, Madero loyalists made at least eight other raids on IRC property. In response, forces loyal to the Mexican government of Porfirio Díaz launched counterattacks that again cost the IRC several hundred dollars in damages. Then in May 1911, revolutionaries under the command of Emilio Madero, Francisco's brother, attacked the city of Torreón, where the IRC owned an important guayule rubber–processing factory. Targeting examples of foreigners' infiltration of the Mexican economy, rebel troops and their civilian supporters ransacked the shops, ranches, and factories of foreign business owners. Insurgents cut railroad and telegraph lines; confiscated weapons, horses, saddles, food, and currency; and brought rubber production to a halt. A few weeks later, *Maderista* forces seized thirty tons of baled guayule, worth over $3,400. Within just a few weeks, these groups of lightly armed rebels had cut off the United States from one of its most important rubber sources.[2]

The situation remained chaotic and intense for several years. Forces loyal to Madero, Díaz, Victoriano Heurta, Venustiano Carranza, Pancho Villa, and others battled for control of northern Mexico. On at least five hundred occasions, Mexican rebels attacked property that belonged to the IRC and its subsidiaries. Remarkably, one company official wrote secretly in 1912 that he would pay five hundred dollars for "the ears and scalp" of the revolutionary who had repeatedly led attacks on the IRC's main properties at Cedros.[3] The IRC eventually filed claims for $260,000 in lost, stolen, and destroyed property, including hundreds of tons of raw guayule ready for processing and sale on the U.S. market. Guayule harvest came to a halt, leaving thousands of agricultural and industrial workers unemployed.[4] The guayule industry, which in 1910 had produced twenty-one million pounds of rubber—over 10 percent of the American rubber consumption—was on the verge of collapse. Any hopes of developing a renewable domestic supply of rubber in North America might have come to an end. But at least according to rubber-industry lore, IRC botanist William B. McCallum packed thousands of tiny guayule seeds into his tobacco tin and smuggled them back to the United States while fleeing Mexico in 1911. From those seeds, the most significant efforts to grow rubber in the United States had their beginning.[5]

These dramatic events illustrate how much America's dependence on imported rubber left its military and economic interests in a dangerously vulnerable position. Standing at the intersection of the vital economic, industrial, political, and military issues of the day, rubber became a vital concern for American political and business leaders. As the American economy became ever more connected with the booming automotive sector, the nation's wealth came to depend increasingly on developments in tropical agriculture. Yet the ever more extended supply lines that separated tropical producers and American consumers remained vulnerable to disruption from war, local rebellion, aggressive pricing policies, and poor weather.

By the time that World War I broke out, leaders of the international powers recognized that rubber had emerged as an industrial and military necessity, yet the potential for war, revolution, and trade restrictions meant that access to rubber supplies stood in a precarious position. One of the few crucial commodities that the United States did not produce, rubber's significance was about to become more clear. In this context, many American leaders began calling for the development of a readily accessible and renewable source of rubber, which in turn enmeshed botanists, agricultural scientists, and other rubber experts in issues that reached into the core of U.S. national security.[6]

Origins of the Guayule Industry

The desert plant guayule fit into these developments as an alternative rubber source. In colonial and postcolonial times, the ranchers who increasingly dominated the economy of northern Mexico viewed the plant as a nuisance because it filled the rumen of grazing cattle with a rubbery mass. Guayule rubber first gained national attention in the United States as part of the Mexican state of Durango's display at the Philadelphia Centennial Exposition of 1876. Various investors sought a means to process the shrub into usable rubber, including Italian and German chemists who developed chemical methods for extraction around the turn of the century. A guayule boom followed between 1900 and 1907, and more than twenty guayule-processing companies sprang up in northern Mexico.[7]

The IRC was the most important. In 1902, Rhode Island senator Nelson W. Aldrich and his son, newspaper publisher Edward B. Aldrich, employed a manufacturing chemist, William Appleton Lawrence, to develop a suitable extraction process. After a year of trials in a makeshift laboratory in their Jamaica, New York, home, Lawrence and his daughter, Clara, also a chemist, announced that they had developed a mechanical process based on grinding the shrub in small stones. Lawrence received a patent for the pebble-mill process in 1903, a development that suggested that guayule could become a commercial rubber source. He promptly assigned the patent to Aldrich's new corporation, the IRC; the company thereafter sent Lawrence an annual payment of five thousand dollars. In subsequent years, the IRC secured several more patents for the processing and purifying of guayule rubber and did not hesitate to bring suit against any firm that seemed to infringe on its patents.[8]

The IRC represented a prototypical example of the networks of American capitalism that thrived at the turn of the century. In addition to the Aldriches, the IRC's principal investors included Wall Street financiers Bernard Baruch, Thomas Fortune Ryan, and Daniel and Sol Guggenheim, each of whom eventually put $925,000 into the enterprise. Other investors included John Regan of Standard Oil, Paul Morton of the Equitable Trust, Charles Sabin of the Guaranty Trust Bank, Harry Payne Whitney, Levi P. Morton, Jacob H. Schiff, and C.K.G. Billings.[9] John D. Rockefeller Jr., Edward Aldrich's brother-in-law, chipped in $25,000.[10] Significantly, these investors recognized the strategic and nationalist implications of their endeavor, as the company's bylaws expressly prohibited selling or transferring any shares "to any but American citizens."[11]

These investors mobilized just as Brazil's *Hevea* rubber trade began to collapse, and just before plantation rubber became established in the East Indies. They understood that the nascent automobile industry would cause the global demand for rubber to explode, and therefore they intended to form a rubber trust modeled on the Rockefeller oil monopoly. In 1904, Baruch traveled to Mexico with the aim of securing control of as much of the plant's native habitat as possible. With the acquiescence of Mexican political leaders, the IRC eventually controlled nearly 3.8 million acres in the center of guayule country. By 1905, Baruch and his IRC colleagues bought and modernized a guayule-processing factory in Torreón in the state of Coahuila. The company soon increased its capitalization, secured two additional factories, installed new technologies and railroad spurs, and announced that it had controlling interest in most of the guayule lands and factories in northern Mexico.[12] The IRC's Mexican investments proved remarkably successful. By 1909, its Torreón factory operated three shifts per day, employed over one thousand workers, and produced up to eight hundred thousand pounds of rubber per month; in the 1909–1910 campaign, net profits exceeded $2.5 million; in its peak year, 1910, Mexican guayule accounted for as much as 19 percent of American rubber consumption.[13]

Meanwhile, the IRC and other firms sought to exploit the guayule native to the Big Bend region of western Texas. By 1906, the company asked for "concessions" from the U.S. government to tap guayule resources on public lands.[14] The IRC also secured contracts with at least four western Texas landowners that allowed the harvest of American rubber resources.[15] This contract called for a minimum harvest of twenty tons per day, an amount that quickly would have decimated the wild rubber resources of the United States. The search for additional rubber lands in Texas continued in 1907. "Captain" William H. Stayton, a well-known promoter of American military preparedness, and his associate L. C. Andrews scouted Texas guayule lands and urged the company to move quickly to obtain leases for guayule that grew on American soil. By 1909, the Texas Rubber Company spent at least sixty thousand dollars for the rights to harvest guayule shrub on public lands. From this remote and uncharted part of the country, crews of immigrant Mexican laborers hauled four thousand tons of shrub by burro and wagon to a factory that operated from 1909 to 1914 at Marathon, Texas. Although costs were too high for the mill to become profitable, it yielded about six hundred tons of crude rubber and thus represented the first production of rubber in the United States from domestic sources.[16]

Despite the guayule industry's impressive growth, two pitfalls lay ahead. The first was the looming exhaustion of wild guayule plants. Despite controlling property that measured roughly sixty-five by forty-five miles square, a 1908 IRC memorandum reported that it would take only six or seven years for the known supply of the desert shrub to be exhausted.[17] With virtually all of the company's assets tied to the fate of this scarce raw material, the writer asked if "it might be well to stop and consider whether we are not grinding up our shrub too rapidly."[18] A year later, Baruch submitted a report that suggested that the supply could be gone in less than three years, rendering the IRC's forty-million-dollars investment worthless. Shrewdly, Baruch sold off his guayule interests in 1910, at about the peak of their value.[19] In this context, the company's remaining investors became involved in a scheme in which King Leopold II offered Americans an interest in the Belgian monarch's controversial rubber operations in the Congo.[20] But efforts in the Congo also required the destruction of native plants and thus provided no long-range supply of raw rubber. As supplies of wild rubber became increasing difficult to secure, the IRC launched the initial stages of its decades-long effort to transform guayule from wild desert shrub into a cultivated American crop.

Early Rubber Crop Research in the United States

Because publicly funded agricultural research tended to focus on widely cultivated crops, the IRC could not readily turn to the U.S. Department of Agriculture (USDA), agricultural experiment stations, or similar avenues for publicly funded agricultural research. Instead, the company initiated its own well-funded research on the guayule plant, a project that it pursued for nearly forty years. First, it tapped into the Carnegie Institution of Washington, which in 1902 had established a Desert Botanical Laboratory near Tucson in Arizona Territory, where research focused on the science of desert plants.[21] To push its research into a more practical and commercial direction, the IRC hired the Carnegie Laboratory's promising young botanist Francis E. Lloyd to bring his research activities to Mexico.[22] As the complexities of cultivating guayule became more apparent, the IRC made further arrangements with about a dozen agricultural scientists, including experts in plant physiology, plant breeding, irrigation techniques, Mexican geography, desert climatology, and other fields.

Research on other rubber crops also took place on American soil. Several investors promoted pinguay, or the Colorado rubber plant, a weedy shrub found in the arid mesas of southwestern Colorado and northern New Mexico; one source suggests that at least one rubber factory operated in southern Colorado by 1907. Tests by B. F. Goodrich and other rubber firms suggested that "Colorado rubber" was of higher quality than guayule.[23] California botanist Francesco Franceschi also experimented with guayule and pinguay, although apparently without connection to commercial developers.[24] Meanwhile, in 1909, Lloyd accepted a post at the Alabama Polytechnic Institute at Auburn and brought guayule seed from Mexico for study under both outdoor and greenhouse conditions. As a result of this research, Lloyd and the Carnegie Institution published a two-hundred-page study of the desert plant that became the bible of guayule studies.[25] To replace Lloyd, the IRC hired a Canadian-born botanist, William B. McCallum, who in 1910 began his thirty-five-year career devoted to study of guayule.[26] At about this time—perhaps presciently, in view of the many hard struggles for a domestic rubber crop—Lloyd abandoned guayule studies to serve as a consultant for the United States Rubber Company, the most prominent American participant in the boom of *Hevea* rubber plantations in the East Indies.

Guayule and the Mexican Revolution

Meanwhile, the guayule industry's second handicap, its vulnerability in Mexico's unstable political climate, was becoming increasingly clear. Many local elites resented foreign dominance in the Mexican guayule territory, and many believed that the Díaz government had instigated policies on irrigation, export duties, and land rights that unfairly favored the interests of foreign investors. Indeed, State Department officials indicated that Díaz himself had informed the IRC that it could "count upon the sympathy of the government and its good will in aiding its affairs, so far as the law permits." Local officials, such as Coahuila Governor Miguel Cárdenas, also were inclined to help, because they made quick profits selling desert lands to the IRC during the guayule boom. Further, the government's devaluation of the peso in 1905 placed investors such as the IRC in a stronger position to buy land and to invest in production facilities.[27]

The guayule industry brought profound tensions to the society and economy of northern Mexico. Demand for guayule sparked aggressive competition for territory that previously had interested only ranchers. Bandits and thieves from rival companies had little difficulty hauling away the increasingly valuable shrub; machete battles and gunfights became commonplace along the burro trails that served as conduits for shrub harvested from the desert. For its part, the IRC tried to demarcate its property by cutting a line of harvested guayule two miles wide around the borders of its vast holdings. Foreign companies also upset local labor markets. The IRC reportedly paid the highest wages in the industry, which forced local elites to pay ever more to compete. Livestock owners felt new pressures as the search for wild guayule sent burros and their drivers deep into the Chihuahuan Desert where both risked death from dehydration. Meanwhile, foreign companies such as the IRC constricted the flow of capital into the local economy, because employees received one half of their pay in scrip redeemable only at the IRC's company store.[28] Overcrowded living conditions overwhelmed the region's primitive infrastructure as new factory workers and guayule harvesters poured into territory that had been sparsely populated. Shantytowns emerged in rural areas (because officials in Torreón drove out vagrants), and rivers and creeks became filled with waste products that brought diseases such as typhoid and smallpox to the new communities.[29]

The situation intensified to the point that guayule was one of the factors that sparked the Mexican Revolution. Critics of the IRC and other American interests found a voice in Francisco Madero. A member of an elite family and a major landowner in northern Mexico, Madero and his relatives had their own reasons to target the IRC. The Maderos were major players in the guayule industry; Francisco served as secretary and his father, Don Evaristo, served as the president and patriarch of the IRC's principal rival, the Compania Explotadora Coahuilense, S.A.[30] "I am not afraid of [the IRC]," Madero declared, and he emerged a hero of those who wished that someone would stand up to the foreign infiltration of Mexico. Undoubtedly funded in part through his own sales of guayule to Western markets, Francisco Madero launched a political career designed to challenge the status quo under Díaz. As the revolution in Mexico unfolded, a new chapter began in the search for a domestic supply of rubber in the United States.[31]

Rebels took direct aim at both of the IRC's most valuable properties: its estate at Cedros and its factory in Torreón. They seized merchandise, corn, arms, carriages, furniture, medicine, cattle, mules, goats, horses, and tackle from the IRC and its subsidiaries.[32] Revolutionaries and their supporters ransacked the shops of foreign merchants in Torreón and massacred over two hundred Chinese residents of the city in a day of violence in May 1911. The U.S. consul stationed in Torreón, G. C. Carothers, worked closely with IRC officials and made protection of their properties a top priority, but he predicted that risks to American investors in Mexico would remain. In a coded message to his superiors in Mexico City and Washington, DC, Carothers stressed that the situation was delicate. "By all means avoid intervention, which would mean destruction to all Americans here," he warned. "This has been an object lesson to the Americans."[33] Heeding this advice, the U.S. government essentially abandoned a rubber industry based only a few hundred miles south of its southern border. The United States would need to look elsewhere for its rubber.

The Mexican Revolution's most ominous threat was to the IRC's seeds and seedlings, the genetic basis for any attempt to turn wild guayule into a cultivated crop. When the revolutionaries attacked in 1911, the company was growing several promising varieties on nurseries and about ten acres of irrigated plots near Cedros, but animals turned loose in the chaos of revolution destroyed much of this investment.[34] In both 1913 and 1914, the Cedros Ranch was "entirely at the mercy of marauding rebels and bandits," while the factory at Torreón was closed between June 1913 and October 1914.[35] Rebels repeatedly burned stockpiles of baled guayule, hundreds of tons in all. In one case, revolutionaries reportedly boarded a train at Cedros, raped the women and girls in front of their husbands and fathers, and then burned the railcars to strand them all. At other times, they smuggled guayule across the border through El Paso and other cities and sold it on the black market for the weapons and hard currency needed to sustain their revolution. Mexican governmental authorities proved unwilling or unable to provide adequate protection against the marauders. The IRC provided higher wages, arms, and horses to employees in an attempt to protect its interests. That effort failed as well, notably after some were forced to stand naked after revolutionaries raided Cedros and seized the workers' clothing and cash.[36]

Rubber as a Strategic War Material

These events, along with rising prices, offered clear warnings of the industrial world's vulnerability to and dependence on imported rubber supplies. William Perkin Jr., son of the discoverer of mauveine, the first important synthetic dye, was especially determined to challenge German chemists in the search for synthetic rubber. Thanks largely to the biochemistry research of his colleague Chaim Weizmann, the Zionist leader and future president of Israel, Perkin announced in 1912 that his team had developed a synthetic rubber derived from alcohol that could be manufactured from renewable plant resources. At the same meeting, the German chemist Carl Duisberg announced his own process, accentuated by his presentation of two synthetic-rubber tires derived from nonrenewable coal derivatives. In sum, the battle lines over the future of the rubber industry had been drawn years before World War I began.[37]

But synthetic rubber remained decades away from commercial viability, and because possible natural-rubber sources in Congo, Peru, Brazil, and Mexico all increasingly proved unviable, U.S. consumers had little choice but to depend on East Asian rubber plantations dominated by British and Dutch firms. As long as these nations and their colonial possessions remained friendly to the United States, the problem of unstable rubber supplies seemed to fade. Indeed, as Thomas Edison declared in September 1913, tropical rubber plantations ensured low prices, ample supplies, and a healthy automobile industry.[38] In an era of world peace, American rubber consumers had little reason to worry. Yet the undeniable reality was that the American rubber market remained in a precarious position, for it had become almost entirely dependent on supplies located halfway around the globe.[39]

The situation changed in August 1914, as the European powers entered into a global war that lasted more than four years. Despite public assertions

that the United States had no part in the early phases of World War I, the
American economy was so intertwined with that of other nations that American business and industrial leaders could not ignore the war and its consequences. For one thing, only 10 percent of American international commerce
traveled on ships registered in the United States; as Germany and Britain
redirected their ships into war service, American ocean commerce came to a
virtual standstill.[40] This had implications for the American rubber industry because the difficulty in transporting the product to the United States meant that
rubber began to pile up on the docks of Singapore. Moreover, modern warfare
proved a logistical nightmare demanding ever more strategic materials. Before
military officers could signal any attack from the trenches, unprecedented tons
of raw material had been gathered from around the globe, processed into fin-
ished goods, transported to the battlefields, and kept ready in storage. The war
also demonstrated the military significance of rubber: gas masks, gaskets, tubing for blood and plasma bags, insulation for telegraph wires, and, of course,
the tires on automobiles, trucks, and ambulances all required imported rubber
in unprecedented quantities. By war's end, rubber had shifted from a rather
exotic mercantile product to an ongoing military necessity. Issues of rubber
supply, war preparedness, and the importance of developing domestic and renewable raw materials thus converged.

Thomas Edison and War Preparedness

The outbreak of war brought Thomas Edison into the debate about the nation's
dangerous dependence on imported commodities. Long before he initiated his
own intensive search for a suitable domestic rubber supply, Edison became one
of the most influential and outspoken critics on the issue of America's dependence on German chemicals. When war broke out, Germany dominated the
world's production of industrial chemicals. German firms produced about 90
percent of America's textile dyes and were nearly as prevalent in pharmaceuti-
cal, fertilizer, and other branches of the chemical industry. Just as the power
of German chemical firms had been a paramount issue in war-preparedness
discussions, efforts to challenge German chemical hegemony were at the core
of a conflict eventually dubbed "The Chemists' War."[41]

Within three weeks of Germany's declaration of war on Britain, Edison
warned that American firms needed quick access to a supply of phenol (also
known as carbolic acid)—a chemical necessary in the manufacture of phonograph records. Edison's company was by far the nation's largest consumer of
phenol. As a result of the British naval blockade of Germany and Britain's decision to place an embargo on the export of strategic chemicals, Edison faced

a serious shortfall. He lobbied Congressman Edward W. Townsend of New Jersey to pressure Britain to release some of its phenol stockpiles to the United States, but this proved unsuccessful. In response, Townsend simply wondered "why this country does not produce such articles as a matter of self-protection to the manufacturers which require them."[42]

Characteristically, Edison pronounced, "This war should teach us to depend on ourselves . . . and to take care of our own needs."[43] Moreover, Edison followed his own advice that "people in the trade [should] act, not talk," and he decided to manufacture phenol himself. Using his customary approach based on trial and error, Edison assigned forty staff members to labor around the clock to test a previously unproved method that he hoped could produce phenol on an industrial scale. Initial results proved encouraging. Within a week, workers began to expand Edison's own chemical plant in Silver Lake, New Jersey. The new facility was in operation by the end of September, and Edison's company soon built a second to meet the sudden demand for American-produced phenol.[44] These operations, however, did not bring closure to the issue. Edison's process relied on benzol, but as supplies of this imported chemical also dried up, the inventor launched a new effort to capture benzol from the coal-tar gases released in the coke ovens of Pennsylvania and Alabama. Edison also began to produce his own aniline, another product essential for his phonographic records but suddenly cut off from German suppliers. Thus by early 1915, Edison successfully had bypassed German dominance and produced his own chemicals derived from domestic sources. No less important, investors who had trusted Edison's judgment received impressive dividends. Before the end of the war, Edison had produced so much phenol that government stockpiles bulged and prices plummeted. Shrewdly, Edison then moved away from the organic chemical business and sold off his interests at a handsome profit to more specialized firms. Indeed, there is evidence that Edison's assistant manipulated the inventor's prominence on the preparedness issue to help the Edison company gain closer ties with the navy and thus more sales for its batteries. In any event, from the public's point of view, the episode demonstrated that Edison had the skill, experience, and desire needed to respond to wartime shortages.[45]

Edison's achievements made him an early authority on the issue of American preparedness for war. In May 1915, the *New York Times* published an important interview with Edison in which the inventor called on Americans to find inexpensive and simple means to produce the technologies necessary in modern warfare. Edison invoked his familiar faith in American ingenuity and inventiveness with a proposal that the United States establish "a great research laboratory" where skilled engineers and scientists could develop the weapons and other technologies that would be necessary should the United States enter the European war. Five weeks later, Secretary of the Navy Josephus Daniels called on Edison to lead such an endeavor, naming him in mid-July 1915 to head the newly formed preparedness organization, the Naval Consulting Board.[46]

The group soon expanded its mandate beyond the navy and spawned other organizations that addressed the broader issues of the nation's industrial capabilities for modern warfare. As automobile executive and Naval Consulting Board member Howard Coffin put it, "Twentieth-century warfare demands that the blood of the soldier must be mingled with from three to four parts of the sweat of the man in the factories, mills, mines, and fields."[47] In 1915 and 1916, Edison, Coffin, and a handful of other prominent American engineers trumpeted the importance of industrial preparedness. Leaders in the preparedness movement organized publicity campaigns to increase public awareness of the issue and conducted industrial inventories that assessed the nation's resources and capabilities. These experts believed that the military setbacks of France and Britain offered Americans evidence that modern warfare demanded greater coordination between government and industry. Moreover, they feared that Germany had done a better job than the United States in integrating its scientific and technological research with its national aims. In brief, Germany seemed to be a rationalized and efficient economic, political, and military power. In the words of one historian, "The vision of a rationalized German society was never far from the minds of American preparedness groups."[48]

The German advantage in chemistry and industrial raw materials received special attention. For instance, Charles Holmes Herty, then head of the American Chemical Society, alerted President Woodrow Wilson and other American leaders of the nation's uneasy dependence on imported chemical compounds. The National Academy of Sciences (NAS), revived in spring 1916, took on the project of securing and stockpiling nitrates from Chile and other countries. The NAS also tapped Columbia University chemist Marston Bogert to investigate dozens of potential problems linked to shortages of strategic chemicals. Bogert responded with a frank report that outlined several serious shortfalls in the American chemical and industrial preparedness; in case after case, he warned that Germany had an immense advantage. His assessment also addressed American dependence on imported rubber and included an inaccurate (but ominous) warning that German chemists had nearly perfected a process to produce synthetic rubber.[49]

The Naval Blockade and Its Impact

On the other hand, American engineers' admiration for German industrial efficiency and chemical expertise came under scrutiny as the impact of the British naval blockade on Germany became clear. Launched in August 1914, the blockade brought economic hardship to Germany by autumn. By the following spring, the blockade had effectively paralyzed sea trade, and much of the German economy had come to a standstill. Government rationing and price-ceiling programs could not bring the economy under control. More important, black-market activities and street disturbances indicated that support for the war had diminished. The "Turnip Winter" of 1916 and 1917 brought especially acute deprivations that sparked food riots and citizen protests that contributed to the eventual collapse of the German government. In all, an estimated seven hundred thousand Germans, mostly innocent civilians, died from the effects of malnutrition during World War I.[50] The widespread suffering demonstrated the imperative for governments thereafter to obtain and distribute the resources necessary to conduct modern warfare.

The blockade also decimated the German rubber industry.[51] German rubber manufacturers' stockpiles had dwindled, skilled laborers joined the military, and transportation networks broke down. The price of crude rubber, rarely above one dollar per pound in the United States, reached nearly twenty dollars per pound in Germany. In April 1915, German officials "absolutely prohibited" the sale of rubber products and launched a scrap-rubber drive. By May, the Prussian government essentially banned all private automobile traffic.[52] Even more ominously, rubber shortages directly hindered the war effort. Military vehicles became useless in the absence of tires, military hospitals had to manage without rubber gloves and drainage tubes, and poorly insulated telephone and telegraph lines caused battlefield breakdowns that contributed to the deaths of thousands of German soldiers.[53]

German chemists and industrialists scrambled to find a solution to their rubber famine. Although World War I generated reports of German ingenuity in developing several substitute, or ersatz, products in response to wartime deprivations, rubber was one commodity that could not easily be replaced. Research on synthetic rubber sped up, but efforts by Carl Duisberg and other chemists fell short of securing a suitable substitute. In four years of war, Germany produced only 2,500 tons of methyl rubber, a hard, leatherlike substitute that could not be used for tires.[54] Germans also launched rather desperate efforts to find natural rubber in plants that would grow in temperate climates. Some of these attempts dated to the 1880s, when Georg Kassner of Breslau issued a report that promoted the sow thistle, *Sonchus oleraceus*, as a sustainable domestic rubber crop. In Kassner's view, the sow thistle could bring industrial, agricultural, and political advantages to the nation. The search became more urgent during World War I, when an Austrian scientist announced in 1916 that he had determined that a wild lettuce, *Lactuca viminea*, had rubber content greater than any other source, including *Hevea* and guayule. The plant grew in the wild in the Danube and Elbe valleys, and there seemed hope that it could be brought under cultivation. Some critics scoffed at schemes to develop German "rubber vegetables," but the editor of *Gummi-Zeitung* defended the project as necessary, because "our enemies hope to cause our starvation through rubber just as they do with grain."[55] In the end, however, wild lettuce and other potential rubber plants contributed nothing to the German war economy.

Rubber and the American War Economy

Long before the United States decided to fight alongside Britain in World War I, rubber severely tested American neutrality. In September 1914, Britain placed rubber and several other strategic goods on a list of contraband items. Two months later, Britain cut off American access to its East Asian rubber supplies, a decision that temporarily left some 250,000 American rubber laborers out of work. Such measures forced American leaders to choose between maintaining access to the products of Britain, its colonies, and its allies, or risk all by continuing trade with the Central Powers. American leaders reluctantly acquiesced to British demands and agreed not to export contraband rubber to Germany or neutral countries that might trade with any of the Central Powers. In the words of one journalist, the experience "placed the United States in a humiliating position" that demonstrated its dependence on imported raw materials.[56]

Meanwhile, British war measures also impelled Americans to seek alternative methods to ensure access to rubber. Some were not ready to give up on the Mexican shrub, now selling for about three times its prewar price. In 1915, a young Texan named Isaac Newton Buckner, a former manager of the IRC's plant at Torreón, snuck into a town where revolutionaries had seized twelve train carloads of guayule. Buckner strolled into the state governor's palace uninvited, a deliberate strategy to get himself arrested and gain an audience with the governor. Buckner persuaded the governor to sell him the bales at half the market price and bribed officials and railroad employees with money and tequila to smuggle the load across the border. Buckner conducted several similar expeditions into Mexico in 1917 and 1918 and thus provided a small supply of Mexican guayule to the Akron rubber companies.[57]

Among the more mainstream efforts, Edgar Davis of the United States Rubber Company led the charge to penetrate the East Indian rubber markets. Even before war broke out, Davis challenged the orthodox view that Americans could trust their British and Dutch suppliers. By 1913, he had acquired vast rubber-producing acreages in the Dutch East Indies, which made his company one of the largest rubber planters in the world. In early 1915, Davis sent his aide David Figart to Sumatra to expand plantation rubber operations and worked with a group of Wall Street financiers, including Baruch, to support these initiatives. In 1916, the Goodyear Tire and Rubber Company acquired twenty thousand acres in Sumatra intended for plantation rubber, and Harvey Firestone began his decades-long campaign to "grow rubber under the American flag if we can."[58] Then president of the Rubber Club of America, Firestone lobbied William Howard Taft, the former U.S. president and former governor of the Philippines, to relax the regulations in that territory that forbade any company to control more than 2,500 acres. Other financiers worked through the American International Corporation, a conglomerate connected with the Rockefellers, Morgans, Guggenheims, and others that funneled American investments into underdeveloped nations. Such initiatives indicated that many Americans recognized the strategic importance of rubber and were prepared to challenge British dominance in the African and Asian colonies.[59]

Despite these warnings, the American public had little awareness of the depth and extent of the economic entanglements that would bring Americans into the conflict. Both presidential candidates in 1916, President Wilson and Republican challenger Charles Evans Hughes, pledged to keep the country out of war and downplayed warnings of the preparedness experts. In August, Congress created a new agency, the Council of National Defense (CND), which in turn created committees that only halfheartedly studied the American dependence on strategic materials like manganese and rubber. This group did not meet formally until March 1917 and its impact was minimal. In the words of one participant, army leaders still were "not even on speaking terms with such words as . . . 'allocation,' 'priority,' and 'quantity production.'"[60]

Rubber politics regained urgency soon after Germany resumed its submarine warfare campaign on 1 February 1917. With the American declaration of war looming, the overlap between the preparedness issue and the rubber issue became more evident. Some Americans predicted that rubber would be the first commodity to be "affected adversely."[61] Baruch played a key role, first as head of the Advisory Commission of the CND, and later as head of the War Industries Board, groups that helped coordinate the war economy. In March 1917, Baruch issued a report that declared rubber one of the essential raw materials that "should be kept in sufficient quantities by the government itself."[62]

After the United States officially declared war on Germany on 6 April 1917, several wartime agencies mobilized to ensure a steady supply of rubber resources. In mid-April, Baruch called on the rubber industry's leaders to be among the first to accept his vision of government leadership in securing and disseminating rubber supplies.[63] Also that month, the Rubber Committee of the CND warned that the nation's manufacturers had as little as thirty days of inventory on hand. Due to the "acuteness of the shipping situation," government officials began to restrict rubber consumption until the flow of raw material from Southeast Asia resumed.[64] Bureaucrats also reiterated their commitment that *"no rubber at all"* should reach the Central Powers. In December 1917, rubber became one of the first commodities to be "licensed," which allowed the War Trade Board to regulate its flow. Importers had to go through the Rubber Association of America (formerly the Rubber Club of America) to have any cargo approved and guarantee that no rubber would go, directly or indirectly, to any country at war with the United States. As a result of these trade negotiations, the United States lessened its dependence on Atlantic shipping lanes for its rubber resources and increased the strategic importance of Pacific trade networks. This shift proved beneficial during World War I but a serious problem in World War II.[65]

The rubber issue also revealed emerging tensions between industry representatives and government policy makers. In 1918, rubber-industry leaders created a War Service Committee, in response to the increased demand for its product and the hardships of government restrictions on production. Industry officials conceded that production of certain rubber goods, like poker chips, bathing caps, and telephone ear pads, could be restricted during the war emergency. Yet their report defined a large percentage of their products as "essential" to the war, including surgical supplies, hot-water bottles, gas masks, conveyor belts, and even the athletic equipment that "contribute largely to National health and physique." Indeed, the war brought a significant increase in the demand for rubber goods, particularly truck tires, insulating materials, and surgical goods. Thus the industry sought greater access to plantation rubber, not less. Rubber-industry officials insisted that an oversupply of rubber lay at hand on East Indian plantations, that American ships could be spared to carry the cargo across the Pacific, and that the United States risked losing market share to other rubber manufacturing nations if it failed to make rubber a higher priority. Industry officials also prepared for the eventual armistice talks and argued that the postwar peace must include assurances that the United States would have access to the rubber under British and Dutch control on "as favorable terms as the manufacturers of any country."[66]

One leader in the field challenged this business-as-usual consensus. Pre-saging his 1920s leadership in the struggle for a domestic rubber crop, Harvey Firestone argued that his colleagues failed to appreciate the seriousness of the nation's dependence on imported rubber. Although he turned down an op-portunity to become the wartime rubber-policy "czar" himself, he did offer (just before departing on a camping trip with his friends Thomas Edison and Henry Ford) to do whatever the government needed to help impose rubber restrictions.[67]

Such restrictions came quickly, as government officials began to get seri-ous about the potential rubber shortages. Concluding that the rubber industry had not "shown a proper spirit of self denial," policy makers began to require "sacrifices on the part of all and serious hardships for some."[68] On the grounds that the nation could not spare cargo space on ships, the War Trade Board al-lowed American rubber producers to import just a fraction of the rubber that they consumed in 1917. The shortage of cargo space emerged as a national crisis, and industrial planners warned that the capacity required to ship sol-diers and war matériel to Europe hampered the ability of American industries to import the raw materials that they needed to keep the war economy hum-ming. As rubber stockpiles were strained, George Peek, commissioner of fin-ished products of the War Industries Board, declared that it was "obvious that something radical" had to be done "at once." Peek instructed rubber manu-facturers to curtail tire and tube production to 50 percent of the 1917 rate, to work vigorously to prevent industries from hoarding rubber resources, and to encourage conservation and recycling of existing supplies. Others mobilized to secure immediately whatever rubber might be available in Brazil, either on its docks or in the wild.[69] Meanwhile, the Priorities Division of the War In-dustries Board issued a directive in September 1918 that placed limits on the production of certain nonessential rubber products and threatened to cut off a manufacturer's access to labor, fuel, and transportation if found in violation.[70] Although still not too serious, the rubber industry and its customers learned a little about what the loss of rubber resources could mean.

Searching for an American Rubber Crop during World War I

Wartime shortages and restrictions encouraged several new attempts to seek a domestic rubber supply. Already decimated by overharvesting, the Mexican Revolution, and the emergence of the East Asian plantation rubber industry, the IRC clearly needed to change its direction. In 1915, the company named an aggressive new president and chair of the board, George H. Carnahan. A native of Ohio, Carnahan had some training in chemistry and geology before following opportunities in the mining industry in Colorado, Ecuador, and Mexico. Carnahan came to know the guayule industry while serving as manager of a copper mine in Mexico, and he had begun working for the IRC's subsidiary at Torreón in 1914. Until his death in 1941, just before World War II finally brought guayule into the public consciousness, Carnahan was the guayule industry's tireless supporter and most vocal spokesman.[71]

Carnahan made his impact on IRC policies felt quickly. First, he aggressively defended company interests in Mexico and fought the new government's efforts to raise taxes on foreign firms. In one document, he suggested that "liberal bribes"—a phrase scratched out and replaced with "unofficial measures"—might be used to influence Mexican policy makers.[72] He also repeatedly lobbied Chandler P. Anderson, a lawyer with close ties to the State Department, in an effort to protect American interests in Mexico, possibly through the overthrow of President Venustiano Carranza.[73] Beyond that, Carnahan hedged his bets by creating a subsidiary that controlled twenty thousand acres on Sumatra suitable for *Hevea* rubber plantations.

Another part of the strategy was to cultivate rubber in the United States, where modest efforts had begun already. After fleeing Mexico in 1911 with his tin of seeds, IRC botanist William McCallum launched a series of sporadic but repeated efforts to establish a renewable rubber industry in the United States. In 1912, McCallum began work at Valley Center near San Diego, California, and had more than one million guayule plants descended from the smuggled Mexican seeds in the field by 1913.[74] IRC executives soon determined that their southern California site lacked the water resources necessary for long-term guayule cultivation. In January 1916, Carnahan, McCallum, and another IRC official strolled unannounced into the Tucson office of G.E.P. Smith, McCallum's former colleague at the University of Arizona, for advice about an alternative site. Within days, the party had decided that property in Pima County, Arizona, about twenty-seven miles south of Tucson, could become the center of a new rubber industry in the United States. Smith quickly conducted surveys to confirm the site's water resources and he filed a lengthy and positive report by telegram to IRC headquarters in New York City on 30 June and 1 July. Carnahan immediately obtained approval from his board of directors to purchase the nine-thousand-acre Arizona tract.[75]

Cognizant of their experiences in Mexico, Carnahan and his colleagues returned to the Arizona ranch on 4 July 1916 with revolvers in hand. Fortunately, southern Arizona was more peaceful than northern Mexico, and IRC officials soon discovered that American political stability offered an important advantage in their pursuit of a sustainable rubber crop. Although IRC officials saw this as a project with wartime implications, they especially envisioned the long-term potential of an American-grown guayule rubber crop; in time, they predicted, the property could yield four million pounds of rubber and profits of five hundred thousand dollars per year.[76] With a borrowed railroad car serving as their temporary office, officials began work right away. The IRC soon hired the former Pima County sheriff to direct a crew of more than ninety Mexican Americans and Yaqui, most of whom lived in tents and corrugated tin shacks, to grub mesquite, build water towers, manufacture adobe bricks, and lay irrigation pipe. Even more important, the IRC quickly directed construction of a nursery, where McCallum applied his "secret treatment" to the precious guayule seed intended to keep viable the genetic basis for the U.S. rubber industry. Workers also built a company town called "Continental" that eventually boasted a school, a research laboratory, a company commissary, and crude housing for 350 employees. In addition, the IRC made arrangements with irrigation experts, plant physiologists, soil chemists, and others from the University of Arizona and the Arizona Agricultural Experiment Station to serve as consultants. Within months, the IRC had transformed untouched desert property into a modern agricultural research facility and a community of American rubber growers.[77]

Although the IRC tried strenuously to hide its activities from the press and public, the Arizona project generated hopes that America's dependence on imported rubber could be solved. Henry C. Pearson, the eminent editor of the trade journal *India Rubber World*, envisioned that the project could lead eventually to the cultivation of guayule on some five hundred million acres— an area the size of Alaska and New Mexico combined. Undaunted by the arid climate and thin soils of the American West, Pearson suggested that the principles of "modern forestry" could create a "great new American industry that would be the source of wealth in time of peace and protection in case of war." The possibilities were "almost unlimited," Pearson gushed, and he predicted that guayule "might easily become one of our greatest agricultural industries."[78] The mass-market magazine *Literary Digest* also carried news of the Arizona operation, adding an editorial comment that it took place on "American territory in a climate ideal for white men."[79]

By November 1917, the company had over five million guayule plants in the Continental nursery. Better still, McCallum's tests on the 1917 harvest from the Valley Center, California, plots suggested "almost incredibly high" rubber percentages. His attempts at selective breeding seemed to work. McCallum wrote that he had "every reason to believe" that the Arizona plants, which started out with even more vigorous growth, could surpass those encouraging results.[80] On the other hand, McCallum was frustrated by the company's difficulties in finding and disciplining reliable labor, as well as the residents' apparent indifference to the ditches of open sewage that ran through the community. In any case, however, the IRC had virtually no chance to expand the guayule industry, because wartime controls had completely suspended production of the acetone needed to process guayule resin.[81]

Other citizens took more haphazard and tentative efforts to find synthetic and alternative natural-rubber sources in the United States. In December 1915, for instance, Congressman John E. Raker of California introduced a bill in Congress that called for the study of sagebrush and greasewood as potential rubber sources, but the bill died in committee when the USDA failed to offer any support.[82] In 1916, a group of western investors promoted the ocotillo shrub (*Fourquieria splendens*) and the saguaro cactus (*Cereus giganteus*) as the next alternative.[83] The National Research Council (NRC) formed a Botanical Raw Products Committee that investigated thousands of species of "economic plants" that could have value as raw materials for American industry. The NRC's Chemistry Committee, with a subcommittee on rubber products headed by David Spence of the Norwalk Tire and Rubber Company, also launched investigations of guayule and other American-grown plants.[84] The Carnegie Institution of Washington expanded its research in this field in 1917 and 1918 with a study of thirty potential rubber-bearing species, four of which offered hope for commercial production.[85]

Harvey Monroe Hall and California Rubber Plants

Meanwhile, in March 1917, a few weeks before the official entry of the United States into the war, officials at the University of California offered the services of its staff to help investigate potential American sources of rubber. By May, two emergency organizations, the Committee on Scientific Research of the State Council of Defense for California and the Committee on Botany to the Pacific Coast Research Conference, arranged to make the search for rubber crops one of their highest priorities. Fearful of an enemy blockade, or a possible "change in the alignment of our allies in present or in future wars," officials decided that "it is not wise for us to take any chance."[86] Thus these groups funded a project that sent University of California botanist Harvey Monroe Hall into the mountains and deserts of southern California and Nevada to find an American rubber plant.[87]

Hall soon declared the rubber rabbitbrush (*Chrysothamnus nauseosus*, sometimes called rabbit bush) to be especially promising. A wildflower common in the foothills and plains of the western United States, rabbitbrush is characterized by a dense collection of bright yellow flowers that bloom much of the year. Its latex is found in the stems and leaves, and rubber can be extracted through mechanical crushing combined with chemical solvents. For nearly a year, Hall's field experience included "hob-nobbing with Indians and teamsters and dogs and chasing thru the hills for the elusive caoutchouc."[88] Hall embraced the mission fully, confident that botanists like him could make a difference in the global conflict. "This work is a great thing for me," he boasted, "in that it makes me think I am doing something against that d___ Kaiser." "Whether we ever use the results of my work or not," Hall explained to his wife, "it is at least done in the spirit of enmity to German Ideas."[89] As Hall tramped through California and Nevada, he studied the insect that seemed vital to pollination of the rabbitbrush, learned tricks of rubber extraction from "Paiute squaws," took samples from rubber plants during both snowstorms in the High Sierra and sandstorms in the southern California desert, and struggled with the taxonomic issues that surrounded the different types, varieties, and species of *Chrysothamnus*.[90] Admitting that his obsessive quest for rubber plants resembled an "absurd hallucination," Hall grew increasingly convinced that the rabbitbrush could solve the nation's shortages. "The possibilities of the investigations are almost limitless," Hall believed. He found "oceans of the plant in Owens Valley," and he scouted territory in western Nevada that might be appropriate for factories that produced rubber from American-grown rubber plants.[91]

Hall also traveled to southern Arizona to meet with McCallum, the IRC's botanist. McCallum's experience helped convince Hall that through selection and careful cultivation, rabbitbrush could easily become a suitable emergency rubber source, and perhaps even an economically viable one in peacetime. "In spite of its low [rubber] content," he explained, "certain other advantages of rabbit bush are so great that I believe it will sometime put guayule entirely in the shade as a cultivated plant."[92] The search for a domestic rubber crop continued in the summer of 1918, as Hall joined the prominent American ecologists Frederic and Edith Clements in a three-month journey across the mountains and deserts of six western states, notable as much for Hall's work with the expedition's countless flat tires, blown radiators, and broken carburetors as for his botanizing.[93] Nevertheless, the overwhelming extent of rabbitbrush astounded Hall. Although rubber yields were poor, and the quality of rubber was of a "low grade," Hall remained enthusiastic because the total amount of rabbitbrush in the western states proved "much greater than we formerly supposed."[94]

Much to Hall's disappointment, however, officials in Washington, DC, remained unconvinced that a domestic rubber crop could address the nation's war needs. With the production of East Indian plantations still on the rise, the director of botanic projects for the NRC suggested that only an "extraordinarily vigorous plant" could alter American dependence on imported *Hevea* rubber.[95] Once the armistice was signed, those who desired a rapid demobilization of the American war economy prevailed, and the rubber trade with the East Indies quickly reopened. As another by-product of the demobilization, the California Defense Council cut off its portion of Hall's research funds in December 1918, one month after Armistice Day. Begging that the council reconsider, Hall did not come home for Christmas that year and continued his research with rubber plants in the cold and windy high plains of western Nevada.[96]

Although ignored at the time, Hall's experiences offered real lessons to policy makers interested in the rubber issue. Hall urged politicians to think in terms of future rubber shortages, rather than the one that had just passed.[97] In a 1919 summary of his results, Hall voiced relief that "removal of the submarine menace" had eliminated any immediate fears of a rubber shortage but warned that the potential for future shortages remained. In all, Hall claimed that his work on rabbitbrush could render "the nation practically independent of all foreign countries if the item of expense and harvesting is not considered."[98] Hall and his Berkeley colleague, Thomas Goodspeed, published a report in 1919 that offered a plain conclusion: "It is eminently desirable that a portion of the rubber consumed in the United States should be produced within our own borders. It is the only important commodity to modern warfare which we have not yet learned to produce."[99] In another report, Goodspeed asserted that they had succeeded in their mission, for he and Hall had definitively located an emergency supply of rubber that grew within the continental United States. Goodspeed concluded, "It can now be said with certainty that a considerable amount of rubber is available in the United States," and the nation could be "practically independent of all foreign countries." According to Goodspeed, the discovery of *Chrysothamnus* "should remove any fear among the American people" of a possible future rubber shortage.[100]

A Failure to Respond

Although Goodspeed's pronouncements were a bit hyperbolic, American officials continued to ignore warnings about the nation's dependence on imported rubber. As a whole, the search for a domestic rubber crop made minimal progress during World War I. In the United States, wartime rubber shortages were modest and short-lived, and only vaguely foreshadowed the potential for a genuine rubber crisis.[101] According to Grosvenor Clarkson, director of the Council of National Defense, "rubber was one of the economic freaks of the war" because wartime circumstances brought an unexpected boon to the American rubber industry.[102] With Germany and Russia unable to obtain East

Indian rubber, and with the maturing rubber plantations of the East Indies still increasing their productivity, British and Dutch producers faced an atypical wartime problem of oversupply and low prices. The war reduced rubber shipments into the eastern seaboard harbors, but with shipping lanes across the Pacific Ocean unimpeded, imports into Pacific ports increased. Indeed, American rubber-industry officials eventually lobbied to have imports from the Far East reduced, because oversupply limited prices and profitability at a time when other industries were operating at full bore. Demand for rubber skyrocketed during the war. In the United States, rubber consumption increased from about 60,000 tons in 1914 to about 240,000 tons in 1919. For the rest of the world, annual consumption increased from about 60,000 tons to about 120,000 tons over the same period. The economies of Akron and Singapore boomed during the war. In all, the combination of low prices for raw materials and worldwide demand for finished goods meant that World War I represented a golden age for the American rubber industry.

Yet even if there were no immediate rubber crisis, the Mexican Revolution and World War I communicated important lessons about the nation's dependence on imported valuable commodities for those willing to listen. Turmoil in Mexico proved that American access to essential raw materials was vulnerable, for even bands of lightly armed revolutionaries could stifle an entire industry and nearly bankrupt a company backed by some of the richest men on Wall Street. With foreign control of resources one of the main reasons for the revolution, U.S. officials were careful to minimize their intervention into the Mexican Revolution. The naval blockades of World War I were even more ominous, for they demonstrated that Germany had been mistaken to count on a brief war or on its ability to develop ersatz replacements for every shortfall. Raw materials and agricultural shortages brought an industrial giant like Germany to its knees. Even the United States, which suffered relatively little threat to its shipping lanes, learned that increased demand for wartime cargo space could hinder seriously the importation of industrial raw materials.

World War I was a turning point in other ways as well. As a total war, the conflict made clear that victory hinged on an unprecedented mastery of the industrial economy, mobilization of the civilian sector, and ability to prepare for the worst. The war demonstrated further that economic leaders in the industrialized nations began to speak a new language that encompassed the terms "war preparedness," "strategic materials," and "synthetic substitutes." In the lingo of the American Chemical Society's wartime president, Charles Herty, it was time to speak of "national self-containedness."[103] Thomas Edison understood these concepts as well as anyone and emerged from the war with an enhanced public reputation as one who had perceived the importance of preparing for the next war. Moreover, his success in developing a domestic source for previously imported benzol and other chemicals brought him additional wealth, another important lesson for those who sought to anticipate future raw material shortages. The war also signaled a turning point in the history of American science, as scientists enhanced their stature as "experts" who could mobilize to solve the nation's economic and military problems. Groups like the American Chemical Society began to argue that support for the sciences was a matter of national defense.[104] In scores of cases, scientists delivered ersatz substitutes and chemical solutions to raw material shortages. Yet for all their successes, the chemists came up short in one important way: synthetic rubber remained out of reach. In contrast to the massive government mobilization of scientists in World War II, efforts during World War I were considerably smaller in scale and scope. Thus the soil remained the first place to look for a domestic rubber solution, as the world's industrial economy became ever more and absolutely dependent on the agricultural products of tropical Asia.

Revolution and war also brought some progress in the search for a domestic rubber crop. For George Carnahan and investors in the guayule industry, turmoil in Mexico and heightened demand for rubber goods sparked the IRC's thirty-year effort to establish guayule as a cultivated row crop that could thrive within the borders of the United States. With its salvaging of seeds in the face of the 1911 Mexican Revolution, and its subsequent investment in nurseries in California and Arizona, the IRC had control of the genetic stock that would spark debates over domestic rubber policy for the next three decades. Several other scientists and institutions used the wartime circumstance as justification for the search for other potential rubber crops, such as ocotillo, rabbitbrush, and milkweed.

Rubber-industry officials also recognized the risks that their business faced. As America's rubber consumption increased, so did its vulnerability to export duties, price controls, and political maneuvering. As the peacetime economy was restored, rubber supplies flowed again, automobile and tire factories hummed, and a false sense of security returned. When British and Dutch rubber planters challenged American rubber producers in 1922, the next episode in the American search for a domestic rubber crop began.

Chapter 2

Domestic Rubber Crops in an Era of Nationalism and Internationalism

In NOVEMBER 1922, eight days after British rubber producers announced a new plan to restrict rubber exports and raise rubber prices, an official in the U.S. Department of War fired off a memo to his colleagues that decried the British action as "ONE OF THE MOST VIOLENT ECONOMIC WARS" the country had ever faced.[1] The British rubber producers' scheme, known as the Stevenson Plan, came to be one of the most significant issues in American foreign and trade policy of the mid-1920s. It also sparked another burst in the American search for a domestic rubber crop, for the episode demonstrated that economic tensions could upset American rubber markets nearly as dramatically as a military conflict. Just as World War I had taught the lesson that the United States had come perilously close to economic disaster because of its dependence on chemicals and other imports from Germany and elsewhere, the Stevenson Plan revealed that even America's allies could disrupt the American rubber supply and thereby imperil the entire consumer economy.

The war also left many Americans convinced that they could solve almost any challenge to the nation's economic leadership, including shortages of imported rubber. As the Stevenson Plan became an important issue in the early 1920s, many responded to the new rallying cry, "America Should Grow Its Own Rubber." Several agricultural scientists, botanists, and chemists did their part to respond to this issue of geopolitical significance. In the economic boom times of the 1920s, however, leadership in this area also came from industrial leaders Thomas Edison, Henry Ford, and Harvey Firestone, men whose influence in public affairs extended far beyond the realms of their business interests. Other Wall Street interest groups, including those who hoped to revive the IRC and the guayule industry, also saw the 1920s as an opportunity to develop an American rubber crop. George Carnahan, for instance, suggested that the IRC's guayule project in Arizona would become "the most important agricultural experiment ever carried under purely American auspices."[2] In the end, however, government officials rebuffed the IRC's proposals to invest in guayule as a new American crop and strategic resource, and the scientific community did not make a commitment to solve the problem. As a result, and despite all the preparedness rhetoric that surrounded the Stevenson Plan, the decade ended with the nation still vulnerable to a potential wartime rubber crisis.

Military and Geopolitical Concerns

Amid the bitterness and instability at the war's end, internationalists lobbied President Wilson in Paris and insisted that the industrial world's access to raw materials was a prerequisite for economic and political stability. American leaders such as Bernard Baruch, William C. Redfield, and Herbert Hoover pushed for a greater degree of free trade, based on the assumption that access to foreign markets would improve the standard of living worldwide and thus increase the demand for American products.[3] Thus they recommended that rubber become a topic at the Paris Peace Conference, where they wanted the United States to insist that American firms would have access to the rubber of Britain and the Netherlands without restriction.[4] The issue of Mexican rubber also arose in the aftermath of war, as some lobbyists traveled to Paris in 1919 to bolster American demands for access to the raw materials of Mexico.[5] During the Harding and Coolidge administrations, an important cadre of economic planners, led particularly by Secretary of Commerce Herbert Hoover, saw an opportunity to adapt the experience of wartime mobilization to the postwar world.[6] Throughout the decade, a wide range of academic institutions, Washington think tanks, and international organizations considered the connections between raw materials and warfare. In the words of one of the scholars who emerged as a leader in this field, Brooks Emeny, the lessons of World War I were clear: military victory was no longer determined by the size of the nation, its population, its national wealth, or size of its armies, but "rather by its capacity for industrialization." Because industrialization was linked to the unequal distribution of raw materials, Emeny maintained, a nation's security depended on the degree of its self-sufficiency and unimpeded access to raw materials.[7]

The U.S. military also understood the implications of World War I and took steps to improve its access to critical raw materials. The National Defense Act of 1920 assigned to the assistant secretary of war the responsibility of planning for the next war. Throughout the interwar years, the Office of the Assistant Secretary of War (OASW) became the center of the government's discussions concerning the threat of rubber shortages. Reformers also created the Army-Navy Munitions Board in 1922, an organization intended to minimize competition between the two military branches for war matériel. Another by-product, the Army Industrial War College, founded in 1924, trained hundreds of American military officers (including such notables as Dwight Eisenhower and Douglas MacArthur) to recognize the innumerable connections between the industrial economy and national defense. In the words of one historian, these reforms signaled a shift from strategies of procurement to concern over "how to use the entire economy to foster national security."[8] In addition, the OASW created a number of commodity committees, including one for rubber, to ensure that military officers continually study the threat to Americans' access to strategic raw materials. For the officers involved, the new infrastructure was aimed to prevent a repeat of the "Battle of Washington," a term that one participant used to describe the chaotic race for resources that emerged in the transition from peacetime to wartime in spring 1917.[9]

As in the private sector, only a small number of government bureaucrats urged continued research on domestic rubber crops in an era of low rubber prices. One example occurred in fall 1921, when officials in both the Department of Commerce's Bureau of Standards and the OASW filed reports in which they pointed to guayule as the best rubber alternative in the case of a war emergency. Experts concluded that synthetic rubber remained unproved and expensive, whereas guayule was both native to American territory and already viable as a commercial product.[10] In February 1922, Quartermaster General H. L. Rogers laid out in plain language that politicians needed to consider the "present weak and dangerous strategic position of this country." With rubber "indispensable" in modern warfare, and with prospects for synthetic rubber "very doubtful," Rogers called for an immediate campaign to secure additional agricultural research funds. In an era when many Americans saw war-preparedness efforts as opportunities for large industries to profit from government contracts, Rogers warned that vigilance with rubber was in bureaucrats' best interests. Foretelling the events of World War II, Rogers suggested that his colleagues might need to demonstrate that they had not neglected their duty to secure strategic raw materials and thereby help protect the nation.[11] William A. Taylor, chief of the USDA's Bureau of Plant Industry, concurred and noted, "For several years we have realized that the rubber situation might become acute for this country on very short notice." Despite these sporadic efforts, attempts to stir Congress and the public on this matter languished and the funds allocated for rubber crop investigations remained minuscule.[12]

Chemists and "National Self-Containedness"

The case histories of rubber and the American chemical industry demonstrate that the accumulation of raw materials mattered as much as the production of weapons, vehicles, tools, and other hardware in the "total war" economy. Like the rubber industry, the American chemical industry managed to keep up with a greatly increased demand during the late war and thus contributed to the emergent faith in the boundless capabilities of America's technical and industrial experts. In the words of former American Chemical Society president Charles Holmes Herty, the war had exposed the "fetish of Teutonic superchemistry" as invalid. Speaking before the Rubber Association of America's "Victory Banquet" of 1919, Secretary of Commerce William C. Redfield praised the rubber industry for validating the American values of unselfishness and industrial might, in contrast to the alleged German values of selfishness and the abuse of science for evil purposes.[13]

Despite such rhetoric, the specter of foreign control loomed ominous for both industries: Germany seemed poised to reassert its domination of the American chemical industry, and the rubber industry remained under the thumb of British and Dutch planters. According to Herty, Americans needed to commit to a doctrine of "national self-containedness" to prepare for the next conflict. Moreover, in era before the hydrocarbons of petroleum became the dominant feedstock for both synthetic chemicals and synthetic rubber, chemists assumed that the carbohydrates produced by American agriculture would provide the raw materials for chemical independence. The trajectories of these two materials soon converged: the renewed and insatiable demand for imported chemicals provided a strong incentive for chemists to develop a synthetic alternative to natural rubber.[14]

American chemists brought in plain language the message of "self-containedness" to a broader audience. Herty and other chemists emerged from the war determined to educate the public about American chemistry; thanks to revenues derived from the seizure of German chemical patents, they soon had the funds to spread their message. Edward E. Slosson, a chemist-turned–scientific journalist, launched this campaign with a series of articles published during the war. He expanded his work in a 1919 text, *Creative Chemistry*. In the introduction, Julius Stieglitz, another former head of the American Chemical Society and professor at the University of Chicago, argued passionately that the American "public should be awakened" and that "the whole nation's welfare demands . . . that our public be enlightened in the matter of chemistry to our national life."[15] The war also revealed another value of skilled chemists: they helped Germany stay in the war despite the Allied blockade through their work on substitutes for cotton, wool, and other basic products. Stieglitz was grateful that the war had forced Americans to become more conscious of their sources of raw materials, but he feared that such momentum would stall unless Congress and other leaders recognized their responsibility to help the American chemical industry. In Stieglitz's view, there could be no doubt that an "unscrupulous competing nation" might easily exploit the weaknesses of America's chemical industry if a false sense of security resumed.[16]

Slosson buttressed this general argument with a chapter that focused on the rubber industry. He called on chemists and agricultural scientists to both solve a problem that had eluded government officials and philanthropists for decades and reduce American dependence on rubber imported from faraway sources. Slosson also warned that British actions twice proved that Americans were still vulnerable to outside pressure: first, during the war, when Britain temporarily cut off American access to East Indian rubber; and again, after the war, when it passed regulations that limited foreign investment in Malayan rubber plantations. In Slosson's view, it was imperative that Americans expand research on three potential alternatives: the development of rubber in America's tropical possessions and allies (such as the Philippines and Haiti, or by the purchase of the Guianas); the extension of the search for guayule and other rubber crops that might grow within the United States; and the development of synthetic rubber.[17] All three strategies, although unsuccessful in the 1920s and 1930s, reemerged as some of the nation's highest priorities amid the rubber crisis of 1942.

In August 1921, Slosson turned to another issue connected with the history of rubber. He proposed a series of twelve "scientific" newspaper and magazine articles that would outline potential issues of political conflict in the Pacific, including trade disputes that would involve rubber.[18] Indeed, in the aftermath of World War I, the United States lessened its dependence on Atlantic shipping lanes for its rubber resources; consequently, the strategic importance of Pacific trade networks increased. By 1922, the popular press was replete with books and articles addressing the "Pacific Problem," a euphemism for Japan's apparent assertiveness in economic and colonial policies. Many writers explained that Japan faced both demographic and economic pressures to expand its territory, and it would be only a matter of time before these pressures impinged on American interests.[19] As tensions rose over economic and naval issues, a few military planners contemplated the possibility of warfare between the United States and Britain. By 1922, the crisis had reached such a stage that President Warren Harding called for an international disarmament conference in Washington, DC, a meeting that eventually diffused tensions by establishing limits on the naval arms race in the Pacific. Soon after the meeting, former assistant secretary of the navy Franklin D. Roosevelt expressed relief that Japan had changed its course, confidently asserting that Japan could be trusted to not threaten American interests.[20]

In any case, for most Americans, rubber was anything but a crisis in the years immediately after World War I. As some wartime officials had correctly predicted, the end of war simply opened the floodgates for bottled-up producers. World rubber production climbed rapidly, as increasingly efficient East Indian plantations spawned overproduction and low prices. No longer luxuries in the American consumer economy, rubber products had become everyday commodities essential in many medical supplies, articles of clothing, telephone and telegraph equipment, and, of course, the booming automobile industry. In this milieu, concern over future supplies faded into the background, and only a few Americans continued the search for a domestic rubber crop.

Domestic Rubber Crops of the Early 1920s

University of California botanist Harvey Monroe Hall was an important exception. Although the wartime crisis had passed, Hall still received funds from the Carnegie Institution of Washington, and he remained optimistic that such research could help the United States develop a native rubber plant that could be cultivated on a commercial scale.[21] Hall also approached rubber magnate Harvey Firestone with a request to bring his science to "some practical account." Aware that scientific research could not be mobilized until the public demanded it, Hall explained that someone with "long vision and very strong patriotic impulses" was needed to lead the effort for a domestic crop.[22] Harvey Firestone proved to be such a person.

Meanwhile, Hall had become "engrossed" in the study of a new favorite plant: the desert milkweed, *Asclepias subulata*. A native of Baja California, Mexico; southern California; and southern Arizona, desert milkweed is usually leafless, characterized by its thick gray-green stems, which contain a milky latex. It can yield up to 8 percent rubber, a bit less than wild guayule. Hall became convinced that selective breeding could surely improve such yields in the future. As Hall confidently wrote to his wife, "I wish you could appreciate the startling importance of my work. It is great fun and will certainly upset the world when the results come in."[23] In 1921, Hall and his colleague Frances L. Long released *Rubber-content of North American Plants*, a lengthy study that went beyond his previous reports on rabbitbrush.[24] Moreover, milkweed by-products offered an attractive raw material for the manufacture of paper. Best of all, and in contrast to guayule and rabbitbrush, it seemed possible to harvest milkweed without uprooting it, thus making the plant the potential basis for a sustainable and renewable domestic rubber industry. Hall brought his results east and made a grand tour that included lectures at the Carnegie Institution and the New York Botanical Garden, as well as visits to the offices of *India Rubber World* and to the laboratories of guayule experts David Spence in Connecticut and Francis Lloyd in Montreal. Hall was especially excited following a visit with officials of the United States Rubber Company. "There is no telling where this will end," he wrote his wife. "It is pleasant to do business with people who have lots of money. They spend about $300,000 a year on what they call research."[25]

The postwar years also represented the high point for another potential American rubber crop, the ocotillo. One of the most distinctive plants of the southwestern deserts, the ocotillo stands up to twenty feet tall with long, woody, spindly branches that appear gray and lifeless for most of the year. Following a rare rainfall, however, hundreds of tiny leaves and brilliant red flowers appear, only to fall off soon after the dry weather returns. Beneath its thick outer layers, the plant holds an intriguing resin that prompted various attempts to find a commercial purpose before and during the war.[26] Early efforts were not very successful, but by 1920, the Ocotillo Products Company of Salome, Arizona, claimed that it had developed extraction methods that yielded products useful for coating hand grenades and ships, as well as providing the material for phonograph records. The company also claimed to have produced two tires from ocotillo rubber that had survived twenty-two thousand miles of use in an environment of "vipers, lizards, Gila monsters, scorpions, tarantulas, stinging flies, and poisonous beetles" and temperatures that "often reached" 132 degrees Fahrenheit. Such boastful language suggests that the promoter may have exaggerated the claim that the factory could process one hundred tons of ocotillo per day, that the plant contained 5 percent of "first-class crude" rubber, and that there was enough ocotillo in the vicinity to keep the plant operating for fifty years. In the end, botanical experts such as Hall demonstrated that ocotillo's resins lacked the qualities of true rubber. Despite attempts to entice Henry Ford to back the idea, the ocotillo factory still had not opened by 1923. Yet another potential rubber crop fell by the wayside.[27]

Rubber and the Florida Land Boom

Long before Edison embarked on his research, other Americans had identified Florida as a potential home for a domestic rubber crop. By the 1890s, Henry Nehrling and several other Florida botanists had become fascinated with the genus *Ficus*, which includes two species native to Florida and a number of others that successfully acclimated there. One species in particular, *Ficus elastica*, had already become a commercial rubber source in the tropics and had the potential to be the same in Florida.[28] The collapse of the Brazilian rubber industry around 1910 sparked further interest in Florida rubber. John Gifford, a well-known promoter of development in the Everglades, issued his own call for rubber crop research. In a 1911 article published in *Everglade Magazine*, Gifford promised that he had "no doubt" that the desert plant guayule would thrive in humid Dade County. Gifford voiced even more enthusiasm for *Cryptostegia grandiflora* and *Cryptostegia madagascariensis*, both vines that the American-born botanist Charles Dolley had systemically studied both in Mexico (while working for the IRC) and at the Nassau Botanical Garden in the Bahamas. Cryptostegia grew easily in southern Florida as an ornamental plant, and Dolley claimed he had nearly achieved a commercial level of rubber production just before World War I. Significantly, Gifford framed his argument in the context of fighting the "colossal graft" of the rubber industry.[29] The famed plant explorer David Fairchild also became enthusiastic for cryptostegia, noting by 1913 that it "seems perfectly at home and should be propagated and tried even in the Everglades." Fairchild described the commercial potential of southern Florida rubber and reported regularly on the progress of the experimental vines that grew near Coral Gables.[30]

Beginning in 1918, railroad and real estate magnate James E. Ingraham began an aggressive study of the rubber possibilities in Dade County. Ingraham reported that a deceased Dade County landowner had left a will in 1897 instructing future settlers on his lands to plant rubber trees (*Ficus elastica*). Surviving a few frosts, the trees had thrived in the twenty years, offering proof, it seemed, of their commercial viability.[31] Ingraham intensified his efforts to develop a southern Florida rubber industry in late 1923, a time that coincided both with the Florida land boom and the consequences of the Stevenson Plan. Ingraham and other promoters stressed that independence from foreign suppliers was a vital issue, particularly in case of war. "This is too important a resource to neglect," he insisted, and he implied it was a patriotic act for Florida residents to give rubber trees a test. Ingraham's real estate associates predicted that rubber in Dade County would be a "howling success."[32]

Some citizens of Fort Myers, Edison and Ford's winter home, also worked to develop a domestic rubber industry. In 1920, for instance, the *Fort Myers News-Press* reported that rubber plants had long been planted in Florida. The unnamed author listed nearly twenty potential species, including two native ficus species and cryptostegia. This article also made the increasingly prevalent "chemurgic" argument that farm wastes would fuel farms of the future. It also expressed concern that rubber industry profits went to millionaire magnates and "smoky-colored natives" rather than southern farmers.[33]

Other Floridians jumped on the bandwagon. Promoters blatantly ignored warnings that a southern Florida rubber industry faced many obstacles. The *Miami Herald*, for instance, published several articles on the topic in 1923 and 1924, editorializing that rubber could be "grown here as successfully, as quickly, and as freely" as anywhere.[34] "We should not permit ourselves to depend upon imported rubber," reported *Manufacturers Record*, an influential southern business journal, "even from our own islands such as the Philippines, if by any possibility we can produce rubber at home." According to an article published in the *Florida Grower*, Floridians could produce enough rubber to make the United States entirely self-sufficient, and Miami would soon become the world's leading rubber port. Senator Duncan Fletcher lobbied for rubber research projects in the state, chambers of commerce touted their towns as low-priced rubber producers, and northern speculators bet a million dollars that they could obtain rubber from the cedar and pine underbrush of the Everglades. In all, by 1925, the peak of the Florida real estate boom, rubber stood not far from center stage.[35]

Emergence of the Stevenson Plan

At about this time, the East Indian rubber industry experienced some significant changes. Since 1915, the area in British colonies under rubber cultivation expanded about 7 percent per year, while production increased an average of 22 percent per year. As a result, postwar crude rubber prices rarely reached 40 cents per pound, hitting a low of 12.5 cents per pound in August 1922. Well below the cost of production, such prices proffered ominous implications for the long-term health of the industry. Of perhaps greater concern, British officials feared that American capitalists would move in and purchase East Indian rubber plantations at "bankrupt prices," thereby reducing Britain's economic clout over a crucial commodity and threatening the entire British Empire.[36] Meanwhile, unstable rubber prices and upheavals among the producing companies brought uncertain levels of production and other risks to rubber buyers and tire manufacturers in the United States. At a time when the demand for rubber products, especially tires, continued to rise, most American manufacturers desperately desired greater stability in the rubber markets.[37]

In this context, British rubber producers sought some form of relief. Although Firestone, Hoover, and other Americans later accused them of a vengeful form of price gouging and market manipulation, the actions taken by British rubber producers can be readily understood. In October 1921, the Rubber

Growers' Association persuaded Winston Churchill, then the secretary of state for the colonies, to form a committee to investigate the rubber situation. Its chairman, Sir James Stevenson, an executive at the John Walker and Sons distillery and a personal adviser to Churchill, became the committee's namesake. By 1922, the committee had several plans on the table, ranging from simple cuts in production to a more revolutionary proposal to replace "the old competitive system" altogether through an internationalist rubber parliament. The plan that prevailed pegged legal exports to specific price targets; exports would be constricted until market prices rebounded. For months, British negotiators tried to persuade the Dutch government to support the restriction scheme, well aware that American consumers might be able to purchase from other suppliers at lower cost. The Dutch declined to participate, recognizing that they could not control native smallholder producers unwilling to submit to European-imposed production quotas. Moreover, planters already had an increasing market share due to the relatively high productivity of their plantations and the high quality of their rubber.[38]

Amid these deliberations, developments in the United States induced the Stevenson committee to act without Dutch support. As expected, few American rubber consumers had any sympathy for the plight of colonial British and Dutch planters who had been quite willing to demand free market prices when supplies were tight.[39] Meanwhile, tensions escalated when American politicians began to increase their pressure on Britain to repay its World War I debts. Some British government officials were shocked that Americans would be so antagonistic and so insensitive to British wartime sacrifices. Among others, Churchill became convinced that Americans were paying too little for rubber. Because the United States consumed 71 percent and British colonies produced 75 percent of the world's crude rubber, higher prices would result in a steady transfer of capital from the United States to Britain and a concomitant decline in Britain's war debt. Reports from industry experts indicated that American demand for rubber would continue to increase even without Dutch support for the scheme. Thus on 13 October 1922, in the last fifteen minutes of the last cabinet meeting before the collapse of Prime Minister Lloyd George's government, Britain approved the Stevenson Plan.[40]

The Stevenson Plan, which took effect on 1 November 1922, offered a rather simple scheme to raise and stabilize prices by linking exports to the price of rubber. In brief, it mandated that British rubber exports would be reduced if the standard crude rubber price fell below twenty-four cents per pound; likewise, exports would be increased if the price rose above thirty-six cents. Those who drafted the Stevenson Plan did not intend to see rubber prices rise excessively, but they failed to anticipate how much American demand would skyrocket throughout the Roaring Twenties. Prices continued to rise, soon far above the target price of thirty-six cents.[41]

Herbert Hoover, Harvey Firestone, and the American Response

The Stevenson Plan raised the ire of the assertive U.S. secretary of commerce, Herbert Hoover, perhaps the most important policy maker during the Harding and Coolidge administrations. Hoover had a fundamental faith that the public and private sectors could work together to create a stronger and more efficient modern state. In particular, Hoover contended that trade associations, think tanks, and scientific experts should provide data and expertise and thus help the business sector work in greater harmony. Government experts would guide and assist, rather than regulate, private industry in the professional management of America's natural resources. Such efficiency would help businesses reduce costs, lower prices for consumers, and increase consumption and prosperity for all. Hoover's agenda also extended into foreign affairs, where he assumed that American values were superior and that America's international influence was destined to spread.[42]

Strategic materials like rubber offered Hoover an opportunity to implement his beliefs. Within weeks of Hoover's appointment, A. L. Viles, general manager of the Rubber Association of America (RAA), offered his group's services in any way that the new secretary of commerce saw fit. Highlighting the RAA's role during World War I, Viles suggested that an atmosphere of even "greater cooperation" between the government and the rubber industry could be cultivated in the future.[43] The RAA's hopes, however, went awry with the Stevenson Plan. The group met in October 1922, just before the Stevenson Plan took effect, and plotted a course of continued cooperation with British suppliers. Although members were not keen to pay higher prices for crude rubber, the majority concluded that price stability and predicable levels of production were worthwhile trade-offs. Hoover, however, questioned this strategy.[44]

Meanwhile, the RAA splintered over the issue. Harvey Firestone, who controlled the fourth largest rubber company in the nation, differed from most other rubber producers. For one thing, his company depended more on British suppliers than did his competitors, who owned or had access to plantations on Dutch territory. Further, and like his friend Henry Ford, Firestone was confident that many automobile interest groups would break from the RAA to support his position. Thus, in a 14 December meeting with his stockholders, Firestone announced a strategy that soon became the company slogan: "America Should Grow Its Own Rubber." Firestone conceived of America's rubber supply as something that could come from the Philippines, Latin America, and the expansion of American-owned operations like Edgar Davis's plantations in Sumatra. Soon thereafter, Firestone traveled to Washington, DC, to lobby President Harding, Secretary Hoover, and members of Congress to challenge the Stevenson Plan and work for American rubber independence. Repeatedly, Firestone made the case that the United States had little choice but to go its own way and try to break free from British control.[45]

Hoover tended to agree with Firestone rather than the majority of the RAA membership. In cabinet meetings, he suggested that Americans should invest in rubber sources in the Western Hemisphere as alternative sources of supply and insurance in case of a war emergency. Secretary of War John W. Weeks and others in the cabinet thought it a higher priority to invest in America's Philippine possessions, in part to deter a possible war in the Pacific. Senator Medill McCormick of Illinois proposed legislation that would encourage investment in South and Central American rubber plantations and thereby gain control of "a national supply" of rubber. With Firestone working behind the scenes, Hoover agreed to support the concept. Meanwhile, the commerce secretary met with a delegation of representatives from the British Rubber Growers' Association in an unsuccessful, last-ditch attempt to persuade them to ease the restrictions of the Stevenson Plan.[46]

Thus Hoover came out in favor of McCormick's proposed legislation, a bill that would grant five hundred thousand dollars for in-depth surveys of potential sources for the nation's rubber needs. Both Hoover and Firestone testified in Congress in favor of the bill; the former invoked rhetoric that linked plants and strategic resources, arguing that "we must prepare some sort of national defense" against such price-control schemes.[47] The plot thickened when Firestone called a meeting in Washington, DC, to publicize his vision of rubber nationalism and voice support for McCormick's bill. Although the RAA sent telegrams that plainly discouraged members from attending, two hundred representatives of industry, agriculture, and government participated in Firestone's event. There, Firestone attacked the British plan as "vicious in principle and unsound economically." In his speech at Firestone's conference, McCormick insisted that investment in the Philippines would never be sufficient. Americans "must grow [rubber] in countries with which our communication will be unbroken, no matter how widespread or immense a war of the future may be."[48] Significantly, Secretary of Agriculture Henry C. Wallace predicted at the conference that the United States soon would produce its own rubber crops, just as American-grown sugar beets had challenged an earlier monopoly of tropical sugarcane producers. "The most desirable thing," Wallace asserted, "is to build up a rubber industry on our own soil, conducted by American methods."[49]

These arguments prevailed. In early March 1923, President Harding, a friend and camping partner of Ford, Edison, and Firestone, signed McCormick's "rubber survey" bill into law. It granted four hundred thousand dollars to the Department of Commerce for the purpose of conducting an extensive survey of potential rubber sources around the globe, plus one hundred thousand dollars to the Department of Agriculture to investigate the possibility of growing rubber on or near American territory. By hiring botanists and other scientists to scrutinize rubber possibilities in countries around the world, the bill signaled a new level of American commitment to the search for biological resources. Tensions remained, however, as Firestone announced his resignation from the RAA, a group that he once led as its president. In the meantime, Churchill declared that Americans should appreciate the price stability that the Stevenson Plan afforded.[50]

Hoover continued to mobilize government involvement in the economics, politics, and science of this strategic agricultural commodity. In March 1923, Hoover wrote to former secretary of commerce William C. Redfield, who then held a position as a lobbyist for Dutch commercial interests, praising the Dutch for staying out of the Stevenson Plan. He warned Redfield, however, that the United States might very well attempt to develop its own rubber industry in "new or nearby areas."[51] Four days later, Hoover approached Wallace in an attempt to coordinate utilization of the new rubber-survey appropriations.[52] Hoover also held a private meeting of the leading rubber producers in New York City on 20 May 1924. Ostensibly, the meeting's purpose was to announce results of the first of the rubber surveys, but Hoover's real aim was to bring the RAA membership over to Firestone's position. Firestone presented his concerns about American dependence on foreign rubber, and Hoover declared that the nation would "undoubtedly" feel shortages by 1928 unless more direct actions were taken. Yet as one historian observed, the meeting proved to be a "dismal" experience for Hoover and Firestone.[53] Except for Firestone, none of the major American rubber producers were willing to challenge British and Dutch growers. While some suggested that Americans might invest in their own plantations in the British and Dutch colonies, William O'Neil, president of the General Tire and Rubber Company, countered that there was "no reason why the British and the Dutch should not continue raising our rubber."[54]

Edison, Ford, Firestone, and Rubber in Florida

While the McCormick bill appropriated funds for government-sponsored agricultural research of domestic rubber crops, Edison, Ford, and Firestone also launched their own efforts at rubber crop research. Aware of his "America

Should Grow Its Own Rubber" campaign, many real estate promoters and amateur scientists approached Firestone with promises of rubber riches in southern Florida. In one case, the developer G. M. Duncan sent three aggressive letters that urged immediate investment in land near Cape Sable purportedly suitable for rubber trees. To be sure, Duncan also increased his price for the land in question. In another, a Louisiana oil broker suggested that Firestone invest in a scheme to plant a series of rubber plant test plots near the Gulf of Mexico, from southern Florida to the Mexican border, each one acre in size and spaced every twenty-five miles apart. Investors would buy vast tracts of land in the areas near the most promising test plots. Although Firestone turned down all proposals along these lines, he nonetheless took them seriously.[55] More to the point, he also began to play a more direct role in the investigation of potential domestic rubber plants. Firestone used his contacts to help import unusual rubber-producing plants from the East Indies for the USDA scientists to investigate, and he brought hundreds of rubber plant specimens from the nearby USDA laboratory to grow and test at his winter home in Miami Beach. Firestone's staff studied soil types, fertilization regimens, plant-quarantine regulations, the cost and supply of labor, and other issues that would determine the potential for a successful domestic rubber industry.[56] Somewhat encouraged, Firestone suggested in March 1925 that Florida "may yet become a garden spot for rubber" and help "relieve America from the firm grip of a foreign monopoly on a commodity vitally essential."[57]

Henry Ford's already-established network of automobile dealers, car salespeople, and industrial scientists made easier his search for a domestic rubber crop. In April 1923, Ford's chemist William H. Smith instructed the head of the firm's operations in Los Angeles to ask the legendary botanist Luther Burbank for the latest in rubber research. Although Burbank expressed little optimism, he mentioned the IRC's operations south of Tucson, Arizona. The Ford dealer in Tucson tried to learn more, but as others had found, the IRC proved quite secretive about its operations because it lacked patent protection and had a stake in guayule's commercial success. The IRC even refused to share results with Arizona Agricultural Experiment Station scientists. Nevertheless, without revealing the purpose of his request, the Tucson dealer managed to obtain some guayule samples that he then sent to Ford's laboratory in Dearborn, Michigan. Meanwhile, a Ford dealer from San Francisco secured an interview with California botanist Harvey Monroe Hall. In August 1923, Smith had the Ford dealer in Gila Bend, Arizona, employ a team of Mexican immigrants to scour the hillsides for fifty miles in every direction from Gila Bend in search of the milkweed *Asclepias subulata*. These explorers managed to gather a few hundred pounds of the plant but warned that it was "practically extinct" and thus not a likely source for American rubber. Ford's staff also asked their dealer in Copenhagen, Denmark, to join the search by investigating progress that German botanists had made in their search for agricultural rubber.[58] In all, Ford's and Firestone's actions illustrate a willingness to use their international clout to augment the agricultural resources of the United States.[59]

Meanwhile, Edison also felt impelled to turn to domestic rubber crop research. He had already received Harvey Monroe Hall's summary of his wartime research, *The Rubber-content of North American Plants*, which he read thoroughly. Tellingly, Edison underlined the passages—six times—that described mechanical harvesting of rubber plants.[60] In January 1923, Firestone sent Edison a copy of a book that he had helped publish, *Rubber: Its History and Development*; within one week, Edison responded that he had "lost no time in going through this book."[61] Also in that year, both Edison and Ford conducted laboratory research with the common milkweed, *Asclepias syriaca*. By July, Edison wrote to Ford's secretary, Ernest Liebold, with news that the milkweed experiments had both encouraging and discouraging implications: "We can get two crops a year," he noted, but also observed "it is going to be a difficult matter to get a commercial process to get the latex out."[62] Ford's staff scientist Smith announced that the milkweed grown in a Dearborn greenhouse could yield 4.2 percent rubber from milkweed, more than the 1 to 2 percent that Hall and Long had predicted in their 1921 study.[63]

Edison also showed an intense interest in guayule. Captain J. Fred Menge, a railroad agent and steamboat operator in Fort Myers and close friend of both Edison and Ford, made a trip to New York City to meet with IRC officials to investigate the potential of guayule in southern Florida. After Menge's return, the *Fort Myers News-Press* reported there was "little doubt" that Lee County could become a center of guayule rubber production and "millions of dollars would be made here."[64] In a letter to Ford, Edison predicted that guayule could yield 680 pounds of rubber per acre. "I think there will be no trouble to reap or plant with machinery," he added, "but you will know all about that." To Firestone, he scratched out a note indicating, "Am having good luck with milkweed & guyule [sic]." By summer 1923, Edison was somewhat desperate for guayule; he asked Firestone, Carnahan, and USDA officials to send guayule seeds, which he intended to plant in both New Jersey and Florida, as well as bales of harvested guayule to be studied in the laboratory. As long as the IRC kept strict proprietary control over its germplasm, however, it was difficult for others to do research on that plant.[65]

Edison, Ford, and Firestone also built relationships with the federal government's researchers on American rubber crops. In late 1923, the funds appropriated in the McCormick bill allowed the USDA to expand its research plots at Chapman Field near Miami. Rubber research became a top priority, and plant explorers supplied *Hevea*, devil tree (*Ficus alstonia*), natal plum (*Carissa macrocarpa*), frangipani (genus *Plumeria*), and other potential tropical rubber species for acclimatization experiments in southern Florida. In December of that year, Edison arranged with Alfred Keys of Chapman Field to have seeds of *Cryptostegia madagascariensis* sent to Fort Myers in time for a spring planting.[66] Meanwhile, Firestone's agents from Liberia and Singapore visited Fort Myers to confer with Edison on how to grow *Hevea* and various other rubber plants, and to keep him supplied with seeds of potential Asian and African species. Edison also saw evidence of the national interest in the project, as newspaper reports of his meeting with Firestone's men prompted several citizens to send the inventor suggestions on how to proceed.[67]

Meanwhile, Ford pursued his own investigations. In spring 1924 Ford sent a representative, W.L.R. Blakely, to interview botanists and nursery specialists throughout Florida, all of whom concurred that cryptostegia could thrive in the southern part of the state.[68] Ford also obtained a sample of the vine from the USDA plant explorer David Fairchild, who suggested that the Ford company's "remarkable powers of organization" were just what was needed to conduct the necessary experiments, develop the machinery, and, notably, to utilize the waste products. "I do trust you will go ahead with this work of commercial rubber production on a Ford basis," Fairchild concluded.[69] Thus Edison, Ford, Firestone, and the USDA formed a collaborative network that shifted attention from the boosters who promoted Florida rubber as part of their real estate schemes to a more systemic search for an American rubber crop.

Ford's Cryptostegia Project in Hendry County

The news that Henry Ford, one of the most admired Americans of his time, was investing in southern-Florida rubber added to the excitement. In May 1924, Ford purchased the Goodno estate, some eight thousand acres about thirty miles east of Fort Myers in remote Hendry County, in part so that he could test potential rubber crops. The media exaggerated all news about the project, some predicting that rubber factories in southern Florida were imminent.[70] Apparently fearing publicity and a concomitant increase in land prices (and perhaps also aware that intellectual property law did not protect improvements in plant research), Ford's associates declared there was "no truth whatever in the report" that their boss had purchased lands for rubber investigations.[71] Yet Ford did indeed begin rubber work that summer, and by late

Figure 2.1. Henry Ford's domestic rubber plants near Goodno at Fort Thompson in Hendry County, Florida, 1925. Courtesy of U.S. Department of the Interior, National Park Service, Edison National Historic Site.

June his employees had transformed a corner of the Goodno property into a small rubber-research facility. Amid orange groves and plantings of pineapple, coconuts, mangoes, avocados, and more, his employees filled twelve plots with 1,300 plants from various species of *Cryptostegia, Euphorbia, Ficus, Clitandra, Iatropa,* and *Landolphia.*[72] The experiments got off to a good start, and Blakeley hoped that they could collect enough shoots from the cryptostegia to have fifteen thousand plants survive the winter. Others suggested that cryptostegia vines could be grafted onto the bark of *Ficus elastica* trees. Alfred Keys, the rubber expert at Chapman Field, visited the site in August and offered several suggestions about possible fertilization and drainage schemes. During these optimistic days, Ford's team believed that cryptostegia could produce two crops per year, which could be mechanically harvested and then placed into huge vats of solvent in order to extract rubber. Edison also kept up with Ford's work in Hendry County.[73] But then a devastating November flood put the experiments under as much as five feet of water. Every plant died—except for sixty-nine specimens of *Cryptostegia grandiflora.*[74]

Despite the failure of the Goodno crop, Ford's secretary, Liebold, met with Florida governor John W. Martin in early 1925 in search of state funds to help drain his property and other potential rubber lands in the Everglades. Keys encouraged Ford's men not to be discouraged and to make a "big planting" of cryptostegia in the spring. Keys, along with Firestone's chief rubber scientist, made another visit to the facility in March 1925. The rubber "plants could not look better," Ford's agent reported in October 1925.[75]

Rumors of Ford's "rubber plantations" continued to spread, and the real estate developers who sold adjacent lands at inflated prices became a real thorn in the sides of Ford's senior staff.[76] At about the same time, Ford purchased twelve thousand acres of land near Savannah, Georgia, amid rumors that he planned to grow rubber crops there as well. The Hendry County research also attracted international attention. In an era that predated cold war tensions, Ford's officials welcomed a delegation of Soviet scientists touring much of the Western Hemisphere in search of rubber plants suitable for southern Russia.[77] Throughout 1926, Ford's employees in Hendry County cooperated extensively with the USDA's rubber researchers, agents in Madagascar, and at Florida nurseries to keep a steady supply of plant material under investigation. In the end, however, Ford's experiments with rubber plants in Hendry County flopped. By spring 1927, most of the plants had fallen victim to a combination of poor soils, drought, hurricane, frosts, and even fires caused by carelessly tossed cigarettes, problems that plagued the project throughout its three-year history.[78] Despite the financial backing of one of the world's richest men, Ford's Florida cryptostegia project could not overcome the risks associated with developing new crops in new environments.

The Stevenson Plan Continued

Public outcry over American dependence on imported rubber peaked again in 1925 and 1926. Despite the Stevenson Plan's promise to curtail production, American demand for rubber goods continued to grow, particularly due to sales of automobiles outfitted with new balloon tires that required even more rubber to produce. Prices rose to a new postwar high of $1.23 per pound in July 1925. Hoover boldly asserted that British planters had "mulcted" Americans of some seven hundred million dollars per year since the Stevenson Plan went into effect. In response, the House of Representatives instructed its Committee on Interstate and Foreign Commerce in December 1925 to investigate the apparent price-fixing by British interests. Firestone and Hoover again seized center stage. In his congressional testimony, Firestone promoted the Bacon bill, a proposal intended to pressure the Philippine government to allow American rubber firms to establish plantation-scale enterprises on Mindanao and other islands. At the same time, Firestone released his new autobiographic treatise, *Men and Rubber*, a book that celebrated the rubber industry and especially Firestone's efforts to secure American rubber independence. "We could recover from the blowing up of New York City and all the big cities of the Atlantic seaboard more quickly," the tire magnate asserted, "than we could recover from the loss of our rubber."[79]

Meanwhile, Hoover told the House committee that the time had come for Americans to fight British rubber monopolists. He named three strategies that would help: conservation and reuse of existing rubber products, the development of synthetic rubber through chemical research, and the search for new sources of supply both at home and abroad.[80] Hoover also attempted to lure American firms like the giant banana producer, the United Fruit Company, to invest in rubber plantations in Central America and to move laborers from banana plantations to potential rubber areas. He also met with officials from the War Department to arrange experimental plantings of rubber crops on military bases in the Panama Canal Zone.[81]

Next Hoover launched a bold publicity campaign that asked Americans to reduce their rubber consumption, conserve tires, and recycle rubber scraps. Just as he had linked conservation with patriotism during World War I (colloquially known as "Hooverizing"), Hoover again used that strategy to take the lead. This program provided another public relations victory when reduced demand seemed to cause a steady decline in the price of rubber. He also worked with Senator Arthur Capper of Kansas to propose a radical piece of legislation that would allow American rubber firms to pool together in a fifty-million-dollar corporation to cooperatively purchase raw rubber at favorable prices. Although this provision would have challenged fundamental American laws against trusts and monopolies, Hoover favored an exception for such an important commodity.[82] Once again, the emergence of the United States as an industrial power brought the issue of strategic tropical resources like rubber to the forefront of public debates.

In concert with changes in East Indian production levels, these maneuvers had their desired effect. From their peak of $1.23 per pound in July 1925, crude rubber prices fell to $0.50 per pound by May 1926. The British rubber monopoly had weakened in spite of the Stevenson Plan. That spring, many American newspapers praised Hoover with headlines similar to one in the *Washington Post*: "Rubber Fight Won by Americans."[83] Regardless of the outcome, both principal protagonists, Hoover and Churchill, had staked out positions in the rubber fight that increased their political stature and positioned them for higher office. Closer to the subject at hand, the rubber-policy debates of the 1920s helped drive Ford, Firestone, Edison, and others to expand the effort for rubber crops in the United States.[84]

Response to the Stevenson Plan at the USDA and Beyond

Developments within the USDA paralleled the politics of the Stevenson Plan. Since the turn of the century, a small army of USDA "plant explorers" had been scouring the world for new crops of potential economic value to the nation. These explorations did not simply reflect the intellectual curiosity of American botanists—they also hoped to utilize new tropical territories of Hawaii, Puerto Rico, the Philippines, and elsewhere to reduce American economic dependence on unfriendly nations. At least three of these bioprospectors, David Fairchild, Walter T. Swingle, and O. F. Cook, focused particularly on plants of tropical and subtropical climates. All three were well acquainted with the agricultural possibilities and the geopolitical implications of rubber, had a genuine interest in developing America's economy through domestically grown tropical and subtropical crops, and had close ties with the state of Florida.[85]

In the 1920s, the flamboyant and outspoken scientist Cook headed USDA research on rubber and other tropical crops.[86] In 1922, he issued a candid analysis of America's potential rubber crisis. In particular, Cook feared the expansion of Japan's power in the Pacific, part of what he called "the general Oriental problem which our civilization must face." The Japanese were so adaptable to new circumstances, Cook predicted, that they might eventually overwhelm the Europeans' Asian colonies and cut off the U.S. rubber supply. As a solution, Cook suggested that the United States aggressively seek to acquire British Honduras (now Belize) and French Pacific island colonies (particularly the Marquesas Islands, still a part of the territory of French Polynesia) as partial repayment for war loans. Speaking before War Department officials, Cook asserted that the people of British Honduras were eager to come under the American flag and begin rubber production. In private, Cook admitted that his racial analysis and diplomatic solutions might not win favor in Washington, DC, so he also called for USDA leaders to begin an immediate search for "all possibilities of rubber production in the United States or in adjacent regions."[87]

Within days of passage of the McCormick bill, which granted one hundred thousand dollars to support USDA rubber research, Cook shifted his attention to Haiti.[88] With plenty of underdeveloped land and a government under the thumb of U.S. occupational forces, Haiti offered the best prospect for Americans to imitate the plantation-style rubber culture of the East Indies. In spring 1924, the USDA established a rubber testing station on the island's northern coast, planting two thousand *Hevea*, five hundred *Funtumia*, and one thousand *Castilloa* trees.[89] Cook explained that rubber investigations elsewhere in the world had limitations—he considered Mexico too politically unstable, the leaders of the Philippines too uncooperative, and the workers of Liberia too unskilled. Instead, he energetically favored increased investment in the USDA projects in Haiti, where real progress was already being made despite the debilitating effects of malaria on its employees.[90]

The McCormick bill also supported efforts to grow rubber within the borders of the United States. In addition to work at Chapman Field, the USDA also expanded rubber work at several plant acclimatization stations. At Bard, Shafter, and Torrey Pines, California, for instance, scientists planted small stands of milkweed, cryptostegia, rabbitbrush, and other potential rubber plants to see if such crops might be suitable for western climates.[91] As Secretary of Agriculture Henry C. Wallace wrote when threatened with the reduction in funding for this research, "We are justified in spending whatever money may be necessary to exhaust the possibilities of rubber production in the continental United States." Cook deemed it "imperative" that the United States prepare for a potential wartime rubber emergency, highlighting progress with desert milkweed as the most promising.[92]

It is significant, however, that this burst of interest in rubber crops in the public sector did not help the cause of guayule enthusiasts in the private sector. For at least two decades, a cat-and-mouse game took place between the IRC and USDA officials. For its part, the IRC remained extremely reluctant to share its intellectual and genetic capital without assurances that its proprietary rights and the value of its stockholders' investments could be protected. Although IRC president Carnahan spent over three decades trying to convince investors that guayule offered a potential solution to American dependence on imported rubber, his message to the government was somewhat different. In a May 1923 meeting with Cook and other USDA officials, for instance, Carnahan indicated that most of the possible domestic rubber-producing plants were "hopeless as commercial prospects." He conceded that guayule had potential, but he was reluctant to share any plant material for research at USDA facilities without assurances of "very discreet handling."[93] In a subsequent meeting with IRC officials in November 1923, Cook explained that USDA policy was to cease investigations once a new crop had been proved to have commercial potential and industrial applications. Because the IRC's researchers had already established guayule's market viability, the USDA assured the company that it was only interested in a few guayule seeds to "complete their display" and provide evidence for published reports on the rubber crop situation. Cook agreed that the IRC deserved some kind of help to protect its shareholders, perhaps including plant patents, and guard their investments in breeding, cultivation, harvesting, and processing technologies.

Because plant patents did not exist in 1923, Cook agreed to shield the company from risk and publicity in more subtle ways. For instance, the USDA agreed to plant and display only the seeds of "ordinary" guayule strains, and it asked for no seeds of the more successful strains that IRC believed would be the basis for their commercial success.[94] Driven by hopes that guayule would one day prevail in the American rubber marketplace, the IRC sought to keep its research secret from the public and its germplasm under proprietary control. Its various appeals for plant patent protection did not get far, so it aggressively sought patents for various aspects of its mechanical operations.[95] As a consequence, from late 1923 until the middle of 1925, amid some the greatest interest in domestic rubber crops in American history, the USDA and the IRC had no contact with each other.[96]

Meanwhile, the Stevenson Plan and its aftermath led the Carnegie Institution of Washington to turn again to University of California botanist Harvey Monroe Hall, America's expert on rabbitbrush, milkweed, and other desert rubber plants. Hall welcomed the opportunity to work with independent funding agencies; as he remarked, it was more difficult to secure agreements with private rubber companies because "their interest goes up and down with the price of rubber!"[97] The Carnegie Institution's ecologist Frederic Clements was especially enthusiastic about the project and traveled to Berkeley in May 1923 to review strategies with Hall. They agreed to send an assistant to an outpost in Fallon, Nevada, to conduct fundamental scientific studies of rabbitbrush and propagate the plant under controlled breeding experiments. Hall also agreed to work as a consultant to the USDA and serve as an adviser for the USDA's milkweed rubber research in Bard, California.[98]

The continued high rubber prices also revived interest in rabbitbrush in other circles. Beginning in August 1925, F. W. Bolzendahl, a representative of the Chrysil (a colloquial name for rabbitbrush rubber) Rubber Association, urged Nevada governor James G. Scrugham to reexamine this plant. Intrigued by the possibility of a new industry for his state, Scrugham traveled to the Omaha, Nebraska, hotel where Bolzendahl based his operations to learn more. Scrugham quickly engaged leaders at the University of Nevada and the Nevada Agricultural Experiment Station to help in the search.[99] Meanwhile, Hall conducted an extensive survey in northern Nevada and eastern Oregon in August 1925 and reported that the area contained even more stands of wild rabbitbrush than he had earlier estimated, nearly ten thousand plants per acre in some places. It seemed that Nevada's rabbitbrush grew well without attention and could be readily hybridized to increase yield. "It appeared possible, even probable," Hall further reported, "that chrysil rubber could be estracted [sic] and marketed with profit." Hall met with Governor Scrugham in September 1925 and urged him to seek federal funding for an experiment station devoted to rubber crop research. As in 1918 and 1919, Hall remained convinced that federal support for rabbitbrush would materialize.[100] According to the experiment station's director Samuel Doten, "Governor Scrugham is so full of chrysil rubber these days that it puts new elasticity into his step and he is beginning to bounce when he goes up and down stairs."[101]

Nevada's research results, however, soon dashed Hall's and Scrugham's hopes. A 1926 report concluded that rabbitbrush offered little chance of agricultural success until plant breeders found ways to make the plant grow more rapidly and become more attractive for investments in mechanical planting and harvesting technologies. Rabbitbrush took longer to mature (five to six years) than guayule (four to five years), its rubber yield was considerably less, and the costs of grubbing and transporting bales of the shrub across the primitive roadways of rural Nevada proved far too costly. In Doten's words, the rubber companies would not be serious until the plant had so much rubber it would "snap back when you pulled it." "I don't say it cannot be done, but me," he added, "I'm gonna send to Montgomery Ward for my tires."[102] Hopes to secure support from the IRC, the USDA, and the United States Rubber Company in the project fizzled, and there seemed no choice but to drop the rabbitbrush project. Nevertheless, Doten recognized that the millions of rabbitbrush plants growing wild in the western deserts might serve as an emergency reserve, if a future war crisis made necessary a one-time harvest of the plants.[103]

The Stevenson Plan also brought guayule back into the public consciousness, especially after rubber prices reached a new peak in 1925. The prospect of high rubber prices revived the industry in western Texas; the factory at Marathon, last operational in 1914, reopened in April 1925. Now under the auspices of the New York–based Border Rubber Company, officials reported that local Mexican and Mexican American laborers were accustomed to the "hardships of camp life" and had a "natural instinct" for finding the shrub amid the rocky hills and sandy deserts of western Texas. At its peak, the plant employed seventy men, operated twenty hours per day, and reportedly produced one ton of American-grown rubber per day. With a cost of production around thirty-five cents per pound, however, the guayule industry would not be profitable if its costs of production rose or if imported *Hevea* rubber prices fell. In fact, both events occurred. Because the Marathon plant made no effort to develop a sustainable and cultivated crop, its owners faced steadily increasing costs to harvest guayule from lands increasingly isolated and distant from the factory. By September 1926, the Marathon plant closed it doors again. It remained silent until World War II once more spurred interest in Texas guayule.[104]

Revival of the Intercontinental Rubber Company

Developments in Texas were minor compared to the opportunities that the Stevenson Plan presented to the nation's leading guayule firm, the IRC. In an atmosphere of high rubber prices and national interest in the political economy of rubber, George Carnahan of the IRC seized the moment to reassert the potential of a domestic guayule industry. In response to schemes to harvest wild guayule, rabbitbrush, ocotillo, and other resources of western deserts as emergency rubber supplies, the IRC proclaimed itself ready to develop guayule as a cultivated and sustainable crop for American farmers.

The IRC had additional reasons to reassert its U.S. endeavors, for revolutionaries continued to attack its property in Mexico. In response, a number of Americans, including Carnahan, had formed the National Association for the Protection of American Rights in Mexico, a group that had lobbied for its interests at the Paris Peace Conference in 1919 and sought restitution for the loss of guayule and other property destroyed amid the Mexican Revolution.[105] Meanwhile, the combination of intermittent violence and low demand for guayule continued to hurt the company's business plan. The Torreón plant closed several times between 1919 and 1922 but reopened in March 1923 as the national response to the Stevenson Plan gelled. The company continued to expand, building three new factories in Mexico in 1923, 1924, and 1925. In spite of this round of political turmoil, production from the company's Mexican operations steadily increased from about 58 tons in 1921 to about 7,400 tons in 1926.[106]

Not surprisingly, the IRC once again turned its attention to producing rubber on American soil. After a lengthy evaluation of pros and cons of IRC operations at Valley Center in southern California, and at Continental in southern Arizona, Carnahan had become convinced that the milder climate of Salinas, California, offered a better chance to transform the desert shrub into a cultivated crop. Early estimates had suggested that the company could obtain as much rubber from one acre of cultivated California guayule as from one hundred acres of the wild, sparsely distributed guayule of northern Mexico. Thus in 1924, the company abandoned its plans to build a rubber industry in southern Arizona and developed a new facility at Salinas that offered a sophisticated and scientific approach to guayule studies, with modern nurseries, chemical laboratories, processing equipment, and more.[107] The company scored another breakthrough in 1925 when it developed a method to treat raw guayule through continuous rather than batch processing. The new method extracted an additional percentage point of rubber and yielded a cleaner, less resinous product that mixed readily with *Hevea* products. Once he secured a

Figure 2.2. View of the Intercontinental Rubber Company's guayule-processing facility at Cedral, Mexico, in 1925, showing walls and watchtowers. Texas Instruments Historical Archives, DeGolyer Library, Southern Methodist University, Dallas, Texas, A2005.0025.

patent for this process, Carnahan controlled the genetic stock, the manufacturing process, and the marketing networks that could become the basis for an entire industry based on an American domestic rubber crop.[108]

Operations in Salinas provided a new business opportunity for the IRC. In 1925 Carnahan alerted his scientist William McCallum, "Confidential: Our principal friends here [in New York City] are again showing a lot of interest in the California work."[109] The IRC's new backers included Hoover's former assistant secretary of commerce Claudius Huston and several Wall Street bankers led by Charles H. Sabin of the Guaranty Trust Company. With profits of nearly one million dollars in 1925, the company also hoped to secure support from Harvey Firestone.[110] Although unsuccessful in that regard, Carnahan did gain an audience with Secretary Hoover in January 1926, where he outlined the comparative advantages of guayule and *Hevea* rubber. Carnahan admitted that wages in the East Indies were only about one-twelfth the level that American workers would expect. That was counterbalanced, however, by the potential to mechanize the planting, cultivating, and harvest of guayule. The result, he predicted, would be that guayule production would require only about one-twelfth as many workers as involved in the tapping of *Hevea* trees. In others words, he claimed that costs of production in the United States and the East Indies would be identical. Thus Carnahan appealed to Hoover's enthusiasm for economic nationalism and human efficiency by explaining that guayule production protected American interests and eliminated through automation the "waste of human effort" associated with a "huge army of hard driven and poorly paid human beings." It was un-American, Carnahan implied, for rubber firms to hire "coolies" when "the production of rubber by machinery manipulated by intelligent, well-paid workers is in line with our national industrial genius." Perhaps hinting at the real point, Carnahan also asked if it might be possible for Congress to grant a seventeen-year patent for the IRC's long investment in guayule seed. With a plant patent bill stalled in Congress, the possibility did not seem out of reach.[111]

After spending over one million dollars on the development of plant breeding, mechanical harvesting, and rubber-extraction technologies, the IRC brought its idea to American investors. Its days of secrecy were over. In February, Carnahan announced that the IRC was now ready, for the first time, to work with American farmers and landowners in the development of an American rubber crop.[112] That month, the company moved beyond experimental work and planted six hundred acres of guayule in the Salinas area for commercial purposes. In advance of the IRC's offering of equity shares for sale to the public in March, two brokerage houses issued favorable reports on American-grown rubber as an investment opportunity. As one might expect, brokers predicted that cultivated American guayule offered a patriotic solution to the rubber problem: "In time of peace this is a matter of National concern; in time of war it might well assume calamitous proportions." One brokerage also quoted Commerce Secretary Hoover's prediction that rubber prices were likely to rise. The real crisis, Hoover suggested, could come by 1928 or 1930 and thus "lay a heavy burden on the American consumer."[113] Meanwhile, Carnahan also approached powerful financiers such as John Raskob, an executive at both General Motors and DuPont, in an effort to demonstrate that tires made from guayule could be superior to those of the highest grades of *Hevea* rubber.[114]

The IRC's publicity campaign continued in the fall. Three IRC experts delivered papers on guayule before the American Chemical Society's annual meeting. McCallum demonstrated the plant's botanical characteristics and described his fourteen years of "painstaking" efforts to overcome the plant's resistance to improved rubber yield. The rubber chemist Spence struck a different tone, emphasizing the IRC's success in developing a "practically continuous" manufacturing process that yielded a consistent and high-quality product. Carnahan described the plant's economic potential and boldly predicted that a commitment of one thousand square miles (1/156th the size of California), with one quarter of the acreage harvested every four years on a rotating basis, could ensure a steady supply of American rubber. If Americans ever found it "necessary or desirable to adopt a Stevenson Plan of their own," he added, this domestic rubber crop would permit it. The IRC stood ready to cooperate with farmers (and investors) willing to help their efforts.[115]

Six hundred chemists heard those talks, which Carnahan described as the "outstanding feature" of the entire meeting.[116] Major newspapers and both of the American rubber trade journals took note of the IRC's new initiatives, and both released special issues that featured the IRC's progress. *Rubber Age* devoted nearly twenty pages to guayule. Its editor suggested that the government offer IRC investors the benefits of tariff protection, for the cost of tariffs might be better than "letting someone else grow this country's rubber at a dictated price." *India Rubber World* celebrated the IRC's "courageous pioneering efforts" and "careful, costly research in botany, chemistry, and engineering problems."[117] The new publicity campaign also featured guayule's potential as a substitute for the distressed cotton farmers of the American South. With the Great Depression already underway in the "Cotton Belt," the IRC and its supporters touted guayule as a crop that could break the perennial grip of cotton monoculture. As guayule gained momentum in 1926, one of McCallum's friends congratulated him for the new dawn in the history of domestic rubber crops: "You've had a long pull of it and it really is very beautiful to have dreams come true."[118]

The Fading Crisis of the Late 1920s

Enthusiasm for guayule was nonetheless beginning to wane elsewhere. In January 1926, Carnahan had convinced Hoover to try to help the IRC gain inexpensive and privileged access to the Gigling Military Reservation in California, a facility that would allow the firm to expand cultivation. On 28 January, Hoover sent a letter to Secretary of War Dwight Davis that highlighted the "prime military importance" of the guayule scheme. War Department officials denied Hoover's request on the grounds that the military needed the land for artillery exercises, and no one else in the Coolidge administration would endorse a proposal that favored the guayule industry over others.[119] Commerce Secretary Hoover's campaigns ran into additional roadblocks. Congressional hearings that year revealed that many fears over the power of international commodity cartels had been exaggerated. In their minority report, Democrats suggested that the Stevenson Plan could be seen as a rational response to Republican tariffs. John Foster Dulles, then a prominent Wall Street attorney, published an article in the internationalist journal *Foreign Affairs* that charged that Hoover's policies were hypocritical and that foreign commodity regulations did not "mulct" Americans any more than American commodity groups

influenced the prices that foreigners paid. Hoover's and Raskob's efforts to create an American conglomerate of rubber buyers and importers fell especially
flat. Members of Congress such as Fiorello La Guardia of New York argued that
the scheme reeked of hypocrisy and that the largest American corporations
would receive most of the benefits while consumers would face even higher
rubber prices. Indeed, congressional testimony revealed that the six largest
American rubber companies made nearly $192 million in net profits during
the first five years of Stevenson Plan price-controls.[120]

Even more significantly, military planners had turned their attention away
from domestic rubber crops. Surveys conducted in 1924 and 1926 determined
that Americans had sufficient rubber stocks on hand to survive eighteen
months without hardship. By the mid-1920s, military officials promoted the
"associationalist" hope that private industry, led by the RAA, would be able
to institute the conservation measures needed to pull the country through any
rubber emergency. In December 1926, amid the IRC's active media blitz, one
War Department official closed the matter by concluding that it was "perhaps
unnecessary to give serious consideration to guayule at present."[121] Similarly,
USDA Secretary William M. Jardine's progress report to Hoover downplayed
research on domestic rubber crops but was more hopeful about developing
Hevea rubber reserves in Haiti or the Canal Zone, territories under U.S. military protection.[122] Other officials noted that the IRC's previous attempts to
grow guayule in Arizona and southern California had not been successful, and
they remained uncertain that the latest guayule excitement was justified. In
opposition to the IRC's program, USDA officials warned farmers that guayule would not be lucrative without a processing factory located nearby. One
suggested that USDA policy should be to remind farmers that Carnahan's
insistence on proprietary control over guayule seed meant that farmers could
not operate independently of the IRC. Presciently, the same official linked the
IRC's campaign to its interest in plant patents and its desire to protect expir-
ing processing patents.[123]

In all, developments of the early 1920s demonstrated that the United
States fell far short of the ideal of "national self-containedness," at least insofar
as rubber was concerned. Despite murmurs of isolationism, American demand
for rubber and other strategic raw materials ensured that the country remained
actively and aggressively involved in world affairs. In contrast to those who
fueled the tempest over rubber prices and the Stevenson Plan, the *New York
Times* editorialized that Americans should accept high rubber prices as the
natural result of market conditions and a fair reward for Britain's "far-sighted
statesmen and empire-builders."[124] Former secretary of commerce William
Redfield joined the refrain in 1926 with a text entitled *Dependent America*.

He warned that Americans should not follow the "cult" and the "false god" of isolationism. With America's "insufficiencies . . . as notable as our resources," Redfield challenged the wisdom of hoping for American-grown solutions to every problem, and he explicitly rejected Firestone's "sentimental" but potentially "dangerous" campaign to develop American-controlled sources of rubber.[125] In contrast, he believed that Americans would do better to admit their dependence on a fragile thread of supplies from faraway territories continually at risk of unstable political circumstances. In case after case, Redfield proved that the materials found in everyday consumer goods—telephones, radios, automobiles, paints, pens, and many more—relied on imported plants and minerals. In brief, he argued that American consumers should recognize and accept their dependence on the men, women, and children around the world who harvested and mined America's strategic resources.[126]

By 1927, nationalist concerns over rubber prices, Japanese aggression, and military preparedness had started to decline and the Stevenson Plan was on its last legs. As Dutch and other producers expanded their plantations and ignored the prices set by British rubber interests, crude rubber prices fell steadily from their peak in July 1925. Moreover, American rubber manufacturers had greatly expanded their utilization of scrap rubber. As a result, the nation's share of the world's raw rubber consumption had fallen from 73 percent in 1922 to 64 percent in 1927. By 1928, with crude rubber selling for $0.28 per pound, the British government officially ended the price-restriction policies of the Stevenson Plan.[127] As the rubber crisis waned, the urgent need to develop an American rubber crop might have faded away forever had not the American inventor Thomas Edison announced in February 1927 that he too was seeking a solution to the next potential rubber crisis. With the backing of his personal friends Ford and Firestone, Edison's campaign marked the next chapter in the American struggle for a domestic rubber crop.

Chapter 3

Thomas Edison and the Challenges
of the New Rubber Crops

I N FEBRUARY AND MARCH 1927, Thomas Edison leaked news to the press that he too had joined in the search for an American rubber crop. A series of reports traced Edison's project, and most indicated there was "no doubt" that he could achieve a successful and viable solution. For his part, Edison asserted that he would "do my bit to see it through, if I have to work twenty-four hours a day until it is an accomplished fact." These early reports suggested that Edison's success hinged on cryptostegia, the imported and invasive vine that thrived in southern Florida. Although this plant did not remain his favorite for long, it seemed to produce good-quality latex, it seemed suitable for mechanical harvesting, and it proved capable of withstanding a light frost. In actuality, Edison had barely begun his research on rubber crops. As he had done in the late 1870s in the early stages of his search for an electric lighting system, Edison's announcement came long before he had actually solved the problem. Nevertheless, Edison's involvement meant that the national spotlight remained focused on the search for domestic rubber crops into the early 1930s, even though American rubber prices had fallen and Britain had lifted the Stevenson Plan's price restrictions. As long as Thomas Edison, the nation's iconic inventor, chemist, and preparedness expert, preached its importance, the effort for a domestic rubber crop remained a vital national issue.[1]

Convinced that strategic plants like rubber lay at the core American military vulnerability—and hopeful to finish one last project before his death—Edison threw himself into this effort with a passion and commitment that belied his eighty years. He masterminded an impressive and systematic study of the problem that occupied his energies until his death in 1931. With financial support from Henry Ford and Harvey Firestone, Edison orchestrated a publicity campaign that generated national enthusiasm for his Fort Myers research; he lobbied politicians in Washington, DC, in order to gain a steady supply of information, seed, and specimens; he plowed through texts on botany, plant physiology, and agricultural science to better understand his subject; and he organized a complex network of plant collectors who brought in seeds and samples of some seventeen thousand plant specimens. He transformed the landscape of his Florida winter home, first by planting hundreds of exotic rubber-bearing trees, vines, and shrubs from the tropics, and later by establishing hundreds of experimental plots of domestic plants that he subjected to a combination of modern methods of scientific agricultural research as well as empirically based, somewhat haphazard methodologies that he had used on many of his other projects.

Edison's work with domestic rubber crops was significant in its own right. His approach to the problem differed from those of IRC, Harvey Monroe Hall, Herbert Hoover, and other rubber crop enthusiasts. From the start, Edison's research focused on a single issue—the search for a domestic source of rubber from plants that could supply the nation during a war emergency. Claiming that he had no interest whatsoever in profit, Edison's sole stated objective was war preparedness. He emphasized repeatedly his goal to find a crop that produced rubber within one year of the beginning of a war emergency. Price was not his major concern; although Americans had been outraged when rubber prices exceeded one dollar per pound due to Stevenson Plan restrictions, Edison set a target price of two dollars per pound. And in view of American labor costs, even this target could be achieved only with a fast-growing crop that could be mechanically sown and reaped.[2]

Edison's Interest in Rubber

Most Edison biographers assert that the inventor's interest in rubber can be linked directly to his visits in 1915 with the plant breeder Luther Burbank in Santa Rosa, California. As Edison did indeed meet with Burbank three times that year (once with Ford and Firestone in attendance), it seems likely that rubber was a topic of conversation. Previous histories suggest that the friends recognized that wartime would bring uncertainty to the global rubber markets; apparently Burbank convinced Edison that new natural sources for rubber could still be found.[3] Edison also was well aware of the chemical shortages that had hindered American industrialists during World War I, and he had taken remarkable steps to develop his own alternative supply. Nevertheless, little evidence survives to demonstrate connections between the

Burbank meetings and Edison's later rubber projects; elements of this story may be apocryphal.[4]

In any case, the story of Edison's study of domestic rubber crops should be placed in a fuller context. As an industrialist, Edison was well aware of the sporadic supply and inconsistent quality of this valuable commodity. Although he had abandoned rubber as an insulator for electric wiring because of its high cost and rapid deterioration, he consumed large quantities of rubber in his phonograph and storage-battery enterprises and a variety of other operations. And because of his appointment to the Naval Consulting Board, Edison had helped coordinate American scientific and technological research in preparation for the nation's potential entry into World War I.[5]

Edison knew of the potential for a domestic rubber solution. In a 1917 interview, the inventor had stated there was "no doubt" that many rubber plants could grow in the United States; the only question was the cost of labor. He recommended some milkweed varieties and highlighted guayule for the "remarkable amount of rubber it yields"; in fact, he attempted to grow some in his New Jersey greenhouses during the war.[6] In 1919, he asked Harvey Firestone Jr., then attending Princeton University, to send to the Edison laboratories at West Orange samples of all of the rubber products that the tire company used. He also approached IRC president George Carnahan for samples of the resin from guayule rubber. As a consequence of these various studies, Harvey Firestone Sr. could report after his 1919 camping trip with Edison and Henry Ford that he was "astounded" by Edison's knowledge of rubber and its properties.[7]

Edison's interest in rubber intensified just as the Stevenson Plan went into effect in late 1922. By this time, Edison's storage battery had emerged as the most successful segment of his business and manufacturing operations. Hard rubber, an inelastic product more like plastic than conventional rubber, was one important component in the battery's design—hard rubber pins were used to insulate the unit and a hard rubber base held the charged plates in place. The "Pontianak," or "pressed Jelutong" grade of rubber, the raw material for Edison's hard rubber, had rapidly increased in price from twelve to fifty cents per pound, despite its poor quality.[8] A dispute with his purveyor, the Joseph Stokes Rubber Company of Trenton, further contributed to Edison's interest in finding a new source for the rubber used in batteries. By December 1923, the inventor's son Charles, now in charge of daily operations at Thomas A. Edison Inc., signed off on a plan for the firm to manufacture its own hard rubber. The company built a new factory in Bloomfield, New Jersey, and production began by spring 1925. In brief, the episode provides another explanation for Thomas Edison's willingness to make rubber his last project.[9]

Edison's Early Fort Myers Experiments

As described in the previous chapter, Edison, Ford, and Firestone had begun to collaborate on rubber crop investigations soon after the Stevenson Plan went into effect. Edison tested milkweed and guayule in West Orange beginning in 1923, Firestone planted rubber trees and shrubs at his Miami Beach winter home, Ford funded extensive experiments with cryptostegia in Hendry County, and all three collaborated with Alfred Keys of the USDA's new plant acclimatization facility at Chapman Field near Miami. Perhaps inevitably, these efforts began to diverge. With their huge corporations dependent on more immediate results, Firestone and Ford now understood that domestic rubber would not meet their immediate needs and they began to look beyond Florida for a solution. Firestone's company launched expeditions to survey *Hevea* rubber prospects in the Philippines and Mexico before settling on Liberia as its target. The departments of Commerce and State and President Coolidge offered enough tacit support for Firestone's endeavors to secure a positive response from Liberia. In the end, the Liberian government granted Firestone the right to lease one million acres of Liberian land for ninety-nine years, in plantation-sized blocks of twenty thousand acres. The Liberian government further guaranteed to issue no export duties above 1 percent of current rubber prices in New York, while promising not to impose import duties on agricultural implements, chemicals, and other plantation supplies. For all practical purposes, the Firestone Tire and Rubber Company controlled Liberia's economy from the mid-1920s through the mid-1950s.[10]

Similarly, Ford negotiated with the government of Brazil to make another attempt to reestablish *Hevea* as a New World commodity. He obtained a tract of 2.5 million acres, almost four times the size of Rhode Island, along the Tapajós River, where he established a vast agro-industrial complex called Fordlandia intended to bring plantation rubber production back to the Americas. By 1934, about 1.5 million rubber trees had been planted. Within a year, however, erosion, labor shortages, illness, and especially the rubber-blight fungus forced Fordlandia to suspend operations. Undaunted, Ford tried again with a new plantation called Belterra, eighty miles downstream, an even larger operation that soon had five million seedlings in the ground. But results there were similar. Until the blight could be managed, and until rubber planters could entice thousands of laborers to move into the underpopulated jungle environments, it seemed that Brazil could not offer a serious alternative to Asian rubber. The two projects proved to be a huge drain on Ford's resources, costing about twenty million dollars, and provided virtually no commercial payoff.[11]

Edison, meanwhile, remained focused on Florida in search of a rubber crop that would serve the nation during a war emergency. Edison first visited the state in 1879; six years later he made a trip to Fort Myers, then a town of about fifty homeowners outnumbered by the cowboys who drove cattle onto ships docked in the Caloosahatchee River. Edison built a home in there in 1886, and he eventually gave the town its slogan: "There's Only One Fort Myers and Ninety Million Will Find It Out." In 1916, his friend Henry Ford built a winter home there as well, only two hundred feet from Edison's. By the time of his rubber research, Edison's property in Fort Myers included a barn, a well-stocked chemical laboratory, a slat house (a small plant nursery with widely spaced boards in the roof, which allowed a mix of sun and shade to bathe seedlings), and several acres suitable for agricultural experiments.[12]

Edison intensified his work with rubber plants in the summer of 1926. That spring and summer, his crew received and planted many rubber species, including *Hevea*, *Funtumia elastica*, and others. Edison's men also expanded their work with *Cryptostegia madagascariensis* in 1926, enough to turn a "corner of the estate into a plantation."[13] Securing more of these seeds from Madagascar was an especially trying ordeal that demonstrated Edison's commitment to the project. Beginning in April 1926, Edison secured the help of Ernest Liebold, Ford's right-hand man; P. W. Grandjeau, secretary of the Ford Motor Company of Canada; and Edwin Mayer, Ford's agent in Tananarive (now Antananarivo), Madagascar, in this effort. Grandjeau reported that he had put pressure on his contacts in Madagascar and "impressed upon them the urgency and importance of this matter," but securing seeds from plants that grew on the far side of the island proved difficult. Mayer responded in coded telegrams explaining his difficulties in finding ripe seed and gaining cooperation of colonial bureaucrats and shipping firms. The crucial seeds finally arrived in Florida in April or May 1927.[14]

Edison's Objectives

Edison emphasized repeatedly that he had no interest in commercial profits; his goal was to find a crop that produced rubber quickly. For this reason, and much to the chagrin of its supporters, Edison often stated in the popular press that guayule could never meet his objective of creating a wartime supply because it required four to five years between planting and harvesting. As his plant selection studies progressed, Edison selected for plants that fit his ever more refined criteria. Although he continued to experiment with vacuum extractors on rubber trees and attempted to breed rubber vines to grow as shrubs, by late 1927 Edison had become convinced that the ideal crop was a perennial herb or weed that could be sown and harvested like a grain. He favored plants that had their rubber in leaves, rather than stems or roots, because these would be easier to harvest and process. Consequently, he preferred plants with numerous and broad leaves. He also sought a plant that could tolerate a light frost, for he hoped to expand production beyond southernmost Florida. Edison's notebooks of early 1929 reveal his plain-language conclusions: "The only possible solution of war rubber is high rubber herb plants produced and extracted by solvents bred to give 80 to 100 lbs rubber per dry ton of leaves. Reaped like wheat baled & shipped to factories in each area. The rubber of such quality as to give inner tubes of fair quality."[15]

Edison continually considered the financial implications of expanding his tests to factory scale production. For instance, he routinely calculated potential output in terms of percentage of rubber and in terms of potential pounds of rubber production per acre. Edison worked out estimates for the total cost of an operational American rubber factory that used a hypothetical ten-thousand-acre field sown with goldenrod, which eventually emerged as his favored crop. He began with the supposition that (if started from "pups" rather than seed), ten thousand acres could produce 3,330 tons of leaves. Because the crop should be harvested within twenty-four days, and a team of two could mow thirty acres per day, he determined that twenty-eight men were needed to work the harvest, each drawing a wage of five dollars per day. Additional men would be needed to hang, de-leaf, bag, and haul the harvest. One ton of dry leaves would produce ninety pounds of rubber; thus the harvest of ten thousand acres would yield 111 tons of rubber.

In separate calculations Edison estimated the costs of production. Assuming that a factory could process twelve tons of plants with 5 percent rubber to produce 1,200 pounds per day, Edison reckoned that the facility would require twenty-four employees earning six dollars per ten-hour day. Additional considerations included expenses for land rent, transportation, and chemical solvents. Elsewhere, Edison calculated the costs, availability, and yield of solvents. Recalling that acetone would be more difficult to obtain during wartime, Edison thus developed extraction methods using benzol and carbon tetrachloride that permitted efficient recovery of the solvents. As a result of these calculations, Edison typically concluded that he could not produce domestic rubber for much less than two dollars per pound, far more than even the highest prices during the height of the Stevenson Plan crisis. In all, these data show that he understood that the rubber crops he favored were intended as an emergency, not as a commercial, rubber product.[16]

Finally, this project also reflected the inventor's interest in farm issues and the social implications of domestic rubber crops. Edison claimed that, at the instigation of Ford, he had hoped to "invent" a scheme to help America's yeoman farmers. Like others interested in reforming southern agriculture, Edison endorsed a plan that involved credits for crops stored in government warehouses and urged farmers to find new crops as alternatives to cotton. "Our object," Edison stated, "is to use some of the most southerly cotton lands to provide rubber so as to prevent overproduction of cotton and to give two different crops to the farmer." Annual rubber crops, he reported, offered a much better chance for the cash-poor southern farmer to get started in the industry than plantation-style rubber trees. Moreover, he stated that local rubber-extraction centers could follow the model of the local cotton gin.[17] Rather remarkably, the inventor's son Charles added that draining the Everglades would not be a difficult engineering task and the region could be converted into a rubber region that rivaled the Amazon valley.[18] In all, Edison's vision of a decentralized emergency rubber economy stood in sharp contrast to the IRC's fierce efforts to centralize and maintain its control of guayule seed and know-how.

Edison's Publicity

As promised in the reports that he leaked to the press, Edison turned his attention wholeheartedly to the rubber project in the spring of 1927. Soon after his eightieth birthday, Edison launched into a flurry of activity that made rubber a passion for the last four years of his life. In typical Edisonian fashion, reminiscent of his self-promotional work with his electric lighting and phonograph inventions, his first step was a widespread publicity campaign. Originating in the Fort Myers press, the story traveled through the Associated Press wire to newspapers across the country. "Edison, Backed by Ford, in South Inventing Machinery to Revolutionize Rubber Industry," claimed the New York Times page 1 headline in its 27 February 1927 edition. The article touted the inventor's latest project as one that "he believes will revolutionize the world's rubber trade." Demand for further information caused the Associated Press to release a follow-up story the next day, and similar articles appeared in many American newspapers well into March 1927.[19]

The stories were quite hyperbolic. They touted Edison's promises of machinery that could reap rubber crops, his confidence that cryptostegia could thrive throughout the South, and his hints that the experiments might solve the problem of American rubber imports. The Fort Myers News-Press published an editorial indicating, "We can look forward to the time when Florida south of the Caloosahatchee can meet the nation's demand for rubber." Reporters descended on Fort Myers seeking further information, but as with his earlier inventions, Edison's next step was to restrict his contact with the press and begin the actual work of innovation.[20]

Reporters were fascinated with Edison's purported work ethic in the search for rubber. "Edison is working way past midnights in his laboratory," the *Times* reported. Another story presented a slightly more plausible account of Edison's day. According to his wife, Mina, the inventor awoke at 6 A.M. and quickly went across the road to view his plantings. He spent most of the midday hours in the laboratory but typically went on short expeditions into the country looking for more specimens. He would not return for dinner until 7 P.M. Edison devoted his evenings to reading about rubber and keeping up with his correspondence.[21] Soon it became obvious that Edison's obsession with rubber plants meant that he was ready to retire from the rest of his industrial empire. In a June 1927 memorandum, son Charles addressed his famous parent: "Father: Dept. of Rubberology, Edison Laboratories. I have come to the conclusion that you really do want to concentrate on rubber and not bother much with the details of the business."[22] Later that year, Charles took over as chair of Thomas A. Edison Inc.

That spring, Edison spent his time in Fort Myers on several major projects. Edison continued to test tropical plants on his property for rubber, and he conducted a Florida tour in search of additional rubber crops. With handfuls of what he claimed were self-produced balls of rubber, Edison would ask complete strangers, "Any rubber trees around here?" as part of his travels in central Florida. He also visited the Naples, Florida, botanist Henry Nehrling, who allowed Edison to study his books, photographs, and manuscripts as part of a crash course on the botany of rubber-bearing plants.[23] Upon his return to New Jersey in May 1927, Edison's enthusiasm for the rubber project continued on multiple fronts. Within days of his arrival, he sent a cryptic telegram to his friend: "Dear Firestone: It would pay you to come over and have a talk on rubber when you come to New York. I am loaded."[24] Meanwhile, Edison studied additional literature on rubber issues, took his own automobile trips in search of New Jersey specimens, and through growing contact with the botanical community arranged for nurseries, plant collectors, and government officials to supply him with additional specimens. Edison's increasingly close relationship with colleagues at the New York Botanic Garden (NYBG) in the Bronx stood at the center of his 1927 initiatives. Such collaboration was essential for the eighty-year-old inventor, just beginning his first serious work in scientific botany, and the NYBG responded with lessons on nomenclature, collection, labeling, mounting, and preparation of botanic specimens.[25] In particular, Edison gained help from John K. Small, the NYBG's head curator and an expert on Florida flora, who personally added the search for rubber plants to the agenda of his multiple field expeditions.[26]

Edison and his associates also made much use of the NYBG library. Edison's notebooks from these library trips provide insights into the inventor's research methodology; his notes were thorough, systematic, and focused completely on the goal of finding a crop suitable for an emergency rubber supply. Beginning with works published at the turn of the century, Edison plowed through writings on the botany of rubber from England's Kew Gardens, the Ceylon Agricultural Institute, the USDA, and other sources. With his help of his assistant Barukh Jonas, who claimed fluency in seven languages, Edison quickly had notes on much of the published literature concerning rubber technology and chemistry, as well as the research on the enzymes, proteids, and other contents of plant protoplasm that could affect rubber yields. Determined to make a contribution to rubber chemistry as well as botany, Edison spent much of the next two years searching for efficient and reliable combinations of solvents suitable for extracting rubber from his new rubber crops. At the end of a quick summer of research, Edison was ready to make a characteristically blunt conclusion: "The general impression I get that nearly the whole of the expmnts [sic] on Latex & rubber solvents are worse than worthless & that every theory is disputed—The chemists who have worked on it . . . were . . . not empirical chemists & not competent to tackle the job."[27]

Edison's visits to the NYBG also served his agenda of seizing the rubber issue for its public relations value. The media followed Edison's repeated visits to the NYBG, allowing the inventor to solidify his image as an "enthusiastic . . . adventurous . . . [and] engrossed" researcher who had no financial interest in the project. Although he claimed to refuse to speak to the press about his rubber project, Edison nevertheless continually leaked word through his secretary William Meadowcroft that he had already identified a considerable number of plants reputed to be capable of producing rubber.[28]

Edison employed an impressive network of amateurs and professionals, colleagues, admirers, and employees in his search for potential rubber-bearing plants. Like his decades-earlier quest for possible filaments for the incandescent light, Edison sought help from virtually anyone who could support his broadly empirical approach to the problem. For instance, Edison also wrote to several American railway companies asking that their agents be on the lookout for potential rubber-bearing plants from the arid regions of the American West. The Union Pacific Railway responded positively, and by August 1927 its agents from Nebraska to California had begun to send samples to Edison's headquarters.[29] As an iconic American hero, Edison also could count on support from admiring citizens. Hundreds of Americans sent letters, many with botanical samples enclosed, suggesting milkweeds, poinsettias, Osage orange, sumac, wild lettuce, and dozens of other relatively common American plants. Edison's staff politely answered virtually all of these letters, tactfully indicating that, in almost all cases, Edison had already considered the suggested plant.[30]

Some of this correspondence, however, sparked Edison's interest. For instance, he asked for further information on the work of A. W. Morrill, an agricultural adviser who claimed that he had established a significant operation growing cryptostegia on the western coast of Mexico. It "would be possible for a rubber company to contract for an almost unlimited amount" of acreage in western Mexico, Morrill reported.[31] Similarly, Edison also had his assistant William Meadowcroft follow up on a letter he received from E. L. Dunbar, who claimed he had obtained some commercial success before World War I with pinguay, the rubber plant found in the arid mesas of western Colorado and northern New Mexico.[32] In another episode, W. Sam Clark of the Clark-dota Fig Company approached the inventor following a July 1927 luncheon with Charles Edison in Los Angeles. This led to Clark's proposal to develop fig (*Ficus*) trees that could produce both rubber and commercially valuable figs. Clark insisted that the potential was enormous, for the yields of twelve to sixteen tons of raw material per acre would eventually "be harmful perhaps to the big interests who own great tracts in South America, but [would] be of tremendous interest to those who purchase rubber and manufacture rubber products." Edison remained skeptical, but he consented to try to work with Clark's fig variety on the Fort Myers property.[33] In these and similar cases, however, it is significant that Edison rejected his suitors' offers to pursue rubber crop research as a profit-making scheme and concentrated instead on his stated goal of developing a rubber crop suited for a war emergency.

Edison's Rubber Research Network, 1927

On a more formal level, Edison hired a team of fourteen field men and plant explorers to find potential rubber-bearing plants.[34] He assigned each to a specific territory and outfitted each with a Ford automobile, camping equipment, and a salary of $100 per month (the equivalent of about $1,100 today) to scour the United States for potential rubber plants.[35] Edison also hoped to collect plants and seeds from Mexico, but significantly, in view of the renewed political tensions of that summer, he did not think he could "dare send a man there."[36] Each field man received Edison's lists of the likely botanical families for which to look. But the inventor also endorsed a very Edisonian and inductive approach to the problem. "Our collectors cut every plant in sight," Edison once wrote, and he had little interest in having his field men determine specimens' official botanical names. These men did not stick to the main roads, and Edison expected them to go deep into the woods and be thorough and persistent in their selection process. Edison later insisted that his plant collectors avoid large towns for their overnight stops, because the hotel expenses were higher and traveling through city streets cost time that could be better spent in the field. To facilitate the collectors' work, he asked railroad companies to provide free transportation between stations, on the grounds that the work was "entirely for the good of the country in general."[37] Speed was the highest priority. Aware of the inherently slow process of agricultural research, he remarked, "It's important I do not lose a year."[38]

As he began to look beyond cryptostegia, Edison's main interests at this time were members of the *Euphorbiaceae* family (includes common spurge), as well as a few from the *Asclepiadaceae* (includes milkweed) and *Apocynaceae* (includes dogbane) families.[39] Edison's aides instructed the field men to carefully identify the date and exact location of their collection and indicate the soil type, root type, and size of the plant. They were then instructed to carefully label, divide, pack, and ship plant materials and seeds by express rail to three separate destinations: West Orange, Fort Myers, and one of Florida's leading commercial nurseries. Edison inspected most of the dozens of plant specimens that arrived daily that summer in West Orange. From there, he sent many of the specimens for transplanting at his Glenmont, New Jersey, estate, and he sent others to the NYBG for professional identification.[40]

By September, as the season for explorations and seed collection in the temperate climates passed, Edison made arrangements with additional plant explorers to cover Texas, Puerto Rico, and Cuba. Scientists and other allies in Italy, Uganda, New Guinea, Morocco, and elsewhere also contributed to the search. Edison's interests quickly became more focused and his instructions and criteria more precise. As political tensions in Mexico ebbed, Edison hired J. N. Rose, associate curator of the Division of Plants at the Smithsonian Institution, to scour southern Texas and to venture into northern Mexico.[41]

There were difficulties inherent in Edison's approach. Efforts to collect plants in South Africa were slowed by the remote terrain, a shortage of skilled plant collectors, and difficulty in transporting the specimens to the United States without spoilage.[42] The American field men also encountered a variety of obstacles that limited their effectiveness. For instance, Myron Shear's work was slowed by difficulties in obtaining an automobile license and the shortage of suitable packing materials in rural Virginia. Howard Barton had to use burros to negotiate steep mountainous terrain in New Mexico, and he was surprised to learn that Edison really did expect him to travel hundreds of miles off the main roads to obtain plants from a distant Navajo reservation.

J. N. Rose found that the ecology in central Texas had already been altered by human encroachment (which made finding new species unlikely) and that renewed political unrest limited his chance to "botanize" in Mexico. Further, Rose made the mistake of becoming rather excited about the rubber potential of poinsettias and guayule, two crops that Edison had already specifically ruled out. In the end, Rose admitted that his "results were chiefly negative."[43]

As Edison's research gained national and international attention, some observers realized its potential to impose on other aspects of government policy and private enterprise. After all, government planners had devoted considerable efforts to investigate the rubber-producing potential of the Philippines, Haiti, and other territories, and had also invested in Chapman Field and other facilities designed to test rubber plants on American soil. Edison's research also challenged leaders of the American rubber industry. Except for Firestone, rubber-industry leaders offered Edison little help and at times issued statements to the press that discounted the entire project. As far as government policy makers were concerned, Edison's desire for the *Euphorbia* seeds from South Africa brought matters to a head. On 26 July 1927 (just three days before the official establishment of the Edison Botanic Research Corporation [EBRC]), Edison traveled to Washington, DC, to meet with officials of the Department of Agriculture, the Department of Commerce, the Army-Navy Munitions Board, and the Smithsonian Institution. It was no coincidence that he received a letter the day before from Dr. Charles Marlatt, chair of the USDA Federal Horticultural Board, which indicated that Edison would not receive special treatment in the government's restrictive plant-quarantine policy. Indeed, Marlatt was well-known for his profound fear of foreign "plant enemies" contaminating American crops, as well as for his lasting and personal rivalries with the plant explorers who saw value in foreign plant life.[44]

During these meetings, it seems evident that Edison promised, or perhaps was requested to promise, that his research objectives would not interfere with the larger diplomatic and trade issues that concerned government officials. On 29 July, Everett G. Holt, chief of the Rubber Division of the Department of Commerce, reiterated his appreciation that Edison's "real purpose" was only research on rubber plants as an emergency and wartime resource. "Press statements which had placed emphasis on the likelihood of our defeating the 'British monopoly' through locally grown rubber had given me and several trade interests a wrong impression," Holt cautiously explained.[45] Yet the Washington, DC, meetings also yielded a victory for Edison. Marlatt's ruling had been overturned, and Edison left the capital laden with seeds, samples, and promises of cooperation from the government agencies.[46] On 29 July, Holt issued a "strictly confidential" memorandum to American representatives in

Johannesburg detailing the kinds of plants that Edison desired and explaining the procedures that had been permitted so that Edison's material could come through "without the delay necessary under plant quarantine regulations."[47] Clearly, Edison's political clout had helped garner support for his search for foreign plant specimens.

The flood of incoming specimens to be cataloged and tested overwhelmed Edison's staff. Assistant William Meadowcroft reported Edison could not keep up with his routine work because "this rubber business seems to reach out to infinity."[48] By 1931, many thousand samples of at least 2,222 different species had arrived in West Orange. In a telling illustration of the tediousness of this work, the staff produced a small rubber stamp stenciled with the words "NO RUBBER" to simplify their labor and more quickly dismiss the failed varieties.[49]

Edison's substantial plantings of potential rubber crops in both Fort Myers and West Orange suggest that he conceived of his work as two separate problems. In New Jersey, he primarily planted native weeds and wildflowers with an unknown potential for producing rubber, his simple goal to find promising varieties for further selection in order to increase their rubber content. The New Jersey gardeners also handled warm-weather crops, like guayule, which they planted in the greenhouse. The Florida staff, on the other hand, mainly planted imported rubber plants, many with known commercial value elsewhere. Edison hoped to acclimatize these species to determine if they could survive and thrive in American soil; he also sought to graft promising varieties onto other specimens with a consistent stature and growth patterns so they could be sown and harvested more easily. At this point, Edison had not yet determined which approach was more promising.

After Edison left in May 1927 for New Jersey, estate manager Frank Stout supervised operations in Fort Myers. Stout handled the steady stream of seeds and seedlings by agents of the Firestone Company, Ford dealers in Madagascar and elsewhere, USDA scientists, and other suppliers from around the world.[50] In April and May 1927, his crew planted 350 specimens of *Cryptostegia madagascariensis*, 100 of the "common" rubber tree (*Ficus elastica*), 150 poinsettias, 25 guayule plants, 6 of *Manihot glaziovii*, and many other potential rubber trees, vines, and shrubs.[51] Much of the laboratory work in 1927 involved initial testing for rubber content in *Cryptostegia madagascariensis* and *grandiflora*. By the end of that year, Edison concluded that the former variety produced a rubber of inadequate quality, and the only redeeming feature of the latter was its exceptionally rapid growth. In general, problems with tropical plants were beginning to push Edison toward a new line of research.[52]

Meanwhile, Edison planted over five hundred plant varieties at his Glenmont estate in New Jersey. Despite hundreds and hundreds of plants that necessitated use of the "NO RUBBER" stamp, Edison's results indicated that some varieties tested at his own home potentially offered more rubber than those obtained through his global search for new plants. In one letter, he counted sixty-two possibilities from the Glenmont trials; in another, he claimed that about 20 percent of his Glenmont trials yielded at least some rubber.[53] "Everything looks very favorable," he wrote to Ford's assistant, Liebold, "to a solution of ample supply of war rubber."[54] While in New Jersey, Edison continued to energetically supervise events in Florida, sending letters almost daily to Frank Stout with specific instructions on the research. Likewise, Edison also actively monitored the activities of the fourteen field men and the flood of plant specimens that they collected; in addition, he sent a flurry of letters to botanists, nurserymen, and political leaders about his project. With a dynamic octogenarian at its center, the American search for a domestic rubber crop had reached a new level of intensity.[55]

Creation of the EBRC

By 1927 it had become clear that the federal government and its agricultural research establishment was not going to make rubber research a high priority. Edison's friends took matters into their own hands in July 1927, when Firestone and Ford agreed to form a new entity to be known either as the Edison Rubber Research Corporation or the Edison Rubber Development Corporation (Firestone said he did not care which, so long as Edison's name remained). Firestone also offered to join Ford and Edison on a camping expedition that focused on the search for rubber plants. After returning from his research at the NYBG, however, Edison declined: "I am sorry to say I cannot go joy riding this year as I am too busily engaged in the rubber investigation."[56]

In lieu of their camping trip, the three famous Americans formally established the Edison Botanic Research Corporation. On 22 July, Liebold sent a check for twenty-five thousand dollars that represented Ford's contribution to the project for 1927. On 29 July, Firestone wrote to Edison with the basic terms of the EBRC: Ford and Firestone pledged to contribute twenty-five thousand dollars annually, while they expected Edison's contribution to be not cash, but in the form of "services he has rendered and will continue to do in the future." There is clear evidence, however, that Edison ignored this latter clause. In a 1928 letter, for instance, Firestone expressed with exasperation that "Mr. Edison has been insisting that he put in the same amount of money in the organization that Mr. Ford and I put in." Edison's brother-in-law John V. Miller oversaw virtually all aspects of management of the EBRC.[57]

Edison and Guayule

Having examined hundreds and then thousands of potential rubber plants, many questioned why Edison did not, in the end, endorse guayule. Edison first contacted George Carnahan of the IRC for guayule rubber samples in 1918, and in both 1917 and 1923 Edison expressed genuine enthusiasm for the shrub's potential. Subsequently, Carnahan regularly kept Edison informed of progress with the California guayule experiments and supplied the inventor with seedlings, fresh rubber samples from the IRC's Mexican factory, and advice on issues of rubber extraction. In turn, Edison sent Carnahan regular updates on the success of the guayule plantings in Florida and progress reports on Edison's other favored crops.[58] Further, the IRC hoped to exploit for its own commercial purposes Edison's well-publicized search for rubber plants. Carnahan visited Edison in Fort Myers in March 1927, soon after Edison's project appeared in the national media spotlight, and he was thrilled that Edison had scratched out this endorsement on a scrap of paper: "The only way is to have private individuals or companies to work up the project with a view of profit that it can succeed."[59]

Yet the relationship between the two deteriorated, marking a fundamental difference about whether perennial plants like goldenrod or shrubs like guayule would be the most suitable source during a rubber emergency. During the summer of 1927, Edison began to turn against the western and desert plants that had dominated previous searches. Edison's tests on one of Harvey Monroe Hall's favored plants, desert milkweed, yielded only one-fifth as much rubber as Hall had reported.[60] "I am now convinced that tropical plants are useless," Edison told J. N. Rose, his plant explorer in Texas in November 1927. "We are finding many strange phenomena."[61] More significantly, Edison came to believe that America's guayule industry could not fill the niche he had in mind. Edison determined that guayule was "not satisfactory for war rubber" due to its high resin content, problems in vulcanization, and the difficulty of planting and harvesting the vast number of plants that would be necessary to meet a wartime emergency. Above all, guayule was hampered by the four-year time period necessary before the shrub reached rubber-bearing maturity. Earlier in September, Edison had met with Carnahan and David Spence of the IRC in his West Orange laboratory, where he "slowly and carefully" explained his decision to regard guayule as unsuitable for war-emergency purposes.[62] For his part, Carnahan came away from this meeting critical of Edison's work. In a note to his ally Bernard Baruch, Carnahan reported that Edison's technique "was far behind that which our own people in California and in a general way the same applies to his methods of rubber analysis which he is applying to plants collected from all over the world."[63]

In subsequent evaluations, Edison continued explicitly to reject guayule. In his private notebooks of February 1929, Edison sketched out a list of handicaps: "Difficult to raise . . . confined to a small area of US . . . the long growing period ties up money & risk is great . . . uncertain if the rubber can be vulcanized to give a reliable inner tube."[64] In January 1930, just before a meeting with President Hoover and other business executives in another effort to secure government support for guayule, Carnahan had asked Edison for a few positive statements about the shrub's potential. Again, Edison's response was negative. Carnahan responded with a five-page rebuttal to Edison's skepticism. Above all, Carnahan stressed that there was no need to wait for war to plant the preparatory crop. For about thirty million dollars, the cost of one battleship, the nation could plant a rubber reserve of 150,000 acres of guayule, a "living warehouse" of 450 million pounds of rubber—about one-half of the nation's annual consumption—ready to be harvested when necessary. Carnahan added that the reserve should be "planted *now* and retained as insurance against a wartime or other emergency such as political circumstances in a field that is ripening for communism." Carnahan also challenged the economics of Edison's scheme, because goldenrod would require 4.5 million acres and as much as nine hundred million dollars to produce the same amount of rubber as the guayule proposal. He also pointed out that Edison's goldenrod plants would pull cotton and wheat lands out of production, lands that would be even more valuable during wartime. Further, Carnahan charged that Edison had not considered the difficulties of storing the seed, keeping it viable, and efficiently distributing it in time for planting in the case of a war emergency. In response, Edison suggested that Carnahan had not learned an important lesson of World War I: that the government was unlikely to spend any significant funds on war-preparedness stockpiles. "I think we will have to solve this problem without their aid," Edison concluded.[65]

Edison also rejected the guayule option because his experiments with guayule rubber tubes and tires in Fort Myers proved the material unsatisfactory. In October 1927, Edison wrote Firestone with a simple query: in the case of war and a rubber embargo, could guayule supply Americans with shoes and inner tubes that offered a "fair mileage"? Within a week, Firestone offered to find out by obtaining guayule from the IRC and producing a few experimental guayule tires and tubes for Edison's Ford Model A truck. By the end of November, Firestone had produced the first tire made entirely of American-grown rubber.[66] Although Edison and Firestone had expressed confidence in these tires, further scrutiny revealed problems. In the words of employee W. A. "Bill" Benney, the guayule tires "all [went] to pieces" after about 4,200 miles. One set of guayule inner tubes went flat before leaving the Fort Myers garage.[67]

In general, guayule rubber tires suffered from impurities of dirt and bark that entered into the final product, causing pinhole leaks and unsatisfactory wear. The issue contributed to the growing tensions between Edison and Carnahan. Carnahan tried to suggest that Firestone's technicians "did not make serious effort," had no desire to see the trial succeed, and took little care in the production process. Edison fired back that his ally had no such motive.[68]

Fort Myers Research at Its Peak

Rubber research activities in Fort Myers reached their peak in the winter of 1927 and 1928. Preparations for this new initiative were complex. Edison had named Bill Benney, a West Orange colleague, to take over as superintendent of the experimental work in Fort Myers. Benney arrived in late November, five weeks before Edison and his entourage, to prepare the grounds and the old botanic laboratory for their new research agenda.[69] In the words of one biographer, Edison's return to Fort Myers in January 1928 was "exhilarating—like Menlo Park all over again."[70] At least six Edison employees from West Orange traveled to Fort Myers for the winter, accompanied by more than one hundred boxes of equipment needed for the research: grinding mills, coffee mills, bread slicers, porcelain balls, pie pans, flasks, solvents, screens, stoppers, plant specimens, books, journal articles, and more. After Edison arrived, he telegraphed New Jersey employees more than a dozen times with instructions to ship ever more materials, mainly chemical laboratory supplies that had been left behind. Edison believed that speed and efficiency were essential if the crew hoped to isolate a useful rubber crop. "We are running 16 extractions per day," Benney reported in January 1928, adding that he had to hire yet another man because "Mr. E. told me to step up to 32 per day next week." Mina Edison reported that her husband was "so absorbed in it [rubber] that I cannot get him to do anything right here in town." "He is just as ambitious as ever," she continued. "He is making progress and that makes him all the keener to keep at it."[71]

Edison also continued to search for new plant specimens. Throughout that winter and spring, John K. Small of the NYBG led crews that scoured southern Florida for potential species. These were not simply strolls through the woods. Based at fourteen different campsites scattered across the fringes of the uncharted Everglades, plant hunters armed with machetes and snake antivenin collected nearly two thousand plant varieties that season.[72] The inventor himself collected plant specimens from his own Fort Myers property and dozens more during automobile trips nearby. Often traveling with his wife, he also explored Lake Okeechobee, Naples, Bonita Springs, Sanibel Island, Captiva Island, Pine Island, and Key Largo.[73]

Edison continued to home in on his favored crops. By April 1928, Edison had concluded that the more than one hundred *Ficus* trees planted at Fort Myers would not be part of the solution "on account of high wages." "Rubber must be obtained from shrubs, perennial, that can be cut by machinery and the rubber extracted by factories."[74] As the inventor turned to other possibilities, he used a characteristically grand approach. In July 1928, Edison's staff sent queries to dozens of chambers of commerce in Alabama, Georgia, South Carolina, North Carolina, and Virginia, asking for information about the northernmost range of four more potential rubber crops: Natal plum (*Carissa grandiflora*), allamanda (*Allamanda cathartica*), flame vine (*Bignonia venusta*), and crepe jasmine (*Tabernaemontana coronaria*).[75]

Meanwhile, two other plants emerged as Edison's leading candidates for the domestic source of rubber. In August, Edison and Miller urged plant collectors to be on the lookout for the "best plant found so far," a blackroot, or *Pterocaulon*, that tested at 5.6 percent rubber.[76] Yet Edison soon made a plant known as "Hullmann Ga. 252" an even higher priority, for it seemed the "best yielder yet."[77] Edison called on various experts to help identify this curious plant; eventually they determined that it was a mugwort or wormwood, perhaps *Artemisia lindheimeriana*. In September, Benney offered farmers in rural central Florida one dollar if they could lead him to the plant, but without success. In October and November, Benney spent days in the Macon, Georgia, area, with imprecise directions, trying to find the same patch of weed that Hullmann had located months earlier. Meanwhile, Edison sent plant explorer William Hand on a search for Hullmann Ga. 252 and other varieties of *Artemisia* that began in the Southeast and continued into Texas and New Mexico. Hand's fieldwork in New Mexico included adventures in snowstorms, crashes with runaway horses, cars stuck in riverbeds, and visits to remote villages where the residents had never heard the name "Edison." By December, they had collected enough to transplant four hundred specimens in Fort Myers. Because the crop was not growing well, Benney decided that he needed to import some Georgia clay as a soil supplement or expand rubber crop research onto Henry Ford's vast landholdings near Savannah.[78]

Edison's work in 1928 also hints at his evolving research design. Edison explained his breeding strategy to Benney, instructing him to take selections from the "high yielders" in each bed of specimens. Edison continued: "These we can bud or graft on the most vigorous plants which give very little rubber. The buds will breed true and always give the same yield as the original plant from which the seeds were taken. . . . By grafting or budding from a good plant on to the most vigorous of the poor plants, *all* will be high yielders."[79] In sum, Edison was following the propagation strategies endorsed by Luther Burbank.

Indeed, Edison made sure that his volumes of Burbank's books followed him whenever he traveled to or from Fort Myers and West Orange, and he once told reporters that "maybe I will have to do little Burbanking with my plants." Edison did not engage in true crossbreeding or hybridization work.[80]

Edison's research design also revealed a deep interest in adapting plants for mechanical planting and harvesting. Edison studied methods of planting and spacing seedlings to learn how "to plant these pieces of root *by machine* in a furrow & at the same time cover them by earth."[81] Beginning in April 1928, Edison sketched out six "schemes" to develop a mechanical technique to harvest flame vine, a plant that he hoped to harvest two times per year if he could find a way to train the plant to climb rather than creep. In one case, the idea was to train the vine to climb a system of five wires strung between fence posts and to design a machine that would cut simultaneously between the three upper wires while allowing the remaining vines to grow anew from their stems attached to the bottom two wires. In another draft, Edison wrote that he might try to build a machine that could lower the vine to allow for cutting and then re-raise it following harvest in order to encourage rapid growth.[82]

Edison also sketched out similar schemes involving other plant species. In one note he reminded himself to "start cutting tops off [m]adagascar[i]ensis & testing leaves of each of best plants, also stems, & propagate the cutting = it may be we can find some with high yield & only semi climber."[83] In another notebook entry, Edison remarked, "I have ordered 20 of the vines pruned away so they will train the best to grow up as a tree so it can be tapped. Its only advantage is its enormously rapid growth."[84] In a notation made in February 1929, Edison suggested the following "scheme & make it go"—to grow the flame vine, a creeping plant in combination with a goldenrod plant, using "a good Solidago . . . stem as support . . . and they could be cut together."[85] These various experiments suggest that Edison considered a wide range of possibilities in his search for a domestic rubber crop. Although they fell short of lasting success, his sustained efforts offer strong evidence that the Edisonian inventive process thrived in Fort Myers during the last years of the inventor's life.

During the 1928 crop year, both Benney and Barukh Jonas remained in Fort Myers to direct the increasingly systematic rubber research conducted there. Because Benney's duties also included management of the estate's numerous building projects and the search for specimens, Jonas began the year in charge of the experimental plantings. Although Jonas boasted that he had read nearly "every book on agricultural science" available in his hometown library, it is not clear why Edison entrusted him with this important phase of his experimental agriculture. Indeed, Jonas's real interests lay in the realm of moral and political philosophy and the perceived decline of the intellectual climate of his day. In 1927—while employed by Edison—he released an issue of his self-published, handwritten magazine, *The Intellect*, that satirized the inventor's passion for solving the rubber crisis: he claimed that Edison had gained twenty-five pounds in one week on a diet of *Para*-rubber steak, guayule biscuits, and glasses of *Hevea* latex.[86] In spring 1928, he sent to Henry Ford, of all people, an issue of *The Intellect* that lambasted automobiles for the "trail of poison" and other social problems that they caused.[87] Jonas also promoted a plan to turn Lee City, Florida, into "a fool-less paradise, a small city consisting exclusively of persons with culture and brains . . . an intellectualistic community." Further, Jonas called for a new science of "hedonology" to study what actually makes people happy, but he conceded that the search could not begin until the price of gasoline fell to twenty-three cents per gallon.[88] In any case, he must have been one of the most peculiar citizens of the area.

Meanwhile, Jonas's work with Edison's rubber plants continued. By April 1928, he reported, "Generally speaking a little order is beginning to emerge from the chaos."[89] Jonas brought in new procedures for starting seeds in the slat house, initiated programs of mulching and caring for soil bacteria, and built ditches both to keep out ants and to improve irrigation.[90] Plantings that summer included a large number of new trials with both northern and tropical specimens. Just weeks after Edison departed for New Jersey, Jonas reported, "The work in the garden is advancing remarkably well. You will simply be unable to recognize the place when you see it."[91] Among the new crops he tested that summer, red, white, and especially black mangrove were priorities: "We are going to have quite a plantation of them." Once again, Edison's interests were shifting to several new favorite species.[92]

By the end of 1928, Jonas faded from the scene and Benney emerged as the real leader of Fort Myers research. He had a close friendship with EBRC manager J. V. Miller, and his work was largely free of the problems that typified Jonas's endeavors. Benney complained to Edison that Jonas's methods were so shoddy that "by next year you will not know much more about the variation of plants than at present." Benney also visited Henry Ford's new plantations in Georgia, which he described as "twenty miles from no where . . . certainly an ideal spot for carrying on any experiments which you wanted kept from the outside world." More importantly, Benney advised that "we are a little too south to raise the variety of plants such as you want raised, and the soil we have to work in is too barren." Edison began to reevaluate the whole project.[93]

One of Edison's most vexing problems was the search for a solvent that could efficiently extract rubber from his elected plant varieties. After a frustrating day of work in the Fort Myers laboratory, Edison declared that "it would seem that rubber has no solvent."[94] Edison spent much of the summer of 1928 testing solvents on twenty-pound batches of mangrove leaves that arrived each week from Florida.[95] Edison's "Notebook A," the first systematic attempt to study the issue, included the results of trials on ninety different solvents. In a long entry to himself, dated 5 November 1928, Edison recalled the "vast number of expmts [sic]" that led him to try different combinations of sulfur dioxide, benzol, and acetone. He found that results could vary according to various means of kneading, stirring, and washing of the rubber globules, and that variation in temperature, light exposure, and humidity could damage the rubber product through oxidation. Later that month, however, Edison fussed, "I am always defeated by the tenacity of the solvents remaining in the rubber extract. This interferes with all reactions I make."[96] Finally on December 7, Edison found a method using sulfur dioxide saturated in water that seemed to work: "Cost 6 cents per lb. of rubber from mangrove, and all can be worked by machinery & get good outputs—*!!!HURRAH!!!*"[97] After "all the thousands of experiments," Edison proclaimed with confidence the discovery of a method that was reliable, efficient, and not susceptible to rapid oxidation.[98]

In January 1929, Edison and his staff geared up for another season of active rubber work in Fort Myers. As before, Edison's notebooks reveal the inventor's ongoing reassessment of his progress.[99] At three separate times in January 1929, Edison compiled a list of his most likely prospects. One list included twenty-one "good plants," limited only to herbs; another included seventeen that tested at over 4 percent rubber; the third, denoted the "best in my index," listed of forty-two plants that had a rubber yield exceeding 3 percent. In another series of notes, he assessed the strengths and weaknesses of numerous other possibilities, eliminating those with poor quality rubber and favoring those likely to produce good yields in terms of pounds per acre.[100]

Edison again inspected the Fort Myers plots on 1 February 1929 and then returned to self-reflection. Edison admitted mistakenly focusing in the past on plants with a high quality of rubber, when the real issue should have been the quantity. Thanks to his new understanding of rubber solvents, Edison believed he could now obtain "good rubber" from the "viscous stuff" that he had previously overlooked. He concluded that these developments required a new approach to selecting the plants for transplanting: "Shrubs, even if perennial, are not so desirable as herbs which are also perennial. Shrubs in many cases cannot stand 100% pruning, many die and others are weakened; whereas an herb perennial is cut just as the seeds are ripe and the plant stems and tops have started to die."[101]

In his annual birthday interview of February 1929, Edison admitted that he had encountered unexpected hurdles.[102] "The patience of Job has been considerably overrated," Edison told a reporter that winter. "He did not know

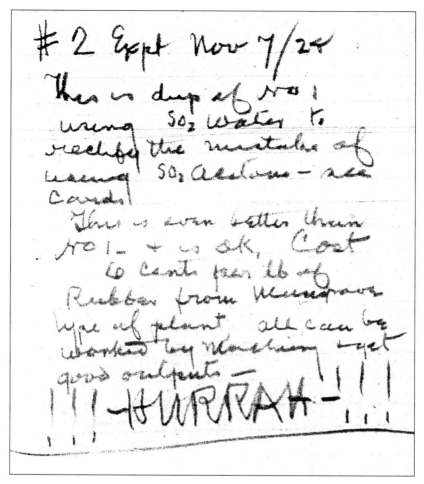

Figure 3.1. Thomas Edison's comments on the rubber-solvent question, 1929. "This is even better than No. 1 + is OK. Lost 6 cents per lb. of Rubber from Mangrove type of plant. All can be worked by machinery & get good outputs—*!!!HURRAH!!!*" Courtesy of U.S. Department of the Interior, National Park Service, Edison National Historic Site.

what patience was." He went on to describe the perplexing and tedious nature of his search.[103] In 1923, he had spoken very highly both of milkweed and especially guayule, two plants that others had also identified as potential rubber crops. In 1925 and 1926, he had spent much time investigating varieties of the ficus tree, also a favorite of several southern Florida real estate promoters, botanists, and nursery specialists. In 1926 and 1927, he had been most interested

Table 3.1
Best in My Index

Latin name	% Rubber	Common name(s)[a]	Type/Remarks[b]
Allamanda cathartica	4.00	Allamanda or Golden Trumpet	Vine
Annona glabra	3.50	Custard Apple or Pond Apple	Tree; very elastic; no good for us; cannot use
Apocynum cannabinum	4.76	Indian Hemp or Hemp Dogbane	
Artemisia lind-heimeriana or lindhermeriana	6.91	Wormwood? or Sagebrush?	Perennial; no comments found; Hullmann Ga 252
Asclepias amplexicaulis	4.40	Clasping Milkweed	Rubber slightly tacky, lacks coherence, good plant
Asclepias speciosa	4.73	Showy Milkweed	Perennial, good plant
Asclepias syriaca	4.83	Common Milkweed or Orange Milkweed	Perennial, light elastic rubber, good plant
Asclepias tuberosa	4.93	Butterfly Milkweed or Indian Paintbrush	Perennial; other specimens tested much lower; good plant if foliage OK
Asclepias variegata	3.75	Redring Milkweed	
Aster cordifolius	3.27	Bluewood Aster	
Aster laevis	3.38	Smooth Aster	
Avicennia nitida	3 to 4	Black mangrove	Shrub; Hand Fla 443
Bignonia venusta	5.87	Flame vine	Vine; could be grown with solidago stems as support; cut them together
Chiococca alba	5.87	Snowberry or West Indian Milkberry	Large shrub curious sticky stuff; may be good if quality of rubber is OK
Chrysothamnus pinifolius	4.13	Rubber Rabbitbrush	Shrub, good plant
Conradina grandiflora	3.43	False Rosemary	Shrub
Dianthera Crassifolia	5.26		Perennial herb, good plant if it grows bushy enough to give good % foliage
Echites umbellata	3.49	Devil's potato	Vine; not good for us
Echitites hieracifolia	3.43	Same as below	Annual herb [same as below]
Erechtites hieracifolia	4.55	Burnweed	Annual herb; Hand Fla 293 & 543

(continued)

Table 3.1. Best in My Index *(continued)*

Latin name	% Rubber	Common name(s)[a]	Type/Remarks[b]
Gibbesia rugelii	4.09		Herb; quite tacky; leaves few and very small; probably cannot use
Gnaphalium obtusifolium	3.17	Rabbit Tobacco	Annual herb; "not good for us"; Hullmann Fla 346
Hamelia patens	3.18	Scarlet Bush	Shrub, NG [= no good]
Jussiaea peruviana or Ludwigia peruviana	3.42	Evening primrose or Peruvian Primrosewillow	Shrub; yellow, dry rubber; may possibly use if pruning does not harm
Laciniaria spicata	4.11	Snakeroot	Perennial herb, stiff white viscous rubber, good plant Hullmann Ga 46
Lonicera fragrantissima	3.47	Sweet Breath of Spring	Shrubby vine
Mascarenhasia elastica	4.17		Small tree; this probably can be cropped as a bush; possibly good
Mesadenia ovata	3.41	Cacalia?	Perennial; fair elastic a little tacky; probably a good plant
Mesadenia atriplicifolia	4.05	Armoglossum	Perennial herb, elastic rubber, good plant
Pterocaulon undulatum	5.60	Blackroot	
Rhabdadenia corallicola	3.12		
Rhabdadenia biflora	3.29	Mangrove Vine	
Rhexia mariana	3.30	Maryland Meadowberry	
Rudbeckia laciniata	3.60	Coneflower or Golden Glow	Very stiff viscous rubber, good plant
Solidago fistulosa	3.00	Pinebarren Goldenrod	Perennial; slight tacky; Not good but foliage may be good
Solidago altissima	3.12	Tall Goldenrod	Perennial; good plant, even reliable; we can get to 4 to 4.5% by selection
Solidago leavenworthii	4.15	Leavenworth's Goldenrod	Perennial; Best is Fla 201; "good plant"
Solidago nemoralis	4.87	Field Goldenrod	Perennial herb, rubber viscous, possibly good for foliage

(continued)

Table 3.1. Best in My Index *(continued)*

Latin name	% Rubber	Common name(s)[a]	Type/Remarks[b]
Solidago mexicana	3.22	Mexican Goldenrod	
Solidago odora	3.07	Sweet Goldenrod or Anisescented Goldenrod	Hullmann Fla 310
Urechites lutea or Pentalinon luteum	3.64	Hammock Viperstail	
Vincetoxicum	3.55	Mosquito plant or Swallowwort	Vine

SOURCE: This is one of three similar tables Edison complied in late January and early February 1929 to assess the status of his rubber experiments before beginning the 1929 crop year. This version is from Notebook (29-01-25), from the ENHS, spelling corrected and rearranged according to alphabetical order.

a. Edison rarely mentioned the common names associated with these plants in his notebooks. These are derived instead from L. H. Bailey, *Standard Cyclopedia of Horticulture*, 3 vols. (New York: Macmillan, 1950); the Hortiplex Plant Database, http://hortiplex.gardenweb.com; and Edison's occasional use of the common names in various notebooks.

b. Most of these are Edison's verbatim remarks, generally indicating the kind of plant, the quality of rubber, and his overall assessment. The table does not include other information noted by Edison, such as plant height, place of origin, and test results of other specimens of the same species. Information in the comments column is taken primarily from Notebook (29-02-01). In a few cases, this column also includes the herbarium identification label used by the EBRC.

in the two varieties of cryptostegia, the vine that Henry Ford too had tested on his southern Florida property. In 1928, Edison had enthusiastically turned to the wormwoods or mugworts, blackroot, flame vine, and black mangrove. In time, however, Edison became disenchanted with all of these.

The Solidago Solution

Finally, by May 1929, after years of experiments on thousands of botanical specimens, Edison had homed his focus on another potential solution: plants in the genus *Solidago*, or goldenrods. He especially favored a variety of *Solidago leavenworthii* that his staffers labeled "Moore Fla. 201." It was a fast grower, occupied a relatively small area, and produced a large volume of leaves per plant. A few days later, Edison wrote that *Solidago*, both the *leavenworthii* and *altissima* varieties, were "good yielders" and "perfect for lbs per acre & for drying." The *leavenworthii* seemed even better because it had fewer dead leaves and it grew considerably taller. "We evidently have lots to find out about the good *Solidagos*," the inventor summarized. Edison estimated that a good goldenrod might yield 1,500 pounds of rubber per acre, and he also noted to keep an eye on a variety that William Hand had collected in New Mexico. "Watch It. PHENOMENON," the inventor wrote.[104] By June 1929, Edison would reveal even in routine correspondence that he had begun to focus on goldenrod.[105]

Edison's enthusiasm for goldenrod had several ramifications. First, his satisfaction with the experiments' progress kept him in Fort Myers until mid-June, the latest stay ever among his two dozen winters in Florida. Secondly, many of Edison's plant-collecting activities came to a halt. After trials on thousands of potential rubber plants, many exotic and from all corners of the globe, it finally seemed that an unremarkable local weed, *Solidago leavenworthii*, could be the answer. Third, the choice of goldenrod suggested that the new U.S. rubber crop might develop in a region far more extensive than southern Florida. In his press leaks about these experiments, Henry Ford hinted that he would plant goldenrod in Georgia and even promoted the use of rubber as a paving surface.[106]

The search for goldenrod took another turn on 17 August 1929. On that day, Hullmann was working near Fort Myers, "4½ miles down the trail, road to the right next to the canal, ½ mile in the woods to left of cypress swamp was in 6 inches of water," when he discovered a specimen soon to change the entire operations in Fort Myers—another variety of *Solidago leavenworthii* labeled as "E.P.C. 573."[107] For years, this plant and its progeny would remain on center stage in Edison's search for a domestic rubber crop. Edison's men propagated this variety again and again until it grew on virtually every plot in Fort Myers, then on Henry Ford's plantation in Bryan County, Georgia, and then on USDA experimental plots near Miami, Charleston, Savannah, and elsewhere. Descendents of the specimen that Hullmann found that day also contributed to American efforts to find a domestic rubber source during World War II. Until synthetic rubber arrived on the scene amid that war, much of the American struggle for a domestic rubber crop centered on E.P.C. 573 and its close relative, E.P.C. 573-A.

Within the next four days, Fort Myers staffers transplanted 336 specimens of this variety in its test plots. Although Edison fell ill in August 1929, he watched their development with ever-greater urgency. EBRC head J. V. Miller instructed Archer to send *leavenworthii* samples to West Orange as soon as they matured. Eventually, Edison lost interest in other *Solidago* species, like *altissima*, that had a substantial percentage of dead or fallen leaves. In a telegram and a letter, both dated 15 September 1929, Miller instructed Hullmann that Edison was "particularly anxious" to see *leavenworthii* harvest results. With thorough detail, Miller added a sketch showing Edison's suggestions on the minutest aspect of the project, including which plants to harvest from each bed.[108] On 21 September, when Hullmann returned from an expedition in which he collected six thousand additional *leavenworthii* plants, he found a telegram on his desk with Edison's instructions on how to prepare the specimens for shipment to New Jersey.[109]

As news of the goldenrod solution reached the press, Edison and his associates milked the discovery to reaffirm his reputation as an indefatigable researcher. Edison granted a formal interview with Edwin Slosson, the science journalist who had repeatedly raised warnings of the American dependence on imported rubber. In this 1929 meeting, Edison described his unconventional research methods in some detail, boasting, "What I like about this research is that there is no money in it." Edison reiterated that his only purpose was to help the nation as a matter of "self-defense" and find a domestic solution to the rubber problem because the "tropics are no place for a white man to live." Also significant, Edison explicitly cited Slosson's book *Creative Chemistry* as an inspiration for his work, and he stressed his chemurgic interest in goldenrod wastes as raw material for paper products.[110]

Industrial and Chemurgic Issues

Edison turned next to preparing for industrial-scale operations. Edison and assistant Fred Ott devoted considerable effort to designing the machinery necessary to bring rubber production up to pilot-scale production. Edison had filed for his patent, "Extraction of Rubber from Plants," in November 1927; he received patent number 1,740,079 in December 1929.[111] Based on the extant mechanical processes used in the guayule industry, Edison's improvements focused on increasing rubber-extraction efficiency from low-grade plants—he was then thinking in terms of plants with only "1% rubber, down to ½ of 1% or less." The key steps included separating the rubbery portions of such crops from those without rubber, and treating the useful product in a "ball mill," a machine that would gently crush the plant material with steel or porcelain balls. The tumbling of these balls would break the woody material into small fibers, suitable as a chemurgic paper resource, while the pith that contained rubber could be further separated from the rest through mesh screens, allowing the rubber to rise to the top of a water bath. Most importantly, the process was small in scale; individual farmers could operate the mill, pressing and drying the rubber produced on their farms before shipment to a central processing plant.[112]

Edison and his crew worked on several other mechanical projects in Fort Myers, and he sketched out ideas on many others. For instance, Fort Myers workers constructed a drying machine that could handle about 225 samples per day when conditions were favorable. In another case, Edison drew a plan for a machine with wheels six feet in diameter and a sharp tool that could harvest poinsettias or similar plants. He also considered methods for stripping leaves from oleander shrubs, calculating the efficiency of such a machine as twenty thousand plants in eight hours. "This is too slow," Edison complained in his notebooks; I "must have 160,000 in 8 hours. 2 acres per man."[113] He also engaged in a long attempt to design a suitable leaf-beating machine. Although most of these machines never worked satisfactorily, their history is another important and overlooked aspect of Edison's role in the search for a domestic answer to a potential wartime rubber emergency.[114]

Meanwhile, Edison took advantage of his connection with industrial chemists to work out the rubber production process. In 1928, he spoke for hours with one of Firestone's chemists on the chemical formulas and procedures useful for testing, extracting, and vulcanizing useful rubber from raw latex.[115] He corresponded frequently in 1929 and 1930 with his chief rubber chemist in West Orange, Fred Schimerka, on the prospects of making the extraction process more efficient and continuous.[116] In 1930, the inventor indicated that his pilot plant in Fort Myers was nearly successful, and he predicted that a larger, perfected factory would be ready at Ford's plantation near Savannah within two years.[117] Edison also considered the problem of the waste products that rubber crops would generate. In 1928, he met with Wheeler McMillen, editor of *Farm and Fireside* and soon to become the leading spokesperson for the chemurgy movement. In that interview, Edison observed, "We overproduce food . . . so the extension of nonfood uses for farm products is a sound, practical and important idea." He readily expressed his support for one of the basic tenets of chemurgy: that the human demand for food is limited by the size of one's stomach, but the human demand for industrial and material products has no limit. As Edison put it on another occasion, pointing to his stomach, "Seems a pity that the farmers' prosperity is bounded by what we can fill in this little space."[118] Similarly, Fred Scheffler, a representative of the Fuller Lehigh Company, visited Edison in Fort Myers in 1930 and claimed that one could expect to generate 6,500–6,800 Btu of energy from each ton—better than some coals—of goldenrod waste.[119] Massachusetts Institute of Technology (MIT) chemists also tested the potential uses of goldenrod wastes. Although their research report was rather negative, MIT President S. W. Stratton's cover letter to Edison offered some optimism: "I feel quite sure that if the material is found to be a rubber source, the pulp will constitute a valuable by-product for building paper, fibre boards, and many other purposes. They would even do for wall-paper or decorative purposes."[120]

In December 1929, as Edison prepared for his earliest departure ever to winter in Fort Myers, his aide William Meadowcroft formally announced to the press that Edison had decided on goldenrod. Yet the tone of this announcement was also cautious, and Meadowcroft denied all rumors that the inventor was close to finding a commercially competitive product.[121] Instead, press reports again stressed Edison's tireless work schedule and his promise to deliver a "present" of a new rubber source to the U.S. government.[122] By the 1930 crop year, however, the declining health of both Thomas Edison and the American economy hindered progress of the rubber research project. In late 1929 and early 1930, Edison hired new plant explorers who spent much of the year crossing Florida and other southern states looking mainly for wild varieties of *Solidago*. Good plants, however, were becoming more elusive. Plant explorer S. T. Moore complained in December 1929, "It just seem like the dame frost and hard raines [sic] was playing tag with me. . . . I am afraide [sic] we mist some good plants do to the frost."[123] In January, Moore nervously reported, "I am working hard but I cannot show anything for my hard work. Each day I think I will find the plant that will make up for my bad luck."[124] Despite frustrations, however, reasons for optimism remained. Several very large varieties of goldenrod—some over ten feet tall—seemed to thrive in Fort Myers, with thirty different plants testing at over 6 percent rubber. Edison confidently concluded, "It looks as if we are careful we will get next year some plants that will give 10% rubber."[125]

Often confined to a wheelchair by 1930, Edison's pace of work on the rubber issue slowed considerably. Mina Edison linked the decline in her husband's health to the rubber experiments: "It comes from nerves as he is not getting rubber results to his satisfaction." She wished that the effort had not received so much attention in the press, for it put added pressure on him.[126] In any case, Edison's enthusiasm and optimism for the rubber project could not be sustained. In his annual birthday interview in 1931, the inventor revealed that he still needed another two years to initiate a pilot plant on Ford's Georgia plantation. That summer, Miller and Archer worked to cut costs by laying off one field hand and cutting the salaries and wages by one-eleventh of all men working on the rubber project.[127] Press reports presented a generally positive description of the situation, although in view of Edison's poor health, there was little sign that he would achieve success during his lifetime. Yet in a press interview, Charles Edison confidently asserted that his father's experiments had passed the "preliminary stage and are in a state of being made

Figure 3.2. Thomas Edison examines an especially large goldenrod specimen, Fort Myers, Florida, January 1931. Courtesy of Edison-Ford Winter Estates, Fort Myers, Florida.

commercially practical." In July 1931, the *New York Times* provided another update on the project, with a notice that Edison was conducting much of his rubber research at his home rather than the West Orange offices. Poignantly, Edison's staff reportedly solved one of the last problems of vulcanization of goldenrod rubber in October 1931. They took a piece of the vulcanized rubber to Edison on his deathbed, an artifact that reportedly brought a glimmer of life to the dying inventor. On 28 October, a leading chemist at Firestone reported that these samples tested "analytically identical" to natural rubber.[128] Edison had seemingly emerged triumphant in his valiant search for a potential rubber crop. But by then it was too late, for Edison had died ten days earlier, on 18 October 1931.[129]

Assessment

Edison's rubber plant research illustrates that even for America's best-known inventor, agriculture was the first place to look for a solution to one of the nation's most pressing industrial issues. Although Edison had little experience in agriculture or botany, he had quickly dismissed synthetic rubber as a viable solution and had promptly assembled an extensive network of plant collectors, botanists, chemists, and even a linguist to help find a domestic source of rubber.

Edison's work with rubber crops also contributed to a significant change in American patent law. In February 1930, Senator John Townsend of Delaware, himself the owner of one hundred thousand acres of orchard property, sponsored a bill that would provide patent protection for plant inventors on par with the protection offered to industrial inventors. Political pressure on President Hoover to do something about the agricultural depression also impelled the legislation, which accorded with Hoover's conviction that opening opportunities in private enterprise was the best hope. Proponents of the bill argued, "We must look to the plant breeder for an acceptable substitute for rubber," and they linked new plant varieties with "public health, prosperity . . . public safety and the national defense."[130] Townsend appealed to Edison for support for his legislation on the grounds that Luther Burbank would have benefited, not mentioning that Edison himself could benefit if his goldenrod varieties were protected. Edison promptly agreed in a widely quoted telegram from Fort Myers: "Nothing that Congress could do to help farming would be of greater value. . . . This [bill] will, I feel sure, give us many Burbanks."[131] The bill offered the standard seventeen years of protection to plant breeders, although only for plants asexually or vegetatively reproduced, such as by cuttings, grafts, and buds. It excluded tubers and any plants reproduced by seed. The bill became law in May 1930 and soon had important implications for the profits of the larger and more successful nurseries. Although Edison never applied for a plant patent of his own, the enactment of this legislation created important new definitions of intellectual property.[132]

Edison's failures were also significant. He struggled for years in a field of scientific inquiry that had high barriers to success and did not yield immediate results. Agricultural research remains an especially expensive and time-consuming process, for virtually every experiment requires an entire growing season before success can be assessed. Countless problems can interfere with that success: difficulty in obtaining and transporting specimens, deterioration of seeds in storage, poor germination rates, unpredictable climate, ill-timed rainfall (or lack of rain), unbalanced nutrition, pests, fungi, inadequate machinery, and the like. The complexity of rubber chemistry added to Edison's problems, because determining appropriate solvents and developing extraction techniques were difficult issues as well. In the words of Harry Ukkelberg, the scientist who attempted to continue the research where Edison had left off, the inventor's understanding of agricultural science was too rudimentary for success: "He started from scratch, you might say, with a plant about which practically nothing was known and in a field in which he had very little information."[133] After years of effort, hundreds of thousands of dollars invested, and tens of thousands of experiments, Edison could not bring American rubber much closer to a possibility of interest for either entrepreneurs or military-preparedness experts.

Despite his efforts, Edison's work also failed to reshape debates over the political economy of rubber crops. The inventor's struggle for a domestic rubber crop culminated at a time when few Americans had potential rubber emergencies on their minds. When Edison died, the nation's unemployment rate stood around 20 percent and the price of rubber stood at a record low of about five cents per pound.[134] Also in 1931, several news reports hinted at progress in the search for a synthetic rubber. As the Edison Botanical Research Corporation admitted in its annual report of 1931, such a discovery would mean that the "need of an emergency supply from Golden Rod will end."[135] For the time being, the urgency that had characterized Edison's search for a domestic rubber crop had quietly evaporated.

Yet Edison's successes were also significant. He made tremendous progress in the last four years of his life, testing some seventeen thousand plant specimens before quickly narrowing the search down to a handful of possibilities. He identified several plants that had the potential to solve the problem as he had framed it, and his premodern selection methods resulted in steady improvements in rubber yields. Through the process, Edison displayed his remarkable powers as a publicist, his talents as a motivator of allies in the private and public spheres, and his mastery of some concepts in plant physiology and experimental agriculture. Edison had kept the search for a domestic rubber crop in the public eye for years. Perhaps most significantly, his warnings that the United States would face a rubber emergency in the next war could not have been more accurate.

Chapter 4

The Nadir of Rubber Crop
Research, 1928–1941

In the spring of 1930, U.S. Army major Dwight Eisenhower embarked on a monthlong expedition that took him from his Washington, DC, desk job through the IRC's guayule operations in California, Texas, and Mexico. In his diary entries from that five-thousand-mile journey, Eisenhower described his encounters with seedy hotels, surly border guards, seemingly endless hot and dusty roads, and memorable "swarms" of Mexican women and children selling tortillas, tamales, and enchiladas.[1] Eisenhower's diary also tells of his repeated meetings with IRC officials and his growing conviction that guayule had to be part of any plan for American war preparedness. The confidential report that Eisenhower and his traveling partner, Major Gilbert Van B. Wilkes, submitted on 6 June 1930 (exactly fourteen years before D-Day) endorsed guayule almost unequivocally: it offered potential employment to thousands of needy Americans; it could provide an alternative crop for American farmers whose overproduction of cotton and grain crops commonly brought low prices; it could help consumers through an overall reduction in the demand for imported rubber; and it could become a permanent addition to the rural economy in semiarid regions of the United States. Eisenhower and Wilkes further speculated that the United States could withstand a war that cut off rubber imports for twelve to eighteen months but warned that any crisis beyond that would require additional preparation. Because guayule required about four years to reach maturity, the time had already arrived for the United States to establish a domestically grown rubber reserve. Thus they endorsed a government commitment to support the maintenance of 400,000 acres (about 625 miles square) of guayule. If harvested on a rotating basis every four years, the scheme could contribute about 160 million pounds annually, or about one-fifth of the nation's annual consumption. Eisenhower and Wilkes accepted and echoed the scientific data and political considerations that IRC officials threw at them, including the idea that the government should support a minimum "reasonable" price for all rubber raised and milled in the United States. "*Under real encouragement,*" Eisenhower and Wilkes boldly concluded, "*the production of guayule would develop rapidly into an important industry in the United States.*"[2]

Yet despite these arguments, the memories of World War I, and the public attention that surrounded Thomas Edison's search for an emergency source of rubber, the War Department did not act on Eisenhower and Wilkes's proposals. Evidently, the report gathered dust in government filing cabinets, its existence forgotten or ignored until 1943, when the nation was in the depths of a real rubber crisis.[3] In retrospect, the 1930s can be seen as the nadir in the quest for an American rubber crop, as other issues took center stage in the American political economy. Above all, alternative rubber crops had no chance whatsoever to compete with the depressed prices of imported *Hevea* rubber. From a postwar peak of $1.23 per pound in 1925, the price of rubber fell to a low of 4.6 cents per pound in 1931.[4] Efforts to develop alternative rubber crops fell by the wayside in the 1930s, as the technological and agricultural limitations of goldenrod, milkweed, guayule, and other possibilities became ever more apparent. Edison's goldenrod experiments continued for a few years after his death, but only in two small projects near Savannah. The IRC became increasingly desperate in efforts to protect its investments in guayule: it suspended guayule production in California and eventually even turned to Fascist Italy as a last hope to find a market for its product. Private- and public-sector experiments with other crops led nowhere. And even though synthetic rubber remained only a laboratory experiment for most American rubber producers, industry leaders presumed that it could be scaled up quickly should the need arise. Most Americans, however, could not even contemplate that the days of cheap and plentiful rubber from the East Indies might come to an end; they had the sense that American rubber sources were secure. By the end of decade, as war approached again, only a few (and often eccentric) enthusiasts still pushed for domestic and natural alternatives to the American dependence on imported rubber.

The overall failure to prepare for an impending military conflict also shaped the neglect of domestic rubber crops during the 1930s. Understandably, American policy makers made domestic politics a higher priority during the decade. In terms of foreign policy, Herbert Hoover's eagerness to challenge the rubber-producing nations gave way to Franklin D. Roosevelt's desire to expand trade agreements, particularly with potential rubber-producing nations in Latin America. Well-publicized projects such as Harvey Firestone's establishment of rubber plantations in Liberia (and Henry Ford's attempts to do the same in Brazil) also created the sense that the United States had alternative sources of rubber, even though these projects had only minimal impact. A few government and industry experts raised their voices to warn that Latin American rubber could not be a solution in the near future, that progress in synthetic rubber research remained slow, and that imminent war in the Pacific necessitated stockpiling Asian rubber and other strategic materials. Yet in the United States, coordinated efforts to prepare for a rubber shortfall remained virtually nonexistent. Meanwhile, Nazi Germany and the Soviet Union made important advances in the preparation for war and its attendant rubber shortages. The autarkic goals and centralized nature of these governments, however, contrasted sharply with the attitudes that prevailed in the United States.

In this context, studying goldenrod, guayule, and other domestic rubber crops became a very low priority in the 1930s. Just days after Edison's death in October 1931, USDA rubber expert O. F. Cook summarized the prospects for an American rubber crop as "precarious." Cook's colleague at the USDA, Karl Kellerman, shared his exasperation: "It is amazing to me that political leaders of this country do not see the desirability of extending research."[5] By April 1932, USDA bureaucrats officially curtailed rubber crop investigations as an economy measure.[6] The search for an alternative rubber source continued through the 1930s, but low rubber prices and other pressing concerns pushed it to the back burner.

The Euphorbia intisy *Episode*

The case of *Euphorbia intisy*, a rubber plant native to Madagascar, represented a good example of a short-lived search for a rubber crop that could grow in America. Madagascar had long been a prime destination for rubber-plant explorers; a 1914 booklet, for instance, mentions about a dozen plants that were harvested for rubber.[7] *Euphorbia intisy* became the most important and, beginning in the 1890s, fueled a significant industry that produced high-quality and easily harvested rubber in Madagascar. The French colony exported about fifty tons of intisy rubber by the beginning of World War I. This boom was short-lived, however, for the race to harvest intisy brought violence among native groups, and rubber collectors traveled in armed bands. Serious overharvesting, which some westerners blamed on the natives' "cupidity," soon rendered the plant virtually extinct.[8]

Interest in this plant had been renewed in 1928, when botanists found that living specimens remained, apparently brought back to life through a root structure hidden beneath the soil. The USDA's Bureau of Plant Industry sent Charles F. Swingle (half brother of Walter T. Swingle, another USDA botanist and tropical-plant expert), who planned a clandestine mission to secure intisy seeds and specimens. Because the rubber trade had died down, officials believed it "should be an easy matter" to pay "substantial bonuses" and get natives to reveal the location of remaining intisy specimens.[9] Government officials made it clear that incidental expenses were not important; Swingle simply had to get live specimens of *Euphorbia intisy* back to the United States. Even more significant, Swingle did not reveal his true objective to French or colonial officials.[10]

Swingle's Madagascar adventure began in France in June 1928, where he joined the French botanist Henri Humbert, already an expert on Madagascar's flora. As Swingle and Humbert's party traveled through northern and central Madagascar, the group observed and collected several potential rubber plants, including cryptostegia, but their real mission was to collect from the drier regions of Madagascar's southwestern corner. The party had difficulty convincing natives to join the expedition, in part because of the locals' awareness of the dangers that lay ahead, and also because of their desire to keep foreigners away from their botanical resources. Swingle and Humbert eventually conscripted a group of forty-one porters after threatening them with a fifteen-day jail term. In order to "conserve their strength," the scientists traveled in the native sedan chair, the *filanzana*, each carried on the backs of four natives. The group ventured into the desert with heavy loads and without enough water. One by one five of the porters fell by the wayside, too weak to continue. The party pressed forward nevertheless, now with Swingle and Humbert down from their *filanzanas*. According to Swingle's "strictly confidential" report, the group eventually managed to deliver some water to the stricken porters, and none "actually died."[11]

On the sixteenth day of their journey, Swingle party's finally discovered a small stand of *Euphorbia intisy*. Natives were reluctant to harvest the plant from its difficult and rocky terrain, but when Swingle offered them forty cents per day, almost three times the normal wage, they came through and Swingle declared the expedition a success. Swingle later met with a group of Norwegian American Lutheran missionaries based in the southeastern part of the island and arranged with them to secure intisy seed without directly informing French colonial officials of the plan. Six months after the journey began, Swingle and his cuttings reached New York in November 1928. By spring 1929, USDA officials had intisy plantings growing at plots in Florida and California. As in so many other cases in the American search for a sustainable domestic rubber industry, press accounts of the Swingle adventure repeated the familiar tale that a "revolution in the American rubber industry" was on the horizon.[12]

Initial enthusiasm for the new potential rubber crop was not sustained. Of the fifty-eight plants that arrived in Washington in November 1928, only twenty were in a condition suitable for propagation. Through considerable trial and error, Swingle developed a method to propagate the plant from cuttings. His best results came through use of an electrically heated box of soil, designed to replicate the high-temperature and low-humidity conditions of the Madagascan desert. Although inconsistent, his results convinced him that the experiment could prove a success.[13]

But suddenly the intisy project came to an end. On 16 June 1930, just as Swingle presented a report that outlined his modest successes, he received word that the USDA would appropriate no funds to continue the project. Swingle fired off another report later that day. "If there ever was any justification for me to risk my life getting this material, or for the thousands of dollars already spent on it," he argued, then the project should be continued. They had every reason to think that the plant would "prove an important American accession." In spite of these arguments, by 1934 only four intisy plants remained in the ground at the test facility in Bard, California. In a decade of economic hardship in the United States and low rubber prices in the East Indies, the goal of a domestic American rubber crop seemed very distant indeed.[14]

Troubles for Guayule and the Intercontinental Rubber Company

A similar fate befell the IRC's twenty-year effort to establish guayule as an American rubber crop. Although the IRC had generated considerable publicity and interest in the guayule possibility during the years of the Stevenson Plan, things unraveled for the company soon after. Virtually every idea that IRC president George Carnahan and his colleagues put forth failed to materialize. Over time the company became increasingly desperate in its search for a sustainable source of income. In spring 1927, for example, Carnahan convinced preparedness guru Bernard Baruch, who maintained ties with the IRC, to present a scheme that would make guayule a major player in the international rubber market. More to the point, such a proposal had the potential to bring immense profits to IRC stockholders. As Carnahan admitted, guayule's promoters "were not and could not be influenced by any altruistic or nationalistic considerations whatever [sic]."[15] In brief, Carnahan wished to create an international syndicate that included the largest rubber companies, in which major firms would promise to purchase specified portions of the IRC annual production at guaranteed prices. The IRC would charge 30 cents per pound for guayule rubber whenever the price for *para* rubber (the highest grade) was below 37.5 cents, charging only 80 percent of the market value of *para* rubber if prices rose above that amount. In a confidential note to Baruch, Carnahan boasted that the company's cost of production was as little as 18 cents per pound, which would yield a gross profit of at least 67 percent.[16]

Carnahan promoted similar schemes through Claudius H. Huston, a close confidant of Herbert Hoover. With a guarantee of a minimum price, American-grown rubber could compete with the imported product; the IRC could sign contracts with California growers, in turn guaranteeing them a minimum price for their efforts. Carnahan hoped that the scheme would mandate that the large rubber firms purchase up to 15 percent of their annual needs as guayule, and the IRC could be assured of annual market of some sixty-three thousand tons.[17] The IRC did not intend to wait for a war emergency to establish a scheme that mimicked the Stevenson Plan's efforts to manipulate rubber prices. Unsurprisingly, however, Carnahan failed to get American rubber companies to commit to this arrangement in an era of low rubber prices and political reluctance to prop up the fortunes of one company in particular. Although the Stevenson Plan had taught that economic warfare could threaten American rubber supplies as much as military conflict could, Carnahan had asked for too much.[18]

In January 1928, the IRC attempted to seize the spotlight again with the development of its "Ampar" brand of automobile tires, a product derived entirely from American-grown guayule rubber.[19] The IRC also produced a new film that touted guayule successes; the company arranged a screening with Thomas Edison in West Orange, New Jersey, and at Ford Motor Company headquarters in Dearborn, Michigan.[20] Tests on an initial run of thirty-one guayule tires suggested that the product was "equal in mileage value and general service quality" to the standard automobile tire. According to the editor of *India Rubber World*, "This sounds like the real answer to our national crude rubber problem."[21] The *Oakland Tribune*'s editorial cartoonist reacted to this development with a sketch that depicted a tire made from American-grown rubber chasing down an overweight and overdressed capitalist, who represented the foreign rubber monopoly. Yet troubles for the IRC emerged within days of their announcement, for production problems forced the company to suspend operations. In an urgent telegram, Carnahan instructed his California colleagues to take no more orders for Ampar tires. There is no sign that Ampar tire production ever amounted to much more than a publicity campaign.[22]

The company's hopes to develop a guayule industry in the American South also ran into difficulties. As it had announced in its publicity blitz of 1926, the IRC hoped that guayule could be developed in southern states eager for an alternative to cotton monoculture. Beginning in 1926, the IRC arranged with agricultural experiment station and railroad company officials to run field tests at four sites in South Carolina, Georgia, and Mississippi. On his way to visit Thomas Edison in Florida in 1927, Carnahan toured each of these experimental fields but found diseased plants and other troubles at each. By June 1928, Carnahan concluded that the desert plant simply was too susceptible to root rot in the wet conditions of the Southeast, and the IRC soon suspended these tests.[23]

Increasingly desperate efforts to transform guayule waste products into fertilizers, artificial fireplace logs, paper products, and the raw materials used in explosives all led to nothing.[24] Likewise, a proposal to develop a guayule industry in Eritrea proved unfeasible due to low and sporadic rainfall.[25] The IRC also turned to chewing gum companies, because the raw materials for their product were in short supply. Removing the poisonous acetone used in the rubber extraction proved difficult, however, and guayule chewing gum also had an annoying "squeak" that made it unsuitable for commercial development, even for the "penny gum" market. Even more problematic was the "objectionable" and "characteristically bitter" "taste of linseed oil or putty" that guayule imparted to chewing gum. Despite extensive research to find a way to remove the offensive taste, representatives of the Wrigley Company complained that guayule gum was "of no interest to us and is without promise." Carnahan then tried Beech-Nut. After months of efforts, Beech-Nut officials firmly told Carnahan that "we are quite sure" the product was not satisfactory. The very next day, Carnahan closed the IRC's research laboratory and laid off his main chemist.[26] Moreover, the violence in Mexico hit home, as "bandits" kidnapped Carnahan's brother, a mining engineer based in Zacatecas. The IRC was on the verge of collapse.[27]

In this context the IRC developed another plan. Just days after the stock market crash, Carnahan appealed for a meeting with President Hoover and offered a screening of his guayule films in the White House. The president declined, on the grounds that his schedule was busy dealing with the nation's "business problems," but Carnahan had his White House meeting by December 1929, during which Hoover suggested that they develop a plan working through the War Department rather than the USDA. Hoover's ally Claudius Huston was "thoroughly sold" on the idea and made arrangements for Carnahan to push for a living reserve of perhaps one hundred thousand acres of guayule rubber.[28] Meanwhile, Carnahan again approached Thomas Edison to see if the inventor had any new discoveries to report, promising to share Edison's successes with leaders of the automotive industry and with President Hoover. Edison declined to help, however, repeating his preference for gold-enrod because it could be made ready more quickly during a war emergency.

Undeterred, Carnahan proceeded with efforts to find supporters in the Department of War. He hoped for high-ranking officials to make a tour of western guayule sites but was pleased to have assurances that Major Dwight Eisenhower was "very competent." The game plan, he told his California col-leagues, was to show Salinas facilities in the most favorable light, and the IRC's Mexican facilities less favorably, in order to impress the need for a domestic rubber supply.[29] For his part, Eisenhower found guayule's prospects as a strate-gic crop intriguing, although he expressed surprise that Carnahan "emphasized over and over [its] confidential nature."[30] Eisenhower and Wilkes returned from their tour of IRC properties with a report that endorsed the company's position in every possible way. As noted earlier, however, the War Department ignored the report, increasingly confident that stockpiling, collecting scrap rubber, and developing synthetic alternatives would be adequate.[31]

Even worse for the IRC, its efforts to develop a California-based guayule industry came to fruition as the global Depression worsened. Its first major plantings in the Salinas Valley began in 1927, just as market forces combined with Secretary Hoover's anti-British campaign to weaken the Stevenson Plan and bring a reduction in rubber prices. Long-term prospects for an industry based on American rubber crops had already begun to diminish. Nevertheless, the company proceeded with the development of machinery custom designed for the guayule industry, as well as pushing ahead with its plantings of guayule in the Salinas Valley. But rubber prices had fallen well below the IRC's cost of production, which forced it to suspend operations at three of its four factories in Mexico and delay opening its long-heralded factory at Salinas.

By January 1931, rubber prices were less than a dime a pound. The IRC's new rubber-extraction mill in Salinas opened with much fanfare, and the company could boast about its $150,000 investment in the endeavor, its wide range of custom-made machinery, and promises of income for the farmers in the valley (whose guayule fields would be harvested only once every four years, on a rotating basis). Yet few found reason to believe the company's claim that it soon would process one hundred million pounds of rubber annually, or 10 percent of the nation's rubber demand.[32] By the end of that year, Carnahan despondently described the "deplorable" situation in the industry. With large supplies and low demand, the future looked grim, not only for guayule, but for the rubber business as a whole.[33] When the Depression brought a dramatic drop in demand for automobiles and other consumer goods, rubber companies

Figure 4.1. The Intercontinental Rubber Company's new guayule rubber processing facility near Salinas, California, about 1931. Intercontinental Rubber Company Records, DeGolyer Library, Southern Methodist University, Dallas, Texas, A1999.2225.

were forced to sell tires and other products for less than the cost of production. For its part, the IRC suffered net losses of $304,000 in 1930, $352,000 in 1931, and $472,000 in 1932. The company substantially reduced its plantings in California and operated its factories in both Mexico and California on a "maintenance basis" only. Several farmers who had contracted with the IRC gave up on guayule altogether; eventually, many dug up and burned the shrub in pyres that ceremoniously marked the defeat of guayule production in California.[34]

On its last legs, the company continued to appeal to the federal government to salvage America's only homegrown rubber crop. In 1931, Carnahan met with President Hoover, Bernard Baruch, and Senator Carter Glass of Virginia in an attempt to secure legislation to increase tariffs on imported rubber. Justified as a means to raise revenue during the Depression, a rubber tariff also would help the struggling guayule industry. The IRC lobbied to have imported guayule rubber from Mexico exempted in order to "preserve its virility for prompt service during a wartime emergency."[35] At another White House meeting with Hoover, Carnahan appealed explicitly for price supports, asking the president to try to persuade Henry Ford to invest in California rubber. Hoover endorsed guayule in general but said that there was nothing he could do in the prevailing economic climate.[36] In another act of desperation, the IRC spread rumors in 1932 that Bolsheviks were trying to steal Mexican plant resources, thus hindering leading Soviet scientists' efforts to gain a visa for botanical travels through Mexico.[37]

Efforts to salvage the IRC proved even more difficult under the Roosevelt administration. In May 1933, Congress considered two appeals that had the blessing of California's state legislature. One asked in general terms for "proper" tariff support for a domestic rubber crop as a tool of unemployment relief. Another included strident warnings about the threat of a war emergency and called for legislation to fund the planting of the guayule seedlings that sat in IRC nurseries. The California resolution also called for federal legislation in which "all government contracts requiring the consumption or use of rubber" would mandate that the rubber must be grown within the territorial limits of the United States or its possessions. Of course, Carnahan and his allies justified such proposals, which would provide funds only to his company and its stockholders, as beneficial to the nation as a whole. As before, guayule supporters justified federal support for several reasons: it offered insurance of strategic rubber supplies in case of war; it developed a new crop well suited for marginal lands; and it promoted a favorable balance of trade in the face of shrinking farm exports.[38]

None of the IRC's proposals got very far. Accustomed to meetings with Republican cabinet officers and in the White House, Carnahan was frustrated by the New Dealers, who left him "camping on . . . the ante-room chairs" of lower-ranked bureaucrats. Nevertheless, he and his allies presented various ideas to connect guayule with the New Deal. Perhaps guayule work could provide employment for the Civilian Conservation Corps' army of underemployed Americans. Perhaps guayule could become part of a plan to restructure southern agriculture and remove marginal lands from production. Rather absurdly, some USDA officials seriously proposed that the very slow-growing guayule might be used as an erosion-control crop.[39]

Meanwhile, the IRC prepared in September 1934 to harvest two thousand acres of the guayule that it had planted in the Salinas Valley about four years before. The IRC mobilized its machinery and turned a cultivated American crop into dozens of hundred-pound rectangular blocks of processed guayule rubber. Although the initial production, six tons per day, amounted to a mere one half of 1 percent of the nation's daily rubber consumption, it did represent the first production of a cultivated American rubber crop on a commercial scale.[40] The IRC used this harvest in a last-gasp effort to mobilize interest in Washington, DC. Carnahan outlined a proposal to Secretary of Agriculture Henry A. Wallace that, as before, explained the value of a "large rubber reserve in the living tissue" of California plants. Carnahan enlisted Bernard Baruch to lobby the president on guayule's behalf, and Carnahan begged USDA officials to make a trip to Salinas to see the operation with their own eyes. This never came to pass, however, and anyone associated with Secretary Wallace declined the invitation.[41]

Instead, the Roosevelt administration continued to seek rubber security through diplomatic initiatives. In April 1934, the United States supported the creation of the International Rubber Restriction Committee, a body that comprised virtually all the leading rubber producers and consumers and helped stabilize the rubber markets. In contrast to the Stevenson Plan, no single nation stood to benefit at the expense of others. Although ignoring the interests of smaller native rubber planters, the scheme attempted to balance the interests of the main rubber producers and the major rubber consumers. Further, American diplomats began to foster relations with the potential rubber-producing nations of the Western Hemisphere. In the aftermath, Carnahan complained that USDA officials had "sold American rubber 'down the river'" because they considered it acceptable for Americans to import rubber as a means to create foreign buying power for America's surplus farm crops.[42] A final blow came in Secretary Wallace's widely read book *New Frontiers*, which explicitly rejected calls for a tariff to support the guayule industry. Indeed, Wallace suggested that if such a tariff were enacted, those in charge of a domestic rubber industry—implying Carnahan and the IRC—would "weep continuously" for additional protections.[43] Carnahan responded with charges that the New Dealers' "Conspiracy of Ignorance" seemed determined to waste money on abstract theories and foolish philosophies, while a vital strategic resource remained untapped in the California soil.[44]

After twenty years of nationalistic rhetoric that defended its enterprise, the IRC began to look elsewhere for a more guayule-friendly nation. With no support from its own government, and with continued hassles from the Mexican government (which still used tax and labor policies to pressure foreign firms), the IRC instead turned to the autarkic economies of Europe. A firm in Nazi Germany became the IRC's best customer, even though it involved converting Germany's "aski marks" (which had no value in the United States) into U.S. dollars through Mexico's currency.[45] Carnahan toured Germany, the Soviet Union, and Spain before determining that Mussolini's Italy offered the best hope for a partnership.[46] Indeed, Italy's war ministry had launched missions to investigate rubber plants in the Soviet Union and hosted conferences on botanic rubber in Rome in an explicit effort to challenge the monopoly of Dutch and British "market bosses."[47] The IRC praised Italy for its "incentive for economic self-containment," a pointed challenge to the Roosevelt administration's internationalist economic policies that had left guayule in the lurch. After several years of preliminary cooperation, the IRC and the Fascist Confederation for Agriculture formalized agreements in 1937, soon after the Italian conquest of Ethiopia. The deal promised that an Italian subsidiary would reimburse the IRC for its seeds, machinery, and technical expertise, and the IRC was offered a royalty of 12.5 percent for any guayule produced over the next seventeen years. Having invested in the IRC's product, the Italians soon were planting guayule in Libya, Sicily, and elsewhere. By early 1941, as the United States was at peace but German and Italian armies waged war against nearly twenty nations, over thirty million guayule plants and seedlings grew in Fascist Italy. Even after the Italians officially joined the war on the Nazi side in June 1940, Carnahan eagerly sought to send another 105 drums of guayule seed to his Italian partners, complaining that only the British blockade prevented the IRC from doing so.[48]

Goldenrod after the Death of Edison

During the 1930s, Edison's goldenrod project suffered a similar fate. Despite the enthusiasm of the late 1920s and the promising results of the early 1930s, prospects for an American-grown goldenrod rubber crop waned in the face of low rubber prices and other governmental priorities. Fewer and fewer Americans believed that Edison's final project could be the answer. Yet the goldenrod project did not end quickly. The Edison Botanic Research Corporation showed no immediate signs of slowing down following Edison's death in October 1931. Results from the fields in Fort Myers were promising that summer—rubber yields increased on 309 of the 433 plots for which data could be compared with the year 1930.[49] Among the 1,500 tests made in 1931, Edison's employees found thirty plants containing 8 percent rubber, five containing 9 percent, and one containing 10 percent. Moreover, goldenrod rubber had been successfully vulcanized just weeks after Edison's death, perpetuating hope that rubber from goldenrod could be "as good as ordinary rubber." Although news reports of recent success in the development of synthetic rubber gave the company some pause, it seemed evident that Edison's work was about to pay off. Thus the EBRC pushed ahead in 1932 without serious interruption. Ford and Firestone continued to support the company financially, John V. Miller continued to manage the EBRC's daily affairs, and Edison company employees continued to conduct botanical trials in Fort Myers and chemical experiments in West Orange.[50]

The 1932 crop year, however, was far less successful. A great many seedlings transplanted from the nursery died in the field. Irrigation and fertilizing strategies did not seem to work. Tensions emerged over what experimental data to record and what those data meant. The final straw came in the summer of 1932 with an outbreak of a plant fungus that the Fort Myers staff could not combat.[51] Firestone company officials complained that the EBRC's scientists lacked real expertise in rubber chemistry, and tellingly, they suggested that guayule expert David Spence, formerly of the IRC, might be out of work and willing to take on the goldenrod project.[52]

The problems impelled Miller to bring on someone with more experience in experimental botany. He went into a July 1932 meeting with Harvey Firestone Jr. determined to continue the work "in an efficient and scientific way." Firestone prompted Miller to write the University of Minnesota's prominent professor of plant pathology E. C. Stakman in search of a person to come to the EBRC's aid. Miller described the job as one with a promising future, and he suggested, "Later on we may want to establish one or two other gardens in Florida or Georgia, and also interest the Government Experiment Stations to raise some of these plants."[53]

Stakman recommended Harry G. Ukkelberg, a thirty-four-year-old Minnesotan who had just completed his master's degree in plant pathology. With a résumé that included over forty courses in the agricultural sciences and several years of practical experience, Ukkelberg was well qualified for the post. In his formal offer of the job and its salary of two hundred dollars per month, Miller explained that "up to this year progress has been very good." Without using the term, Miller urged that Ukkelberg also approach the problem with an eye on its chemurgic potential: "It certainly would prove a commercial success if we could develop some byproducts from the stems or other extracts from the leaves."[54]

En route to Fort Myers, Ukkelberg stopped in Washington, DC, to meet with O. F. Cook, the USDA's expert on rubber plants. It is likely that the two also discussed Luther Burbank, for Cook had recently published a passionate essay that analyzed the strengths and weaknesses of Burbank's approach to plant breeding.[55] Indeed, beginning in about 1910, when Burbank was near the peak of his public acclaim and popularity, American agricultural scientists began to challenge Burbank for his apparent lack of rigorous breeding methods, shoddy record keeping, and lack of interest in finding or applying fundamental scientific laws of plant breeding and genetics. As historian Paolo Palladino has shown, scientific specialists challenged Burbank, the "Wizard of Santa Rosa," precisely because his old-fashioned and empirical methods resembled those of Thomas Edison, the "Wizard of Menlo Park." Although both Edison and Burbank remained in the highest possible public esteem, Ukkelberg found himself—in his first professional job—in a position where he would challenge both American icons.[56]

He went to work immediately. Taking advantage of the flowers of *Solidago leavenworthii* in bloom when he arrived in September 1932, Ukkelberg started with a series of self-fertilization experiments; a few weeks later, he began work on cross-fertilizations. There is little evidence that Edison and his colleagues had seriously considered these issues. Under pressure from Miller, Ford, and the Firestones—the EBRC leaders who expected immediate results—Ukkelberg stressed that success in plant genetics did not come quickly. It was simply too early to know if the tests were having any effect at all. One first had to see if a seed was created through the process, wait to see if it germinated, and then wait months to see if the offspring possessed any improved characteristics.[57]

As a plant pathologist, Ukkelberg was especially interested in the threats of plant disease on the goldenrod experiments. He feared that his predecessors' enthusiasm for using root cuttings as a means of propagation increased the threat of fungal root diseases.[58] Within two weeks of his arrival, Ukkelberg had assistant Walter Hullmann construct a crude soil-sterilizing device. Intended to be a stopgap solution, it was constructed from a 550-gallon gasoline tank cut in half, mounted onto a homemade cradle built from three Model T frames, then packed and clayed to hold the heat. The device was also designed to kill off weed seeds and soil fungi before the soil was used for goldenrod transplants. The failed effort to secure a better soil sterilizer signaled the fate of Edison's goldenrod experiments. Miller initially offered to help pay for "an A1" soil sterilizer, because he did not "want to jeopardize ultimate results by using inferior apparatus." In subsequent efforts to find a more appropriate machine for the task, however, Miller rejected the notion of paying full price and hoped to find one at "some secondhand outfit in Newark, [New Jersey]." Although Ukkelberg explained that the homemade system probably was not effective in killing all the dangerous bacteria and fungi, Miller deemed the cost of three hundred to four hundred dollars "out of line."[59] Ukkelberg also recognized that he would need to improve the quality and quantity of water available to the site. With a new tank to store rainwater, a new pit well that quickly hit surface water, and a new pump, it became possible to irrigate the experimental gardens twice daily.[60]

At this time the EBRC also began to strengthen its ties with the USDA. In contrast to Edison's determination to lead an independent team of researchers and a network of allies, Ford and Firestone were now ready to consult outside experts. William Taylor, chief of the Bureau of Plant Industry, soon offered to test goldenrod at several USDA facilities. In November 1932, Ukkelberg sent 150 goldenrod cuttings to the USDA plant inspection station near Washington, DC, the beginning of what soon would become a flood of Fort Myers plant materials.[61] Also, as part of Henry Ford's interest in expanding goldenrod experimentation, Ukkelberg made arrangements to send Fort Myers cuttings to universities and agricultural experiment facilities in Michigan, Ohio, Iowa, and Tennessee.[62]

Despite the many difficulties that the EBRC had encountered at Fort Myers that summer, and the abrupt transition to a new scheme under Harry Ukkelberg, the 1932 crop year showed continued progress in the experimental work. Ukkelberg's quarterly report of 19 January 1933 had plenty of good news for Miller and the EBRC's directors. Summarizing his results, Ukkelberg concluded with enthusiasm that the "maximum percentage of rubber has not yet been obtained."[63] In outlining the plans for 1933, Ukkelberg reiterated his belief that rubber yields would soon reach a plateau unless breeding replaced selection as the primary strategy. Thus Ukkelberg planted a large number of the plots in the variety E.P.C. 573-A, and he filled many of the remaining plots with other species of *Solidago*—*mirabilis, fistulosa, stricta, nashii, altissima, tortiflora,* and *elliottii*—probably reflecting his thinking that he could hybridize desirable qualities from these species with the rubber-producing characteristic of *leavenworthii.*[64] Ukkelberg implemented a systematic set of agricultural experiments designed to isolate the factors responsible for maximum rubber production. By mid-summer when almost all the plants stood six to eight feet tall, a sea of goldenrod covered the plots around the laboratory in Fort Myers. A domestic rubber crop still seemed a genuine possibility.

A year later, however, results from Fort Myers were quite disappointing. In 1932, 49 percent of the E.P.C. 573-A plants tested showed a yield of at least 7 percent. In contrast, only 9 percent of the E.P.C. 573-A plants tested in 1933 yielded more than 7 percent rubber.[65] In attempting to explain these disappointing results to Miller, Ukkelberg again stressed that he had never been satisfied with the selection methods that Edison and his staff had employed. As before, Ukkelberg hinted that Edison and Burbank had not really deserved their reputations as "wizards," at least not in terms of contemporary agricultural science. Indeed, he could not see how Edison's results had seemed so successful in the first place, because "when selecting plants on the basis of percentage of rubber alone it is easy to see how wrong selections can be made." Ukkelberg reiterated this criticism two weeks later: "I have reached some definite conclusion: That we are placing too much emphasis on the percentage of rubber without considering under what conditions this percentage has been obtained." He went on to say, "I feel we have a good start in making the golden rod a rubber producing plant but that much improvement can still be made." In Ukkelberg's analysis, the primary problem was that "selection within a pure clonal line is of no value or at best of doubtful value." As far as he was concerned, only open and controlled pollination tests would bring real

Figure 4.2. Edison Botanic Research Corporation scientist Harry G. Ukkelberg amid a sea of goldenrod, Fort Myers, Florida, August 1933. Courtesy of Edison-Ford Winter Estates, Fort Myers, Florida.

improvements. Ukkelberg urged Miller to read selected pages of a recent text-book, *Genetics in Relation to Plant Breeding*, that openly criticized Burbank's methods and taught the plain lesson' that purely vegetative propagations could never yield systemically higher-yielding plants.[66] Ukkelberg carefully and repeatedly explained the limitations of Edison's reliance on clonal lines, or the use of root cuttings and pups as the material for most of the goldenrod propagations.

Naturally, officers of the EBRC also faced questions about the expense and long-term direction of rubber research in Fort Myers. In April 1933, Harvey Firestone Jr. hinted that, in view of the "difficult times," it might be time for the EBRC to reduce its commitment to the project. Firestone's chemist Earl Babcock visited both Chapman Field and Fort Myers in January 1934 and concluded that, in the long run, the USDA facility was more likely to make significant research discoveries than the EBRC.[67] Thus in early 1934, Miller offered four proposals to the EBRC's stockholders: (a) continue another season of work similar to that of 1933, for a cost of about thirty thousand dollars; (b) continue and broaden the project to develop machinery for planting, harvesting, and drying, which would require an additional investment to hire a mechanical, a chemical engineer, or both; (c) abandon all crop work beyond keeping a line of valuable goldenrod strains alive until needed for a war emergency; or (d) to turn the entire experiment over to the government and allow its scientists to continue or discontinue the project as they saw fit. Miller seemed to tip his hand in this memo in favor of the first option: "It would be most unfortunate to stop the experiments at this time," because another season of research would surely contribute to the overall value of the project.[68]

John V. Miller and principal stockholders Ford, Firestone, and Mina Edison met in Fort Myers to discuss these alternatives. The group praised the success of the EBRC work, declaring that the results "fully proved that acceptable rubber could be obtained from goldenrod. . . . Mr. Edison's original purpose . . . had been accomplished." But they conceded that progress on the extracting process and associated machinery was several years away; on that basis they decided that the better option was to induce the federal government to take over the project. To that end, the EBRC officers sent Miller and Charles Edison to Washington, DC, to present their proposal. Privately, Harvey Firestone admitted that he was willing to "close it up" altogether if government help did not materialize.[69]

On 5 April 1934, Charles Edison and Miller met with Secretary of Agriculture Wallace and Associate Chief of the Bureau of Plant Industry F. D. Richey. Edison brought along photographs of the Fort Myers activities and small but impressive samples of rubber produced from Fort Myers goldenrod. Wallace and Richey cautiously explained, however, and that funds were tight. The government's budget for all rubber crop research at several facilities in the United States and beyond was only sixty thousand dollars, only double what the EBRC spent annually on its modest operations in Fort Myers and West Orange. Yet, Richey and Wallace said they were "considerably interested" in the Edison research; they sent USDA scientist Loren G. Polhamus to Fort Myers to learn more.

Although Mina Edison was reluctant to agree, the USDA issued a formal offer in late April to take over the chemical testing operations and begin experimental work on goldenrod at its plant introduction station near Savannah.[70] The reasons for the USDA's decision to locate its goldenrod research near Savannah are not immediately obvious, especially in view of its steadfast rejection of all overtures that came from the IRC in 1933 and 1934. In any case, the USDA station in Savannah had established itself by the early 1930s as a national center for bamboo research. In addition, the property had about 2,500 of varieties of foreign plants under cultivation, including chayote from Guatemala, dwarf lemon from China, flowering apricot from Japan, and juniper from North Africa.[71] Although the planting of goldenrod, a native American weed, represented a departure for the station's agenda, it potentially had more practical consequences. With a home just a few miles away, Henry Ford may well have used his political influence to secure the transfer to Savannah. In any case, the cooperative work with USDA scientists kept the project alive while providing government scientists access to the man Ford would soon hire to oversee experimentation on his Georgia properties, Harry Ukkelberg.

Despite Mina Edison's objections and the uncertain federal commitment to take over the project, officials of the EBRC accepted the terms suggested by the USDA.[72] In May 1934, Miller ordered Ukkelberg to prepare immediately for the project's transfer to Savannah. By July, Ukkelberg had sent 3,500 plants to Savannah; by November, the total was over 22,000. In spring 1935, Ukkelberg shipped another 19,830 specimens of the Fort Myers plants to the government investigators.[73] The USDA also took over from the Edison family the task of publicizing progress with goldenrod. In 1934, for example, it issued a report that asserted "goldenrod is expected to develop as the most promising commercially adaptable [domestic] rubber source."[74] That day, however, lay in the future. After another round of rubber extraction and testing in early 1934, the EBRC possessed a total of seven pounds of goldenrod rubber.[75]

Halfhearted discussions in Washington, DC, over the fate of the goldenrod project help explain how the United States failed to adequately prepare for the rubber crisis of 1942. In January 1935, USDA secretary Wallace informed President Roosevelt that Edison's work "would be crippled" without a last-minute appeal for budget adjustments.[76] The president responded, however, that he did not consider the issue of rubber crops of "sufficient urgency" to make a special appropriation.[77] The government's reluctance to pursue the project came as a blow to the Edison family. In a letter to her son Charles, Mina Edison expressed regret that "the Botanic has been taken away from us."[78] Charles Edison made another appeal for government support for his father's goldenrod research in 1936, this time enlisting the support of Eleanor Roosevelt, who in turn lobbied Secretary Wallace. As in the previous year, Wallace then appealed to President Roosevelt, explaining that Bureau of Plant Industry scientists were working on the project even to the extent of "drastic curtailment of other important lines of rubber investigation." USDA researchers had determined, however, that Edison's oft-claimed result of 13 percent rubber from one strain of goldenrod could not be replicated: they deemed "unpromising" the likelihood of real progress. Although skeptical, Wallace concluded with a direct appeal for further funding for domestic rubber crop investigations, because it was "conceivable that any one of several conditions" might cut entirely American access to East Indian rubber, "which would be a real catastrophe." Once again, President Roosevelt declined to expand the American search for a domestic rubber crop. Later that year, the Department of Agriculture turned its attention to studying the blight disease that affected *Hevea*, and the Department of State intensified efforts to develop rubber plantings in allied nations of Central and South America.[79]

If the USDA's goldenrod research in Savannah continued on a shoestring, the EBRC's search for a domestic rubber crop continued in Fort Myers on a thread. In 1934, workers planted 220 beds with goldenrod, 64 percent fewer than in the year before.[80] In spring 1935, Ukkelberg planted only 60 beds.[81] Nevertheless, the EBRC managed to remain in operation, hanging on because of Mina Edison's unwillingness to see the project end and Miller's conviction that Ukkelberg's research remained of value.

Finally in May 1936, Edison's widow (recently married to Edward Everett Hughes) and her son Charles finally consented dissolve the EBRC and close the Fort Myers project "as soon as possible." Miller met with his sister and nephew several times that spring before all agreed on 29 May 1936 to the closing of the company.[82] A meeting held in West Orange on 1 July 1936 made the dissolution of the EBRC complete. Declaring that Thomas Edison's plans "had been realized as it has been proven without a doubt that certain varieties of goldenrod (*Solidago*) yield excellent raw rubber," Miller, Mina Edison Hughes, and Charles Edison voted to reward Ukkelberg "for his conscientious and excellent work," to sell as much equipment to Henry Ford as he might want for his own agricultural investigations, and to assign Edison's patent for rubber extraction to the USDA.[83] Thus by the mid-1930s, Edison's goldenrod project, the IRC's investments in guayule, and the USDA's enthusiasm for milkweed, intisy, and other rubber crops had all diminished to almost minuscule levels. The American search for a domestic rubber crop had reached its nadir.

Chemurgy, Economic Nationalism, and Rubber in the Savannah Area

At the same time, however, agriculture's prominence in the nexus of industry, science, and geopolitics remained in full view. Known as "chemurgy," the idea that Americans should develop agricultural sources for important industrial raw materials began to reach a broader audience in the mid-1930s. Wheeler McMillen, a leading chemurgist and associate editor of *Farm and Fireside*, interviewed Ford and Edison in 1928 on the rubber research in Fort Myers. Ford subsequently kept in touch with McMillen and often invited him to Dearborn, Michigan, to further discuss the industrial utilization of farm crops. For his part, McMillen became convinced that chemurgy would grow after garnering the support of the "three men whom I regarded as the three greatest Americans then living . . . Edison, Ford, and Hoover."[84]

The movement entered the national spotlight in 1935, when Ford and William Hale, an executive with the Dow Chemical Company, resolved to create a national organization to support the chemurgic idea. They invited several hundred representatives of agriculture, science, and industry to the founding meeting in Dearborn of what became the National Farm Chemurgic Council (NFCC), with Henry and Edsel Ford as cohosts. Surrounded by mementos from Edison's career, Abraham Lincoln's desk and chairs, and within the replica of Independence Hall that Ford had erected at Greenfield Village, the chemurgists issued with pomp and circumstance a dramatic "Declaration of Dependence upon the Soil." "When in the course of the life of a Nation, its people become neglectful of the laws of nature," the chemurgists declared, "necessity impels them to turn to the soil" and seek a "new frontier" in agricultural crop research. The chemurgy movement was unabashedly political. Leading chemurgists scorned the Roosevelt administration's call for crop reductions and artificial scarcity, ridiculed the New Deal's programs of welfare and work relief as a poor substitute for "honest work," and warned that national leaders had failed to appreciate U.S. vulnerability to foreign nations' control of strategic materials. Germany and Japan, they predicted, could not be depended on, and they challenged the Roosevelt administration's eagerness to expand bilateral trade relations with other nations.[85]

Meanwhile, the threads that connected Ford, Edison, and the struggle for domestic sources of American industrial raw materials became increasingly intertwined. In May 1936, Henry Ford hired Ukkelberg, recently released from his work in Fort Myers, to direct experimental agricultural work at Ford Farms in Ways Station, Georgia.[86] Although only four miles from the USDA facility, Ford's agricultural-research operation remained separate and distinct from the USDA plant introduction station. The record shows conflicting accounts of what kind of work was accomplished at Ford's Georgia retreat in the late 1930s. According to Ukkelberg, his boss "desired to carry on the goldenrod work at Savannah," and Ukkelberg was eager to do so. He made arrangements with Miller to take goldenrod specimens, about ten thousand plants in all, from Fort Myers as a start for a crop on the Ford property, and he brought some of the old EBRC equipment to outfit his laboratory.[87] Ukkelberg planted goldenrod soon after he arrived and asked Ford to supply a drying room, greenhouse, and laboratory sufficient to continue hybridization experiments.[88] As Ukkelberg later recalled, however, Ford told him that first winter that "we won't carry on the goldenrod work any more. Just plant a little for sentimental reasons. I did this for a number of years until I finally knew he wasn't interested any more." In other reminiscences, Ukkelberg suggested that he had realized, as Edison himself might have in due time, that goldenrod simply could not produce enough rubber to be the solution for a wartime emergency.[89]

Yet such comments are at odds with surviving documents clearly indicating that goldenrod research continued at Ford's Georgia properties until at least 1944, and with Ukkelberg's involvement. In 1937, for instance, the crew planted fifty-nine different varieties of goldenrod, each in a row of two hundred plants. As in Fort Myers, Ukkelberg kept close track of the percentage of rubber, weight of leaves, height, and success of plants generated from seeds and from cuttings. Also as before, Ukkelberg followed closely the progress of the hybrid crosses that he had bred in Fort Myers, concluding in 1938 that "the ability to produce rubber is an inherited condition," apparently justifying his experiments in goldenrod genetics. In general, however, these tests were small in scale and the results not very encouraging. Rubber yields throughout the late 1930s rarely exceeded 7 percent; the highest was a specimen of 9.13 percent rubber in the 1938 crop.[90]

Just a few miles across the Ogeechee River from Ford's landholdings, the USDA also continued goldenrod research at its plant introduction station near Savannah. Researchers planted and analyzed goldenrod each year, beginning in 1934, and soon homed in on four varieties of *Solidago leavenworthii*. Experimenters tested the goldenrod under different regimens of propagation techniques, soils, pest control, fertilization, harvest strategies, storage methods, and more. Unsurprisingly, this research did not change the fact that goldenrod rubber could not compete economically with imported *Hevea* rubber. Yet rubber yields rose steadily, and it seemed that, as Edison had foreseen, the plant still had some potential value in the case of a real raw materials shortage. When that rubber emergency came in 1942, the data that the Savannah scientists had accumulated made it relatively easy to scale up operations in 1943.[91]

Minor Possibilities

As intisy, guayule, and goldenrod all faded into relative obscurity, a similar fate awaited the other rubber plants that briefly garnered some attention in the 1930s. For instance, the USDA continued the search for rubber crops at the plant acclimatization station in Bard, California, a facility that paralleled the one near Savannah. There, the search for a plant suitable for arid lands focused on *Asclepias erosa*, a milkweed species that the California botanist Harvey Monroe Hall had championed following his original research of the rubber rabbitbrush. From 1931 to 1934, government scientists at the Bard station planted several test plots of milkweed in California and Arizona. Extensive research revealed optimal strategies for its propagation, culture, and harvest. The plant germinated readily, grew rapidly, and its harvested leaves could be stored for months without loss of rubber content. Best of all, the plant yielded as much as 13 percent rubber, with the mean yield a very promising 8.6 percent. According to another report, "Milkweed farms should not be difficult to establish" should *Hevea* ever lose its competitive economic advantage.[92]

Another attempt to develop an American rubber crop centered on poinsettia, *Euphorbia pulcherrima*. Countless citizens had suggested to Edison that the plant's gummy sap was reminiscent of rubber, although the inventor quickly rejected poinsettia as a possibility. Nevertheless, Edgar B. Davis, the magnate who once controlled the largest of American rubber investments in the Dutch East Indies, embraced the plant as an attractive alternative to cotton monoculture for the American South. Late in his career, Davis had created a million-dollar philanthropic organization, the Luling Foundation, which eagerly tested guayule and poinsettia in the mid-1930s as potential rubber crops for southern Texas.[93] Similarly, Orlando businessman Chester Kennison persuaded Florida's governor and agricultural leaders to conduct research on this plant. Florida's experiment station officials were skeptical. Indeed, both Kennison and Davis soon found that poinsettia offered little potential, so yet another potential rubber crop fell out of favor.[94] Another aggressive poinsettia promoter, Herman E. Pitman, appeared on the scene in 1934 and claimed that he could supply 60 percent of nation's rubber needs from poinsettia. These exaggerations were quickly exposed. Pitman was arrested for fraud, thereafter seeming like little more than a crackpot.[95] Amid this episode, one frustrated USDA official complained, "If only people knew what a terrific job this chase of an all-American rubber source is!"[96] As in so many other cases, realists who kept their eye on cheap and secure sources of imported *Hevea* rubber had little time for the enthusiasts who continued to call for a domestic rubber crop.

Rubber and Debates over War Preparedness

Geopolitical struggles over rubber ran into yet another formidable barrier in the 1930s. As Germany, Italy, and Japan revealed their aggressive international aims, the political tide in the United States turned toward ever more strident isolationism. Although the chemurgy movement's leaders called for an aggressive strategy of economic nationalism and autarky, many Americans preferred to ignore altogether ominous international developments. Isolationist politicians began arguing that any policy that veered from past practice amounted to a step toward American intervention into the world conflict. Most prominently, Senator Gerald Nye of North Dakota hosted congressional hearings from 1934 to 1936 revealing that government contracts in World War I had been riddled with conflicts of interest among military and industrial leaders. As leftist critics saw it, "merchants of death" deliberately pushed the country into international entanglements that would allow their "war profiteering."[97]

Those who called for war preparedness in the United States found themselves on the defensive. Few could speak about the need to prepare for shortages of strategic materials like rubber. One exception was Charles Edison. In November 1936, President Roosevelt tapped Edison to replace Henry Latrobe Roosevelt as assistant secretary of the navy, the same post that both he and Theodore Roosevelt had used as a stepping-stone in their own political careers.[98] Due to the debilitating illness of Secretary of the Navy Claude Swanson, Edison soon took the lead in preparing the nation's navy for potential war. Outspoken in favor of a strong defense, Edison naturally attracted the attention of Bernard Baruch, the nation's most prominent promoter of war preparedness.[99] For his part, Baruch delivered speeches that warned of Nazi designs on Mexico and other Latin American nations, and he urged a rubber stockpile as part of the nation's strategic response. The internationalist Council of Foreign Relations also joined the debate, highlighting rubber and the lessons of the Stevenson Plan as a warning against shortsighted and isolationist policies.[100] As a whole, however, the isolationist environment of the late 1930s pushed the notion of domestic rubber crops even further into the background.

In this milieu, American policy makers made only minimal efforts to prepare for a potential rubber shortage through alternative strategies. The first of these centered on the assumption that a synthetic-rubber solution remained just around the corner. DuPont's announcement in fall 1931 that it had developed a commercially viable form of synthetic rubber fit this pattern, what one historian has cited as confirmation of a faith that Americans had the ingenuity and resources needed to overcome the economic hardships of the Depression. More broadly, historians have identified an "undercurrent of optimism . . . and American triumphalism" that lay behind the faith that a real rubber crisis would not occur.[101] Executives at the Standard Oil Company of New Jersey (SONJ) contributed to this attitude by assuring reporters that the company had blueprints in place and could begin to build synthetic-rubber factories "immediately" if the government asked it to do so.[102] Such optimism also meshed with the widespread hopes that the United States could avoid entanglement with the deepening world crises. In the words of one isolationist text that touted synthetic rubber, "One thing is certain: if there are any reasons for an uncertain, long-distance war with Japan . . . rubber . . . is not one of them."[103] The widespread assumption that Americans had the know-how to master synthetic rubber yielded little in an era when private industry lacked incentive to make a wholesale change in its infrastructure.[104]

As war in Europe began to loom even more ominously, a few American policy makers turned to a second preparedness strategy: the stockpiling of rubber. Beginning in 1936, a small cadre from the departments of State, Commerce, War, Navy, Treasury, and Interior unofficially created an Interdepartmental Committee on Strategic Materials. Led by State Department economist Herbert Feis, this group continuously warned of the possibility of a crisis with several strategic materials and called for rubber stockpiles as a fundamental element in America's national security strategy. For Feis and his allies, the rise of Japanese power in the Pacific and the decline of Britain's imperial power made Southeast Asia an area of great strategic importance. Tin, quinine, starch, and other commodities were also at risk. Yet President Roosevelt hindered efforts to stockpile rubber and other materials, mainly because he preferred to address the unemployment issue more directly. In general, neither the major rubber companies nor the federal government was ready to invest in complex proposals to stockpile rubber that would disrupt international agreements regulating rubber prices and production.[105]

Rubber stockpiling efforts finally made headway when war appeared imminent. In June 1939, Feis and Joseph Kennedy Sr., the American ambassador to the United Kingdom and father of the future president, hammered out a controversial barter agreement that allowed the United States to exchange about six hundred thousand bales of surplus cotton for about ninety thousand tons of British rubber. The paths of Secretary Wallace and Charles Edison, now the acting secretary of the navy, crossed once again, because both were among those responsible for the transfer, inspection, and storage of the bartered rubber. The two settled the arrangements on 25 August 1939, one week before the German invasion of Poland.[106]

The scheme became increasingly complex once Britain entered the war against Germany. Cargo space became scarce, Britain needed cash more than it needed cotton, and British officials were eager to use rubber to draw the United States closer into an alliance. Finally, by June 1940, leaders of the preparedness movement induced Congress to fund the Rubber Reserve Company (RRC), a government agency charged with purchasing and stockpiling rubber from Southeast Asian suppliers, overcoming problems obtaining cargo space in Pacific ships and transcontinental railroads, and managing warehouses. The RRC's haphazardly developed programs and modest successes offered isolationists another reason to believe that Japanese aggression in Southeast Asia would not harm American interests.[107]

The third preparedness strategy, the hope that Latin American nations could again supply the United States with rubber, accorded well with the Roosevelt administration's hemispheric approach to trade issues. For decades, most attempts to establish a Latin American rubber industry, including Henry Ford's multimillion-dollar investment in Brazilian rubber plantations, had been decimated by endemic rubber blight. One of the best hopes to solve that problem lay in the USDA plant introduction station at Chapman Field near Miami. Established at the height of the Stevenson Plan, this facility had dabbled with research on guayule, cryptostegia, and goldenrod, but its main contribution in the 1930s had been the development of a blight-free strain of *Hevea* that advocates believed might lead to the reestablishment of the Latin American rubber industry. By 1934, thirty thousand *Hevea* rubber trees grew on this Florida property.[108] Wallace and his key scientific advisor, Earl Bressman, invoked geopolitical reasons for their push for Latin American rubber, warning that German, Japanese, and Italian investors had already made headway in Latin America in their own race for alternatives to Asian rubber. According to internationalists like Baruch and Wallace, politicians who opposed American investments abroad stood in the way.[109] Even more frustrated once war had begun in Europe, Wallace complained to Baruch that he had been trying to "drive home with Congressional leaders for two years without success" the notion that investments in Latin American agriculture and forestry were an essential strategy to reduce dependence on Asian resources.[110]

These three initiatives—synthetic rubber, stockpiling, and efforts to jumpstart Latin American rubber production—did not, however, add up to much. Many Americans had little concern about the ominous developments in global affairs, and neither did many of their leaders. As Congressman Hamilton Fish III said in 1939, "I cannot conceive of any war that would cut off these raw materials."[111] Secretary of State Cordell Hull succinctly articulated this attitude when he noted that both domestic rubber crops and synthetic-rubber substitutes would be inferior to and more expensive than imported *Hevea* rubber. From this premise, he concluded that *any* rubber alternative would be detrimental to the American way of life.[112] In a speech delivered before the Army Industrial College in November 1939—after Nazi Germany and the Soviet Union had already conquered and dismembered Poland, and after Japan had brought much of eastern Asia under its control—A. L. Viles, longtime head of the Rubber Manufacturers' Association (formerly the Rubber Association of America), expressed virtually no concern about any potential threat to American rubber resources. Viles dismissed the shortcomings in

Germans' synthetic rubber program as a sign of their "abysmal stupidity," the calls for North American rubber crops as "certainly impractical," and efforts to develop Latin American *Hevea* plantations as too risky in an era when "some chap" in rural Brazil could turn against American corporate interests for political reasons. Seeing no possibility of a disruption in the eastern Asian supply chain, Viles assured the leaders of America's military-preparedness efforts that "our military requirements could be fully met."[113]

Rubber Debates before Pearl Harbor

As German and Japanese armies continued to conquer one nation and colony after another from 1939 to 1941, some important American leaders still actively opposed investment in domestic rubber crops. Opposition to the domestic rubber alternative came from four principal sources: the influence of isolationists who suggested that real war preparedness was unnecessary; faith that modest investments in synthetic rubber and rubber stockpiling would be adequate; belief in an effort to encourage a natural rubber revival in Latin American nations as a long-term strategy; and trust that the Dutch and British could withstand or contain Japanese aggression. Goodyear executive J. A. Seiberling, for instance, indicated in an April 1940 speech that he doubted the Japanese would be so "brash" as to attack "the rubber islands."[114] Even in the fall of 1941, President Roosevelt explained in a letter to his wife that his top priority was to appease the Japanese and avoid anything that could disrupt the trade in tin and rubber.[115] In this context, some experts rejected rubber stockpiling, investment in synthetic rubber factories, and the establishment of rubber plantations in Latin America, because each approach could disrupt and duplicate a "profitable and established industry."[116]

Many assumed that any rubber shortfall would be small in scale and brief in duration, and thus could be answered through a relatively modest program in stockpiling natural rubber and nudging along the synthetic-rubber industry. In a May 1940 press conference, Roosevelt reassured Americans that the situation was not dire and that he had faith in scientists' ability to develop substitutes should that become necessary.[117] Rubber industry leaders opposed a substantial investment in synthetic rubber or alternative natural rubber sources because the results would only exacerbate the oversupply problems that had plagued the industry since the early 1930s. While Nazi Germany and the Soviet Union made a massive commitment to synthetic rubber as vital to their war-preparedness efforts, American leaders eschewed this approach. Instead, in October 1940, Jesse Jones, head of the new RRC, announced plans for a modest investment in synthetic-rubber factories with the potential to produce one hundred thousand tons. Under pressure to not "throw up huge shadow of [synthetic-rubber] plants before they are needed," however, Jones subsequently reduced planned production to forty thousand tons, reducing it again in spring 1941 to a mere ten thousand tons.[118] With the benefit of hindsight, critics were eager to blame Jones for the rubber crisis that emerged in 1942.[119] Yet even the modest stockpiling program made a real difference when the rubber crisis arrived in the aftermath of Pearl Harbor. The United States managed to secure over six hundred thousand tons of rubber by the time Singapore fell to the Japanese in February 1942, an amount that was enough, but barely, to tide the nation over until synthetic rubber began production in 1943.[120]

Meanwhile, USDA secretary Wallace and his allies insisted that the nation's rubber policy should be part of a broader strategy to expand the market for American farm products. These "agricultural internationalists" saw plants like rubber as more than raw materials with industrial and military importance; they also were tools that could be used to leverage broader strategic goals. The solution lay in developing *Hevea* rubber plantations in tropical Latin America, not in experimenting with domestic rubber plants. Investments in Latin American rubber not only would balance the risk of dependence on the Asian trade, but they also would allow American farmers and manufacturers to develop reciprocal trade agreements for goods that Latin Americans demanded.[121] Yet these efforts ignored or downplayed one crucial fact: it would take at least seven years for a new industry in Latin America based on the *Hevea* rubber tree to reach maturity.

Nevertheless, on 10 May 1940 (the very day that German armies invaded the Netherlands and thus exposed the Dutch East Indies to a Japanese invasion), Wallace announced plans for an aggressive strategy to promote the revival of rubber production in Latin America and the Caribbean. In subsequent months, the United States signed trade deals with Brazil, Costa Rica, Nicaragua, Peru, Bolivia, and ten other potential rubber-producing nations to solidify its access to potential Latin American rubber supplies.[122] These agreements typically promised to cover basic expenses of establishing a rubber industry, including a supply of the disease-resistant *Hevea* seeds that had been developed at Chapman Field. Beginning in September 1940, the USDA also pulled together several teams of university botanists and agronomists, who were sent into Brazil, Colombia, Honduras, the Philippines, and elsewhere with instructions to scour the countryside for the genetic materials of promising strains of potential rubber varieties.[123]

In the face of so many sources of opposition, supporters of domestic rubber found little support at the federal level. At the War Department, war-preparedness experts had buried and forgotten Dwight Eisenhower's 1930 recommendation that four hundred thousand acres of guayule be planted as a permanent and living rubber supply available to harvest in the case of a war emergency. In the State Department, officials continued to ignore recommendations from Herbert Feis's small committee that warned of rubber shortages in Southeast Asia. Then in spring 1940, even as German armies swept through western Europe, the House of Representatives tried to remove all appropriations for rubber plant investigations. Convinced that research on goldenrod and similar plants was merely a "plaything of the scientists," Missouri congressman Clarence Cannon wanted to cut the Bureau of Plant Industry's entire appropriation for rubber and other tropical plants. Although the final budget salvaged twenty-three thousand dollars for the rubber research in Savannah and at Chapman Field, it could not be clearer that domestic rubber remained very low among American priorities in the spring of 1940.[124]

Yet there were a few exceptions to the general mood of indifference. For instance, an unsigned USDA report from April 1940 highlighted the "imperative" that the United States do something about its dependence on a "remote" corner of world for rubber supplies. Going over the main areas of progress in rubber crop research, the author argued that having the foresight to plant *Hevea* in southern Florida and guayule in the Southwest might aid the United States if a rubber supply crisis were to arise.[125] In May 1940, Drew Pearson and Robert Allen complained in their nationally syndicated "Washington Merry-Go-Round" column that the U.S. departments of Agriculture and State had shown "amazing nonchalance" on the rubber issue.[126]

Also in the spring of 1940, Salinas, California, officials and their Washington, DC, allies launched an effort to revive the guayule industry. That April, the local chamber of commerce sent a flood of letters that alerted influential journalists, members of Congress, cabinet officials, and President Roosevelt of the guayule alternative. In the House, John Z. Anderson, the congressman who represented the IRC's home base of Salinas, entered the stage as one of guayule's main promoters.[127] In June, Senator Sheridan Downey of California proposed a bill that called for increased funding to support the defense of the United States and its allies in the Western Hemisphere. In congressional hearings over this bill, several rubber and chemical industry experts testified that the synthetic-rubber solution could not be guaranteed. Texas senator Morris Shepard added the dire prediction that "civilization could not go on if we did not have rubber tires" and "the present material civilization of the US could not last a year." For his part, Downey warned colleagues that they would be "criminally neglectful" if they did not prepare for rubber and other material shortages.[128]

Guayule's promoters in 1940 also enlisted the help of war-prepared-ness guru Bernard Baruch. Although he had helped organize the exchange of American cotton for British rubber in 1939, Baruch was troubled by the threats to the American rubber supply and had concluded by spring 1940 that synthetic rubber remained too distant in the future to offer any signifi-cant contribution to a future war effort.[129] As part of his efforts to promote his military-preparedness bill, Senator Downey delivered a speech warning of imminent rubber shortages, the need to establish closer ties with Mexico, and the limitations of the synthetic-rubber program.[130] Baruch saw the news reports and wrote to Downey that very day, "I am sorry I did not know you during the last three years while I was preaching preparedness. We might have made greater headway."[131] Within a month, Baruch brought Carnahan to Washington, DC, to argue for guayule before the National Defense Advisory Commission (NDAC), an agency that quietly worked on war-preparedness issues. Carnahan's testimony indicated that the IRC could produce no more than 550 tons of rubber per year from its Mexican sources—about 0.08 per-cent of the nation's peacetime rubber consumption. For an investment of $112 million, however, the government could produce guayule on 780,000 acres of otherwise uncultivated western lands that could yield 100,000 tons of crude guayule rubber per year—about 15 percent of normal consumption. Realizing that the IRC would not obtain from the U.S. government the kinds of royalty agreements and patent protections that the Italian government had offered, Carnahan put a new proposal on the table: he was willing to sell all of the IRC's assets in the United States (but not its relatively profitable operations in Mexico) to the government. After twenty-eight years of attempts to establish a guayule industry in the United States, the IRC was ready to get out, and the threat of war offered the IRC the opportunity to recoup its investments. "We may hope to cash in," as Carnahan put it, in a way that they never could in a free market. From this point forward, the IRC's real agenda was to sell its domestic rubber assets to the U.S. government.[132]

Most officials, however, remained skeptical about guayule. Even fewer could support a government buyout of a private firm. In the House, Congress-man Everett Dirksen noted that the USDA had spent over four hundred thou-sand dollars in the previous eight years searching for domestic rubber crops yet had produced virtually no tangible results.[133] In July 1940, Secretary of Agriculture Wallace responded to such appeals by pointing to the "sympa-thetic attention" that his department had given to guayule developments over the years. Wallace admitted, "It is true that at present there exists an emer-gency with regard to rubber supplies in the United States," yet he said that he could not "visualize" a situation in which guayule rubber could be competitive with imported sources.[134] Further, Wallace had publicly and explicitly pointed to guayule as the kind of project that the government simply could not support.[135] Carnahan, Baruch, Anderson, and Downey persisted in their efforts to find federal support for guayule, but to no avail.[136]

Most government leaders assumed that no rubber emergency loomed. The NDAC rejected the guayule proposal, convinced that guayule would offer "practically no immediate solution" if the regular supply were cut off.[137] Officials in the Council of National Defense optimistically foresaw no imminent rubber emergency, and they praised an August 1940 issue of *Fortune* magazine that endorsed the government's decision to not invest in synthetic-rubber factories before they were needed.[138] Even Henry Knight, the man in charge of USDA research on the industrial utilization of farm crops, felt no sense of urgency to develop alternative sources of rubber. After a December 1940 meeting with USDA colleagues, Knight wrote in his diary, "My advice to [Henry A. Wallace] was to go ahead with the South American plans but to re-evaluate in about five years."[139]

Efforts to promote guayule intensified in 1941. In April, nationally syndicated columnist Damon Runyon touted the guayule possibility with a column that bluntly accused American leaders of failures in their war-preparedness efforts.[140] Immediately thereafter, Congressman Anderson of Salinas again took to the House floor with a speech that repeatedly criticized government officials for policies that he deemed "unthinkable" and "perfectly ridiculous." Predicting that Pacific rubber supplies would soon be interrupted, Anderson pleaded, "When will we wake up?" Following applause from his congressional colleagues, Anderson concluded with an appeal for the planting of an emergency reserve of guayule, perhaps of 400,000 acres, a living rubber reserve that could yield some 125,000 tons.[141]

Guayule's fate faced additional hurdles after the IRC's longtime president George Carnahan died in March 1941. Recognizing that he might expect "violent political opposition" to their bailout strategy, the company's new leader, Henry G. Atwater, determined that only a "carefully mapped and sustained campaign yet to be created" could succeed. Convinced that IRC stockholders deserved roughly $3.5 million as fair return for their thirty-year investment in a strategic resource, Atwater began lobbying Washington, DC, officials every week. Several officials agreed that the IRC deserved some compensation for its seeds and know-how, but none could recall a precedent or imagine a scheme for such a bailout.[142] Moreover, most USDA officials remained committed to Latin American rubber. In April, Wallace's successor as secretary of agriculture, Claude Wickard, warned President Roosevelt that recent publicity about the guayule solution had been somewhat overblown. Only in the case of an "extreme emergency," resulting in a "complete cut-off" of rubber supplies from Asia, would the United States need to turn to a domestic crop.[143] The president took Wickard's advice to heart.

Once again, guayule supporters altered their strategy. Because skeptics had challenged the feasibility of the 400,000- and 780,000-acre plans, Congressman Anderson introduced a new bill that called for the government to buy or lease the American resources of the IRC and to begin guayule production on only 45,000 acres.[144] Amid the many discussions on the guayule possibility, several USDA officials presented their opinions. On 6 June 1941, Elmer W. Brandes offered a sober analysis of the plant's pros and cons, which included high praise for the IRC's advanced research and "ingenious machinery." If a Pacific war lasted until 1947, Brandes predicted, the United Sates could hope to obtain over 60 percent of its rubber from guayule but only 12 percent from synthetic sources.[145] Loren Polhamus, the department's rubber crop expert, concluded that guayule offered a "definite measure of insurance" in the case of a rubber emergency, but that the government had "no plans" to invest in it or to offer financial assistance to others who wished to do so.[146] In contrast, another USDA official suggested that the overcapacity of rubber in East Indies meant that no one would ever invest seriously in guayule. Even if East Indies were cut off, an "almost unlimited quantity" of rubber was possible from both synthetic and Latin American sources.[147]

Anderson's bill, however, stalled. Despite his weekly lobbying efforts, the IRC's Atwater could not persuade war-preparedness experts in the Office of Price Administration, the Department of the Navy, the RRC, or other agencies to commit to the bailout. He often found bureaucrats to be on vacation, evidently not worried that the Nazi invasion of the Soviet Union or Japanese advances into Indochina could threaten American interests.[148] Several members of Congress and Bureau of the Budget officials questioned why the IRC would want to sell its assets in California but not those in Mexico. Others debated the concept and future implications of a government bailout of a private American company like the IRC.[149] Firestone and most other large rubber companies joined the federal officials in expressing little interest in guayule rubber.[150] Jesse Jones's Reconstruction Finance Corporation (RFC) returned an evaluation that was "not sympathetic."[151] In September 1941, another government agency studied the issue and concluded that an investment in guayule would be far more cost effective than one in synthetic rubber, but it downplayed both possibilities because it saw no need or justification for government intervention into the private rubber industry.[152]

Only a few experts beyond the Salinas Valley spoke up in favor of guayule and other alternative crops. One was Ernst Hauser, associate professor of chemical engineering at the Massachusetts Institute of Technology. Hauser had been warning Americans about the potential rubber crisis since 1935, the year he immigrated to the United States.[153] In a June 1941 speech, Hauser directly challenged the politicians' confidence in the rubber situation. Surveying the alternatives, Hauser concluded that guayule might be considered the "ideal home supply." Although the chemical engineer Hauser was inclined to support synthetic rubber—and did so in 1942—he admitted that the time required to develop both synthetic rubber and rubber from Latin American *Hevea* plantations was even longer than that needed for guayule. Hauser scolded those who still doubted the guayule alternative: "We have waited too long in our systematic studies of home-made rubber; let's not wait too long in our endeavor toward home-grown rubber."[154]

Guayule found another influential and vocal supporter in William O'Neil, president of the General Tire and Rubber Company. "Why copy Hitler with costly synthetic rubber?" O'Neil asked, when guayule promised a cheaper, more reliable, and homegrown solution.[155] Thus in late 1941, even before Pearl Harbor, O'Neil and his company's publicist, W. H. Mason, toured the country trying to drum up interest in guayule. O'Neil and Mason predicted that a rubber emergency was imminent and that investments in guayule would cost less, divert fewer important materials from war production, and offer better long-range benefits for the American economy than any investment in synthetic rubber.[156]

Despite the warnings from Baruch, Hauser, O'Neil, and others, few American politicians, business leaders, or common citizens concerned themselves with any impending threat to the rubber supply. Rubber was not yet front-page news, and the reports that did appear offered an optimistic and complacent message. The *Dallas Morning News*, for instance, endorsed the government's efforts to develop rubber projects in Latin American nations as signs that "Uncle Sam . . . is a canny investor." That newspaper also touted synthetic rubber derived from petroleum sources as a growth industry for Texas, thus revealing another motivation among those who had more faith in synthetics rather than experimental crops like guayule.[157] Popular press articles also stressed the simplicity, speed, and reliability of converting petroleum into rubber and assured motorists that tire and rubber supplies would be adequate.[158] The trade journal *Chemical and Metallurgical Engineering* predicted in late 1941 that synthetic-rubber production in 1942 could reach ninety thousand tons, about one-eighth of the nation's needs.[159] Meanwhile, Goodyear published a series of advertisements headlined "A Report on the Rubber Situation" that praised government programs and progress with synthetic rubber, and predicted there would be enough tires for everyone.[160] And the students who had heard A. L. Viles's 1939 prediction of ample rubber resources were no longer in the classroom; the Army Industrial College had closed its doors by December 1941, and its students were sent to help with the war effort. In all, the combination of public complacency, the unconventional provisions of the Anderson bill, and active opposition from governmental and industrial leaders thwarted any serious interest in domestic rubber crops in the late 1930s and early 1940s, even as a world war spread across Europe and Asia.

Chapter 5

Crops in War

Rubber Plant Research
on the Grand Scale

On SUNDAY 7 DECEMBER 1941, under the headline "U.S. Grows Own Latex," the *New York Times* published an extensive article that touted guayule as the crop that could make the nation independent of imported rubber.[1] The timing of this article was pure coincidence, as dozens of similar news stories had appeared in American newspapers and magazines from time to time over the previous two decades.[2] The *Times* story hinted at the possibility of genuine rubber shortages linked to a war with Japan, and it implied that Congress should pass the Anderson bill that called for the planting of forty-five thousand acres of guayule under federal government control. The article did not suggest, however, that a rubber crisis, one ominous for the entire Allied war effort, was imminent.

Yet events of that very day proved otherwise. With the attack on Pearl Harbor, the Japanese launched a wave of campaigns that led to seizure of the rubber industry's plantations, warehouses, and ships across the Pacific and Southeast Asia. The fall of Singapore in February 1942 cut off the United States and its Allies from at least 95 percent of their rubber supply. Within a matter of weeks, the United States had no access to its largest agricultural import, a natural commodity vital to its industrial, consumer, and war economy. Despite decades of warnings from Thomas Edison, Henry Ford, Dwight Eisenhower, Bernard Baruch, and many others, the worst-case scenario finally had come true.

Within days of Pearl Harbor, American leaders revived their hopes that a domestic agricultural solution to the imminent rubber shortages could be found. Old and discarded plans and proposals to grow domestic rubber crops suddenly became current again. Most centered on guayule, the one crop that had a track record of feasibility. Other plants also received renewed attention, including goldenrod, milkweed, cryptostegia, and rabbitbrush. In addition, a new candidate appeared on the scene, as kok-sagyz, or the Russian dandelion, seemed to offer an especially promising solution to the crisis. Common citizens presented various other plants as panaceas, including such unlikely possibilities as poinsettia, leafy spurge, wild potatoes, Osage orange, elderberries, and more. And as will be seen in the next chapter, others demonstrated that common grain crops readily could be converted into synthetic rubber, making every American wheat and corn farmer a potential rubber producer. War and agricultural science had seemingly come together at last to unlock the potential of domestic rubber crops.

In contrast to the isolated, small-scale, and ephemeral projects that Edison and other rubber crop enthusiasts had pursued in the 1920s and 1930s, the advent of global war demanded a more aggressive and comprehensive approach. The federal government became far more involved in agricultural and industrial policy than it had been earlier wars, stepping into areas where private industry could not meet the challenge.[3] Whereas World War I had brought an ambiguous relationship between science and the war effort, previous barriers had been torn down by the early 1940s. Many Americans now placed their faith in the scientific community's ability to find a domestic agricultural solution to the rubber crisis, which simultaneously enhanced the career prospects of thousands of scientists. As John Collyer, president of the B. F. Goodrich Company, put it during the depths of the rubber crisis, university scientists and engineers were the "military Commandos and Rangers" needed in effort to win the Second World War.[4] In the end, synthetic rubber derived from petroleum, not domestic rubber crops, brought the most significant changes in the political economy of rubber. In today's environment, in which there are searches for high-tech and synthetic solutions to every problem, that result does not seem so surprising. It is worth remembering, however, that World War II was the dawn of much of this thinking. In the aftermath of Pearl Harbor, the soil was the first place to look, and many Americans sought a natural, sustainable, and domestic source of rubber.

A vast national commitment to mobilize science, industry, and agriculture ensued under the rubric of the Emergency Rubber Project (ERP). Created in March 1942, the ERP emerged as something like the Manhattan Project of the plant sciences, comparable to some degree in terms of scale, urgency, and interdisciplinary scope. Between 1942 and 1945, over one thousand American plant pathologists, plant physiologists, geneticists, agronomists, entomologists, foresters, and agricultural engineers worked to find an agricultural solution for America's rubber emergency. Tens of thousands of field hands, construction workers, office staff, and other supporting employees joined the effort as well. In all, the ERP led to some of the best-funded and most intensive agricultural research conducted in the nation's history. Ironically, the ERP and subsequent research made guayule the most thoroughly understood agricultural plant that still has no significant commercial market.[5]

After Pearl Harbor

With synthetic rubber a technological possibility but still far from reality in terms of mass production, many saw little choice but to turn to guayule and other obscure crops. Even before the smoke had cleared from Pearl Harbor, public officials who had virtually ignored the rubber issue for years quickly mobilized. Within three days of the Pearl Harbor bombing, government officials warned that rubber rationing was likely. On 10 December, Senator Sheridan Downey of California introduced S 2152, a new call for the U.S. government to acquire all domestic patents and rights of IRC, to plant forty-five thousand acres of guayule in 1942, and to finance the nurseries, surveys, and other projects necessary to support the guayule industry through the wartime crisis. To prevent profiteering, the Office of Price Administration (OPA) mandated on 13 December that guayule prices be frozen at their level of 6 December 1941. That same day, IRC officials met with Commerce Secretary Jesse Jones in a revived effort to sell their expertise in guayule to the federal government.[6] Meanwhile, USDA and Texas agricultural officials sent agents into western Texas to report on the status of guayule fields that the IRC had abandoned a decade earlier.[7] Others asked scientists based in Savannah to scour the vicinity for enough wild plants in seed that would enable a rapid expansion of the goldenrod project.[8] On 16 December, Undersecretary of Agriculture Paul Appleby finally responded to a June 1941 request for an assessment of Congressman Jack Anderson's proposed guayule legislation, HR 5030. The bill, Appleby suggested, was "desirable" and fit well into a "comprehensive plan for the progressive utilization of natural rubber from several available sources."[9] Anderson lauded the effort to move forward his proposal, which had been "gathering dust on various shelves around the downtown offices."[10] In spite of the new urgency, however, Anderson's bill remained "pigeonholed" in the House while the Senate considered the Downey bill. No guayule proposal emerged from committee before the close of congressional business in 1941.[11]

Skeptics remained. Despite what seemed to be a desperate situation, some government officials remained lukewarm to the idea, and congressional funding for domestic rubber crops was still months away. USDA official Elmer Brandes testified that guayule was certainly a proven crop in agronomic terms, but he embraced the administration's preference to develop a Latin American rubber industry. Vice President Henry A. Wallace (former USDA secretary) advised Commerce Secretary Jesse Jones that guayule was overrated and unlikely to affect the war effort before 1946. For his part, Jones downplayed the severity of the rubber crisis altogether, stating, "I have not felt too uncomfortable about rubber."[12]

Hearings and public debates over the guayule legislation resumed early in 1942. In the House Committee on Agriculture, several congressmen saw the proposed legislation as little more than a government bailout of an unprofitable corporation, which put IRC executives C. L. Baker and Henry G. Atwater on the defensive. Reluctant to state exactly how much money they hoped to receive for their unprofitable American seeds, patents, and other resources, Butler and Atwater nonetheless admitted that they had no plans to sell their profitable assets in Mexico. Forced to concede the botanic reality that slow-growing guayule probably would not contribute much rubber to the war effort before 1946, Baker and Atwater also revealed that the IRC had paid its investors substantial dividends in the past few years. Moreover, despite claims that the IRC had patriotic interests in mind, many beneficiaries of the proposed bailout lived in the Netherlands, for a Dutch holding company held 63 percent of its stock. Several congressmen found it odd that the IRC's executives insisted that the government could do a better job with guayule than a company with thirty-five years' experience in the business. Congressman Adolph Sabath of Illinois added fuel to the fire by charging that no rubber shortage existed at all; rather, he suspected that supplies were "controlled by a few for the purpose of mulcting the American people."[13] In all, the proposal smelled of a classic case of war profiteering.

Debates over domestic rubber crops also bogged down over other issues. Missouri senator Harry Truman, among others, asked about the implications of Edison's goldenrod experiments. Several members of Congress used the platform to blame Roosevelt administration officials for poor preparedness. For instance, Congressman Hampton Fulmer of South Carolina repeatedly bashed Roosevelt's "Good Neighbor" policy as a sign that the State Department preferred "some other country that we might be in love with" over the interests of American farmers.[14] A few weeks later, Fulmer complained that investing in guayule would be pouring money into "a rat hole." He predicted—and events proved him to be perceptive—that powerful interests would call for its end once the emergency had passed.[15]

Nevertheless, the need to do something about the crisis pressured Congress into pushing through a guayule bill. MIT rubber chemist Ernst Hauser again stood at the center of the debate. In a frank letter that supported the Anderson bill, Hauser scoffed at Jesse Jones's optimism and willingness to invest several hundred million dollars in the synthetic-rubber gamble. Even Germany, the nation that had best mastered production, recognized that synthetic rubber was no substitute for natural rubber in many military applications; indeed, the Germans had just initiated a broad effort to obtain natural rubber from its Japanese allies. The United States had an important advantage, Hauser believed, because it could develop a proven alternative source within its own borders. Failure to invest in guayule would be "committing a crime" that might prolong the war or even cost the United States a final victory, Hauser declared.[16] In addition, Hauser predicted that Japanese tactics in Southeast Asia would lead to postwar disruptions in the natural-rubber industry, and he foresaw guayule as part of a long-term solution.[17]

Congressional debates highlighted both the urgency of the problem and the costly and dangerous precedent of bailing out the IRC. Congressman John Flannagan of Virginia conceded that the IRC was "taking advantage of our situation" but considered it "criminally neglectful of the welfare of our country" not to pass legislation immediately and "push production of guayule rubber as hard as we can." Invoking the image of U.S. Marines who had died courageously defending Wake Island in December 1941, Flannagan called for his peers to pass the bill even though "no monuments" would be built for the politicians who corrected mistakes of those who had failed to prepare the country for war.[18] The terms of this bill were significant. In contrast to the IRC's original request for about $3.5 million for its American assets, the final legislation limited the government's expenditure to $2 million. After further negotiations, the company had to settle for about $1.7 million; IRC vice president Atwater urged his board members to accept the offer and to admit that they were lucky to be "out of a very bad holding."[19] In early February, the full Congress finally passed a bill that authorized the USDA to lease seventy-five thousand acres for the immediate planting of guayule within the borders of the United States. With several USDA officials already in Salinas to prepare for the 1942 crop year, the local newspaper reported it "almost a foregone conclusion" that the president would sign the legislation.[20]

Yet President Roosevelt vetoed the bill. In an action that reveals the administration's interpretation of the rubber crisis, Roosevelt responded that the country must promote any possible crop, "regardless of whether within or without the United States." Because of the clause limiting rubber crop research to the borders of the United States, Roosevelt saw the bill as a slap in the face of his own diplomatic efforts with Latin American nations.[21] The veto greatly frustrated Senator Downey, who complained that the "processes of democracy" threatened to delay the entire project.[22] As the deadline for planting the seedlings of the 1942 crop year approached, Downey warned that the whole rubber mess could have "consequences to our national existence and national safety that might be so incalculably tragic as to be beyond description."[23] Nevertheless Downey and Anderson introduced a new version of the bill, with the offending clause removed, which passed the Congress on 28 February.[24]

On 5 March 1942, an extra edition of the *Salinas Index-Journal* blared the headline "GUAYULE BILL SIGNED." The bill formally created the ERP, authorized the government to purchase all U.S. assets of the IRC, and transferred the firm's patents, property, machinery, employees, and, most significantly, its 22,867 pounds of guayule seed to the U.S. Forest Service (USFS). By noon of that same day, government representatives already stationed in Salinas had planted thirty-seven thousand seedlings from the IRC's nursery.[25]

The news that Roosevelt had signed the guayule bill that very morning made Salinas's Guayule Rubber Day—an event already planned for 5 March—a more robust celebration. Local politicians, merchants, builders, hotel operators, and other chamber of commerce members heaped praise on General Tire's William O'Neil and the local members of Congress who had successfully lobbied for the bill. Celebrations continued at the evening banquet, where rubber expert David Spence explained that through thick seeding and intense irrigation, it might be possible to greatly accelerate guayule rubber production and reduce the nation's emergency rubber needs as soon as 1944. Others asserted that the guayule rubber was likely to become a permanent cash crop in the valley.[26] Within weeks, Congress had also settled on the budgetary issues necessary to support the ERP: in contrast to the twenty-three thousand dollars spent in fiscal year 1941 for domestic rubber research, it appropriated over twenty-two million dollars to support the ERP in fiscal years 1942 and 1943.[27] Eighty days after the attack on Pearl Harbor, and after decades of indifference to the IRC's continual pleas for government support, the future of California's guayule rubber industry looked bright. Just a year or so after interest in domestic rubber crops had been at its nadir, America's domestic rubber industry reached a new zenith.

A National Mobilization for Domestic Rubber Crops

National war mobilization also revived the search for alternative rubber sources at the grassroots level. For instance, Hallie Leyda of Shreve, Ohio, urged USDA secretary Claude Wickard to invest in guayule rather than imported or synthetic rubber. "Why not" allow American farmers to receive some revenue from rubber, she asked, "instead of giving it to the rubber barons of Malaya?"[28] Elizabeth Hobson of Great Notch, New Jersey, assured Secretary of Commerce Jesse Jones that "all the kids and rest of us" would send him two bushels of milkweed to contribute to the war effort.[29] Eleven-year-old Marlene Leda of Milwaukee wrote President Roosevelt that she had observed a rubbery substance at the bottom of the barrel when her father manufactured elderberry wine. Government bureaucrats politely responded that their own tests on elderberries had not proven promising.[30]

Promoters of chemurgy and other forms of agricultural development also stressed the potential of domestic rubber crops. *Dallas Morning News* columnist Victor Schoffelmeyer wrote in December 1941 that it was "high time" for Americans to take the chemurgists' message seriously, reminding readers of his own endorsement of guayule nearly twenty years earlier.[31] In subsequent essays, Schoffelmeyer reviewed the state's several experiments with the crop over the decades and urged Texans to "bestir themselves" to gain access to a portion of the guayule program that seemed destined for the Salinas, California, area.[32] DeWitt Hicks, mayor of Waco, Texas, and a dealer in rubber products himself, concluded after a tour of the guayule habitat in remote western Texas that the plant could thrive in the Waco area. At their annual meeting in Waco in early January 1942, the Texas Agricultural Workers sent a resolution drafted by Schoffelmeyer to Secretary Wickard asking that Texans be offered a "proper share" of the effort to meet the war emergency through guayule production.[33]

Officials in other states also hoped for a portion of the new American rubber bonanza. Congressman James Scrugham, who had tried to develop a rabbitbrush rubber industry while serving as Nevada's governor in the 1920s, now led a new campaign to quickly harvest the plant.[34] A radio broadcast distributed throughout a network of western stations touted the plant as one linked to the future of American freedom. "When this war is finally won," the broadcasters asserted, "we may find that a good portion of the credit for victory goes to the science of botany."[35] Thomas H. Goodspeed, the California botanist and Harvey Monroe Hall's colleague in the search for rabbitbrush during World War I, again lobbied for the quick harvest of the millions of pounds of rubber already available in America's western deserts.[36] Samuel Doten, head of the Nevada Agricultural Experiment Station since the previous wave of enthusiasm for rabbitbrush in 1925–1926, argued that the time had come to harvest Nevada's living rubber reserve once and for all.[37] In the end, the rabbitbrush plants were left in place because of the lack of processing facilities. But thanks to Scrugham's lobbying, the phrase "and other crops" had been added to the federal guayule bill, which empowered the ERP to conduct research on goldenrod, rabbitbrush, kok-sagyz, cryptostegia, and other potential rubber crops during the war emergency.

Rubber crop enthusiasm also spread to Minnesota, another unlikely rubber-producing state. There, agricultural scientists recalled a letter that they had received in 1940 from Carl Pfaender, a Civilian Conservation Corps (CCC) worker from New Ulm, who had become enamored of the rubber potential of the leafy spurge, *Euphorbia esula*. In fact, university scientists had concluded that the plant did produce rubber and that its quality was reasonably high. Yet the topic was hardly a high priority in 1940 and 1941, and Pfaender realized that leafy spurge would not become important unless a rubber "catastrophe" hit the nation. Other agricultural officials in Minnesota had classified leafy spurge as a weed and worked hard for its eradication. Indeed, in fall 1941, CCC officials forced Pfaender himself to apply herbicide to several acres of the very plants that he had been tending as a potential rubber crop.[38]

All of this changed as botanists and other plant experts began mobilizing for the war effort. Ross A. Gortner, head of the Division of Biochemistry at the University of Minnesota, initiated a search for colleagues who might be interested in propagating, not eradicating, the state's remaining stands of leafy spurge. At a meeting on 20 January 1942, Gortner suggested the weed might become "a new crop for a permanent agriculture" on the marginal and "cut-over" lands of northern Minnesota and North Dakota.[39] Scientists predicted that leafy spurge might yield 150 to 200 *pounds* of rubber per acre. When the *Minneapolis Morning Tribune* spread word three days later that Gortner had hinted at 200 to 300 *tons* of rubber per acre, a flood of Minnesota citizens responded by offering their land and labor to turn leafy spurge into the state's new agricultural crop. The whole episode soon proved an insurmountable embarrassment for Minnesota's rubber crop enthusiasts when evidence of low yields and concerns about spreading the weed surfaced.[40]

One of the more important examples of wartime domestic rubber mobilization occurred at Cornell University. On 9 March 1942, the B. F. Goodrich Company offered the university a ten-thousand-dollar grant for the investigation of plants that could produce rubber on American soil. Only three days later, Cornell hosted a "Conference on Rubber Producing Plants," where several Goodrich officials discussed rubber plant issues with many of the university's taxonomists, cytologists, biochemists, physiologists, and cytogeneticists. Within a week, Dr. Lewis Knudson and the entire department of botany committed to the project; indeed, Knudson announced in mid-semester that his seminar in plant physiology would drop its normal topics and focus instead on the physiology of rubber plants. In early April, Knudson and his colleagues traveled to the USDA offices in Washington, DC, to collect data from previous studies (especially from Edison's notes), which they needed in order to launch their own research program. To stimulate allies at the grassroots level, Cornell officials called on local 4-H chapters and Future Farmers of America clubs to collect seeds from other potential rubber-producing plants of upstate New York.[41] The search for rubber crops spread across other campuses as well. Also at Cornell, Herbert Whetzel, one of America's most prominent plant patholo-gists, decided to study a relative of the potato, *Apios tuberosa*, as a suitable rubber source. He consulted with both a Princeton University scholar and an "old Seneca medicine woman" from the nearby Oneida reservation before concluding that the tuber could not solve the crisis.[42] By early 1943, Cornell scientists had investigated roughly 10,000 individual plants of 1,200 different species for their rubber content.[43]

At Clemson College, scientist J. H. Mitchell surveyed plants found in upstate South Carolina; like Edison, he deemed goldenrod the most promising possibility. At Virginia Polytechnic Institute, chemical engineering student William Jones carefully studied the *Solidago* species native to rural Virginia and West Virginia, concluding that none offered a feasible or economical solu-tion.[44] Agronomists in South Dakota and Oregon tested guayule, milkweeds, moleweed (*Euphorbia lathyrus*), snow bush (*Ceanothus velutinus*), prickly let-tuce (*Lactuca scariola*), and other plants suggested by both scientists and com-mon citizens alike.[45] In Utah, the president of Brigham Young University agreed to collaborate with the government of Iran in an effort to evaluate the guayule's potential in both regions.[46] Several researchers claimed that cotton, soybeans, and pine trees could yield American-grown rubber.[47] And in Ha-waii, Governor Ingram Stainback supported a project intended to tap the few thousand rubber trees scattered across the islands, survivors of an effort to es-tablish a Hawaiian rubber industry thirty years earlier. Enlisting convict labor to help complete the task, the governor hoped that Hawaii would provide a modest eight tons of *Hevea* rubber to the war effort.[48]

The wartime crisis also revived interest in the scientist George Washing-ton Carver's comments, first made public during World War I, that the sweet potato could be a rubber source.[49] With memories of Ford's collaboration with the aging Edison still fresh, rumors persisted that Carver had become Ford's new ally in the rubber project. Carver also received many letters from War Production Board (WPB) officials and common citizens who sought his advice on the potential of milkweed, Osage orange, sweet potatoes, and other crops as a domestic source of rubber, enhancing his mythic status an African American hero.[50] These grassroots efforts were a far cry from the model that historians have labeled "Big Science." If the ERP's scale, urgency, and commitment of agricultural and scientific expertise resembled that of the Manhattan Project, these smaller episodes underscore the breadth and depth of the struggle for domestic rubber sources in 1942.

Mobilization at Salinas

Even before the guayule legislation had overcome congressional delays and the presidential veto, dozens of government officials had descended on Salinas to launch the American search for a domestic and agricultural solution to the impending rubber crisis. Given carte blanche to organize the project, Christopher Granger, assistant chief of the USFS, chose fellow forester Major Evan W. Kelley to direct operations in Salinas. Granger and Kelley had both been exposed to war and resource issues while serving in the Tenth Forestry Engineers regiment during World War I. To meet the tremendous logistical needs for wood products, the military sent foresters to France to utilize the timber resources available there. For Kelley, Granger, and others, the experience made clear the urgency of securing raw materials during war.[51]

Kelley arrived in Salinas in February 1942 and immediately sent colleagues out to scout potential nursery sites, procure materials, inventory the IRC properties, and mobilize a workforce unaccustomed to rubber plant research.[52] His most difficult task was to manage the scarce genetic resources—almost twenty-three thousand pounds of guayule seed—that the government intended to purchase from the IRC. A military guard stood sentinel over drums of the precious material, the same seeds that had been destined to further the enemy's cause as part of the IRC's previous agreements with Fascist Italy.[53] As the project developed, Kelley and other ERP officials repeatedly sought the advice of outside engineers and management experts, and they came to believe that government officials could manage the project with greater expediency and skill than could private enterprise. Even as the official appropriation for the project still awaited congressional and presidential approval, Kelley mobilized local business owners to find the help needed to prepare the spring plantings and secure housing for the influx of new rubber employees.

Better known as "America's Salad Bowl," Salinas became a rubber boomtown in the spring of 1942. With little regard for expense, and at a pace that reflected the emergency nature of the project, Salinas's hundreds of new residents constructed scores of cookhouses, nurseries, roads, offices, and experimental facilities. Construction also included new electric, water, and sewage systems; 150 miles of irrigation lines; and a network of over 9,000 miles of "duckboards" (to support the workers and machinery that would travel over soft ground). By April, about 875 acres of guayule seeds were planted in fields, a first step toward the 75,000 acres authorized by Congress. Perhaps more

Figure 5.1. Salinas, California, as a rubber boomtown, showing the mobilization of the Emergency Rubber Project in March 1942. Photograph by Carl A. Taylor. National Archives.

important, virtually all the remaining seeds soon were planted in nursery beds to prepare for the next crop year, because transplants from nurseries seemed to do better than those started from seed. Amid this buzz of activity, the local chamber of commerce prepared materials that presented the guayule nurseries and fields as an attraction for visitors to witness an important manifestation of wartime mobilization.[54]

The labor question was an especially pressing problem. The ERP needed 1,900 additional workers to help with plantings and transplantings by May. As on other defense projects, the wartime emergency brought great pressure on the local workforce. Circumstances were exacerbated in the Salinas area, because March 1942 also marked the beginning of the deportation of local Japanese Americans, including many farm laborers, to the various internment camps that were quickly being built in isolated regions away from the coast. Thus, at the same moment that workers were hastily constructing barracks to house tens of thousands of displaced Japanese Americans at Manzanar and other internment camps, government officials authorized similar barracks in Salinas and other California towns in order to house thousands of newly arrived scientists, construction workers, and field hands associated with the ERP.

Labor demands intensified in May as weeds appeared amid the young guayule seedlings. After a two-day trial, ERP officials determined that "perhaps women can handle" the job and soon put hundreds of women from local high schools and colleges to work the fields. Unsurprisingly, local officials used gendered language in their praise for the women's "rapid and delicate finger work" and for the patriotic eagerness to work on their hands and knees for fifty cents per hour in the California sun. Newspaper accounts described these women as if they were Soviet-era shock workers, touting the girls' new records of productivity weeding and seeding the guayule beds. In all, nearly three thousand women and girls helped weed the Salinas area rubber plots in 1942. Meanwhile, ERP officials found different kinds of work for high school boys, mainly in indoor positions at the guayule nurseries. The next year, officials boasted that a shift to chemical and incendiary technologies to combat weeds meant that the whole project could be completed with only three hundred workers. The women who were so crucial to the nation's rubber project in 1942 were deemed expendable within the year.[55]

Meanwhile, because agreements with local growers stipulated that no more than three hundred workers could be drawn from the local workforce,

Figure 5.2. Guayule seedlings in an irrigated nursery, Salinas, California, 1942. National Archives.

the ERP turned to labor recruited from urban centers and braceros, the Mexican nationals who were hastily recruited as field workers during the war. Thus the demand for guayule rubber workers in Salinas played a significant but somewhat overlooked role in the history of California's increasing dependence on the work of Mexican nationals and their descendents. The sudden influx of Mexican workers into the guayule fields brought considerable social upheaval, ranging from the pressure on mess hall cooks to learn Mexican cuisine, to National Guardsmen who kept the immigrants from eating fruit from the citrus trees, to more serious and overt cases of racial tensions and suspicions. Forty percent of the ERP's 4,300 workers in 1943 were Mexican nationals, demonstrating that the rubber project contributed to a new social landscape in the Salinas Valley and beyond.[56]

Rubber and Manzanar

The political nature of the nation's search for domestic rubber crops became especially apparent as a number of California scientists and humanitarians mobilized for the guayule project. Robert Millikan, winner of the 1923 Nobel Prize in Physics and the president of the California Institute of Technology (Caltech), was most prominent among them. Biologist James F. Bonner, botanist Frits W. Went, and plant physiologist Robert Emerson—all from Caltech as well—also joined the cause. Soon after the European war broke out in 1939, Went and Bonner met with George Carnahan of the IRC and signed contracts promising to investigate the biochemistry, genetics, and physiology of guayule.[57] The Caltech scientists saw that guayule presented an unusual and fascinating "test case" in the history of agriculture, one where scientists' expertise on a relatively untested plant could remain ahead of the knowledge of farm practitioners.[58] Went also believed that guayule offered a more promising answer to a rubber crisis than synthetic rubber, and that this research should be recognized as an essential part of national defense efforts.[59] Their work began in October 1940; in September 1941, it moved to field trials, in cooperation with the IRC in Salinas. When the ERP took over the IRC's facilities in March 1942, officials argued that contracts with Caltech were now null and void. In any case, they believed that the top priority had to be rubber production, not research in plant physiology.

But this chapter in the dramatic search for a domestic rubber had just begun. As soon as Caltech officials received the letter officially terminating their access to the ERP, they launched their own project. Robert Emerson took the lead; in addition to being an expert on process of photosynthesis, Emerson was a committed Quaker and pacifist. Along with Hugh H. Anderson, another

Pasadena Quaker and activist who also had close with the Caltech scientists, Emerson responded quickly to the forced internment of more than one hundred thousand of Japanese Americans who lived near the Pacific coast. Emerson and Anderson both criticized inequities within the capitalist system, and Anderson passionately promoted rural and consumer cooperatives as an alternative. Both also recognized that many Japanese Americans had succeeded in the California nursery business; indeed, groups such as the White American Nurserymen of Los Angeles had lobbied for the internment of Japanese Americans largely because of their reputations as skilled and hard-working gardeners.[60] Millikan too had opposed the internments; as he stated in another context, he did not wish to see talented Japanese American scientists "locked up in [the Manzanar] Concentration Camp [with] nothing to do but wash dishes." Looking beyond the immediate issue, Millikan also saw opportunities to develop the economic potential of American deserts, and he and Bonner both warned that the rise of a synthetic-rubber industry based on nonrenewable petroleum was detrimental to the nation's long-term interests.[61]

Thus Millikan lobbied ERP officials for help, arguing that his colleagues' "foresight" to work on the rubber issue "should not be curbed just at the moment that it has become the most valuable." By late April, ERP director Kelley conceded that helping the Caltech researchers caused no problems, and Emerson traveled to Manzanar to find ways in which Caltech could sponsor the Japanese Americans' scientific work.[62] Beginning in April 1942, Emerson enlisted Anderson and others to help transport fourteen gunnysacks full of cuttings—the tops of some twenty-five thousand guayule seedlings—to Manzanar.[63]

Within days, the project moved into full swing. After painstakingly obtaining permission from various military and government officials, Emerson eventually solicited a minuscule budget of a few hundred dollars to pay for farm equipment, glassware, gasoline, and, somewhat ironically, spare tires.[64] There were difficulties: high winds destroyed one of the hastily constructed lath houses; rats, ants, and insects attacked the nurseries; and jackrabbits feasted on the young plants. Emerson brought in a few greyhounds from Los Angeles racetracks to hunt the rabbits down.[65] Nevertheless, by June 1942, the Japanese Americans at Manzanar had planted some 169,000 guayule plants grown from cuttings. According to one sympathetic source, "Japanese ingenuity came to the rescue" at the Poston War Relocation Center in Arizona as well, despite the "soaring temperatures . . . [and] continuous dust storms" of the western Arizona summer.[66]

Emerson also mobilized the talented scientists and nurserymen who had been assigned to Manzanar and other internment camps. Significantly, he

Figure 5.3. Dr. Shimpe Nishimura, an interned Japanese American scientist who directed guayule research at the Manzanar War Relocation Center in Owens Valley, California. Courtesy of county of Inyo, Eastern California Museum.

persuaded one of his former graduate students, Dr. Shimpe Nishimura, who was on the verge of returning to Japan rather than endure the camps, to stay in the United States and take over leadership of the Manzanar research. Emerson also arranged the transfer of a chemist with experience in synthetic-rubber chemistry from the Santa Anita assembly center to Manzanar. Emerson and his allies also persuaded sympathetic scholars from California universities to deliver talks behind the Manzanar barbed wire on guayule genetics, rubber chemistry, and related subjects. Eventually, chemists Kenzie Nozaki and Frank Hirosawa, cytologist Dr. Masuo Kodani, nurserymen Walter T. Watanabe and George Yokomizo, horticulturalists Tomoichi "Green Thumb" Hata and Frank Akira Kageyama, mechanical engineers Homer Kimura and Joe Iwamasa, statistician Shuichi Ogura, and several others formed a team of guayule researchers.[67]

Evidence that this small group of Japanese American scientists, working on shoestring budgets in makeshift laboratories and on tiny research plots, had more success than the large and well-funded research teams of the USDA brought another politically charged dynamic into debates over the American search for a domestic rubber crop.[68] Within weeks of the internees' arrival, *Science News Letter* published a brief article that touted the skillful, innovative, and patriotic work of the men "camped in California's dry interior."[69] In June 1942, the photographer Dorothea Lange, already well known for her sympathies for downtrodden Americans, arrived to document the Manzanar experience for the War Relocation Authority (WRA). In one especially evocative photograph, Lange captured a Japanese American worker tending the guayule seedlings under a latticed shed. The resultant image, depicting streaks of light and shadow across the evacuee's body, gave the impression that the interned researcher was wearing a prison uniform. Because these photographs did not capture the more positive image that government officials were hoping for, most were impounded in government files for decades.[70] Meanwhile, the journalist and Quaker activist Grace Nichols prepared an extensive article that fully praised the scientists' work as a "humble" demonstration of their patriotism and willingness to "be of a service to the country which has denied them

Figure 5.4. Japanese American evacuee shown in the lath house at the Manzanar War Relocation Center sorting guayule seedlings for transplanting, 1942. Photograph by Dorothea Lange. National Archives.

the rights of free men."[71] Then on 9 September 1942, the *Washington Post* published a lengthy article under the headline "Two Jap Nurserymen Experiment to Solve Rubber Shortage by Guayule Production." Reporter Neil Naiden extensively praised the Japanese American scientists' "unbelievable patience," "exceptional skill," and desire to prove loyalty to the United States. In view of their success in experiments that other researchers had declared impossible, the article concluded that "never before have such formidable forces joined in a concentrated attack similar to the experiment at Manzanar."[72]

This news threatened to embarrass anyone associated with the ERP's official guayule research conducted at Salinas. Fred McCargar, secretary of the Salinas Chamber of Commerce and a longtime promoter of guayule projects there, contacted Federal Bureau of Investigation director J. Edgar Hoover to voice his strenuous objection to the *Washington Post* story. According to McCargar, an article that described the Manzanar research in such positive terms could be dismissed as mere "propaganda" but in wartime might prove far more dangerous than that. Indeed, McCargar claimed that his colleagues in Salinas long had feared that allies of the Japanese government had been trying to get their hands on guayule seeds, although he was confident that none had slipped into the enemy's hands. McCargar also explained to Hoover how much important guayule research the IRC had done in the past, highlighting the progress that had occurred since the ERP assumed control in March 1942. McCargar simply doubted that a few Japanese men could do as much as the hundreds of USDA scientists and technicians working on guayule issues, suggesting a ploy to allow Japanese Americans to reclaim their former property despite the "almost unanamous [sic] feeling" that they should "never" be allowed to return to coastal California.[73]

This pressure had the desired effect: government cooperation with guayule research at the Japanese American internment camps quickly evaporated. At Manzanar, officials shut off the water that helped irrigate the guayule experiments. Unbeknownst to the camp director, Kageyama and Ogura kept the plants alive for several weeks by watering them at night. Meanwhile, WRA officials instructed journalist Grace Nichols that she could not publicize her account of the Manzanar research, on the grounds that it would create a "distorted picture" in comparison with the larger work of the Salinas researchers. Her censored article never appeared in print.[74]

The government's reaction also affected the Caltech researchers. Dillon Myer, head of the WRA, repeatedly insisted that Emerson had no official authorization to conduct research that might detract from ERP efforts. By November 1942, Emerson had been stripped of his travel authorization, his self-proclaimed title of "consultant without compensation," and permission to set foot on the internment camps in Arizona. One official informed Emerson that he wished the Japanese had never been allowed to enter the United States. Another urged Emerson to apply for a patent so that he, rather than the Japanese Americans, might accrue "whatever credit or financial benefits" might be forthcoming from the new guayule extraction techniques. Another USDA official expressed respect for Emerson's desire to help the Japanese Americans "in their present difficult situation" and help find them postwar jobs in noncompetitive sectors of the agricultural economy. "Such sociological and humanitarian concerns are praiseworthy," but they were not pertinent to the government's wartime mission. For Emerson, the new policies amounted to an "organized effort to reduce the Japanese to slavery" and to keep them as "houseboys, maidservants, and stoop labor."[75]

Reevaluations of Rubber Crop Research

Meanwhile, the rubber crisis remained at the center of the nation's daily news cycle. The hastily prepared book *Rationed Rubber and What to Do About It* reflected the anxieties of the times. Cowritten by journalist Williams Haynes and MIT chemical engineering professor Ernst Hauser, *Rationed Rubber* warned, "Our present rubber situation is serious, deadly serious, far more serious than many of us have believed."[76] Although Hauser still thought that guayule might offer a long-term solution for American war preparedness and offered a "good hedge against a postwar squeeze," he and Haynes believed that the crisis had reached a turning point. Only synthetic rubber could solve the immediate problem. Above all, Haynes and Hauser made clear that neither the American public nor its political leaders had prepared for the imminent rubber emergency.[77]

President Roosevelt turned to the elder statesman Bernard Baruch to bring some order to the rubber mess. From early August to mid-September 1942, Baruch, MIT president Karl Compton, Harvard University president James Conant, and others met day and night to study rubber from every angle. They called many expert witnesses, secured scores of reports, and speedily retraced the discoveries of Henry Ford and Thomas Edison, the IRC, the Truman committee (discussed in more detail in chapter 6), and others who had investigated various processes for producing rubber from natural and synthetic sources.[78] The Baruch committee focused mainly on synthetic rubber, including its possible production from sustainable agricultural sources, but also looked closely at other possible rubber crops. MIT's Hauser provided testimony on the potential of guayule and cryptostegia. The committee heard positive reports on the USDA's work in Salinas and hints of research success at Manzanar, as well as warnings that the guayule project could become "another public scandal" if bureaucratic bottlenecks remained in place. Behind the scenes, Baruch worked to remove such bottlenecks.[79]

The final Baruch report, issued on 10 September 1942, is deservedly well remembered for its frank assessment of the rubber crisis. A classic iteration of the significance of strategic crops in wartime, the report asserted that the rubber shortage presented a genuine "threat to the safety of our Nation and the success of the Allied cause." It minced no words: "If we fail to secure quickly a large new rubber supply, our war effort and our domestic economy will collapse." In its urgent tone and frank assessment of the status of synthetic-rubber endeavors, the report convinced American consumers to take the rubber situation seriously. Further, it labeled reports on the potential of rubber crops as "over-optimistic," and it did not include any natural rubber from domestic sources in its estimates of how the rubber crisis could be ended. Nevertheless, the Baruch report contained positive assessments of both cryptostegia and guayule; in addition, it urged the end to limitations on the acreage and other resources permitted for the guayule project. MIT's Hauser, for one, was delighted that government agencies had finally received the "prodding" that they needed to address the nation's need for natural rubber.[80] Because many citizens and policy makers received the Baruch report as the bible of rubber policy, it is not surprising that its recommendations had an immediate effect. Senator Downey introduced a bill that increased the permissible guayule acreage from seventy-five thousand to five hundred thousand acres for the 1943 crop year. That legislation easily passed both houses. In the words of one member of Congress, the United States had learned a valuable lesson about the "asininity of depending on countries on the other side of the world" for strategic materials. By November, Congress had appropriated an additional nineteen million dollars to the ERP for fiscal year 1943.[81]

The urgency of the Baruch report triggered an immense expansion in the geographic and physical scope of the ERP's guayule project. Project directors immediately had to prepare enough seedlings for the 1943 crop year. The ERP quickly established ten new nurseries in California, Arizona, New Mexico, and Texas, which required a substantial investment in new buildings, irrigation lines, fences, and farm equipment. Hundreds of new workers (including armed guards) had to be housed and fed at these sites, and officials made plans to hire fifty thousand people for the 1943 harvest. The ERP also became considerably more aggressive in finding property, enticing owners and operators of prison yards, grapefruit farms, cattle ranches, experiment stations, botanic gardens, and hundreds of other properties to plant guayule.[82]

The most significant change, however, came when ERP officials targeted the more valuable and irrigated acreage in California's agricultural heartland.[83] ERP bureaucrats no longer treated guayule as an experimental crop for farmers willing to dabble with something new on a few idle acres; it now required a serious commitment from farmers, who were expected to take a chance on a government program that probably would not yield any rubber for four years. Rubber also competed for space, labor, and water resources among some of the state's most influential cotton, fruit, vegetable, and sugar beet growers. As will be seen, many landowners remained unconvinced that the guayule project was in their long-term interest. Because many Californians questioned the long-term prognosis for guayule agriculture, the overzealous expansion of the ERP probably hastened its decline.

Kok-sagyz Fever

The history of the Soviet Union's experiments with kok-sagyz, also known as the Russian dandelion, presented another embarrassing example of Americans' failure to prepare for a rubber crisis. A related plant had been discovered in 1929, when one of the explorers associated with the eminent Soviet botanist N. I. Vavilov's expedition to Central Asia noticed a Kyrgyz highlander chewing the root of "tau-sagyz" in a manner similar to chewing gum. On a follow-up expedition in the Tien Shan Mountains in 1931, another Vavilov colleague discovered kok-sagyz. By 1932, the Soviets claimed to have solved their rubber problem with tau-sagyz and promised that they would soon plant 250,000 acres.[84] By the end of the decade, Soviet efforts had shifted to kok-sagyz, and they aggressively researched and widely planted the crop as part of the third Five-Year Plan of socialist economic development. From Siberia to Uzbekistan to Belorussia, they established a network of experimental farms, state laboratories devoted to extraction technologies, and institutes that built machines for the mechanical planting and harvesting of rubber crops. In the words of Premier Joseph Stalin, "There is everything in our country but rubber. But after a couple of years we will have that too."[85] The Soviets' tendency for hyperbole notwithstanding, the young socialist nation achieved more progress in rubber crop research than did the United States during the 1930s. In the United States, however, tau-sagyz and kok-sagyz were just two more minor rubber plants, known only to specialists.

Pearl Harbor changed that immediately. On 20 December 1941, J. W. Pincus, an engineer who long had represented Soviet interests in the United States, offered the USDA his services in the search for kok-sagyz seeds. The Americans were lukewarm at first, for they considered guayule a more likely source, and previous efforts to obtain kok-sagyz seed had led to nothing.[86] Pressured to take advantage of the new alliance with the Soviet Union, politicians and scientists soon directed considerable attention to the search.

Dr. Paul Kolachov, director of research and development at the Seagram's distillery, emerged as America's leading evangelist for kok-sagyz.[87] In a series of speeches and press releases, many of which were connected to the National Farm Chemurgic Council, Kolachov announced that a solution to the rubber crisis had been found.[88] According to Kolachov, the crop's main advantage was the ability to reach a rubber-producing stage within one year rather than the four or five required for guayule, or the seven or more needed for *Hevea* plantations. "You can plant this dandelion in April and have tires in October," he promised. He suggested further that the United States eventually could fulfill its entire rubber demand through a renewable domestic crop of kok-sagyz. Supporters promised that kok-sagyz offered another important advantage: unlike most rubber substitutes, it could be grown in a wide range of temperate climates, which attracted U.S. and Canadian officials eager for a new crop for the underdeveloped areas of the Great Plains and Rocky Mountain West. Moreover, it appeared relatively adaptable to American agricultural practices, because the seeding, cultivating, and harvesting machinery used for sugar beets and potatoes seemed suitable for the Russian dandelion.[89]

In early 1942, the desperate effort to bring kok-sagyz seeds to America became part of a highly publicized drama that tested the newly formed Soviet-American alliance. Advocates for kok-sagyz linked their campaign with an organization called the Seed Committee for Russian War Relief, a group that appealed for American aid for Soviet citizens left hungry by advancing German armies. American agricultural scientists did their part, offering certain strains of American oat, corn, and soybean seeds to their Soviet colleagues in a not-so-subtle attempt to secure kok-sagyz seeds in exchange.[90] Meanwhile, the highest officials in the departments of State, Commerce, and Agriculture and other diplomats all negotiated for the seed with Soviet officials.[91] Finally, on 9 May, two precious jute bags of seed—inconspicuously marked "R1" and "R2"—arrived in the United States after a dramatic air journey from Kuibyshev, the USSR's temporary seat of government, via Teheran, Cairo, Khartoum, and Lagos. Several more seed bags reached the United States via a ship that traveled from Basra to Bombay, where the priceless cargo was carefully subdivided into separate loads sent via several more ships, in case any were lost at sea.[92] The seeds' sojourn entered the public discourse as evidence of how America's new ally contributed to the war effort. On 22 June 1942—"National Aid to Russia Day"—Americans publicly thanked the Soviets for their important contribution to the search for a domestic rubber crop.[93]

The geopolitics of rubber affected the Soviet Union just as it affected the United States, and the socialist nation mobilized its agricultural, botanic, and industrial experts in an effort to find an alternative. Soviet experts had visited Ford's cryptostegia research plots in Florida in 1925, and they had kept up with goldenrod research before and after Edison's death. Soviet research on guayule had begun in 1927 with test plots beyond the Caucasus in Azerbaijan and intense genetic research at Vavilov's institute in Leningrad. USDA scientists Walter Swingle, Elmer Brandes, and others monitored Soviet experiments with rubber crops, and Brandes made a trip to inspect Soviet progress with rubber plants in 1937. (Brandes also came away from this trip with an "instructive" experience with Soviet synthetic-rubber tires: "Starting with five new tires . . . we experienced five blowouts and finally spent the night in the desert surrounded by howling jackals.")[94] The Soviets also kept abreast of developments with the IRC's guayule project in Salinas, USDA intisy introductions in Arizona, and other rubber crop research in the United States.[95] As with most others who asked, however, the IRC refused to share its precious guayule seed with Soviet experts.[96]

Events took yet another turn when the American rubber crisis coincided with Trofim D. Lysenko's emergence as the Soviet Union's leading plant breeder. Kok-sagyz fit into this drama because it offered Lysenko the chance to promote his unconventional doctrines of plant physiology. Lysenko demonstrated that kok-sagyz accorded with his theory of "vernalization"—that germination rates could be improved substantially if the seeds were kept moist under ice or snow for a period of fifteen to twenty days before planting. According to Lysenko, seeds treated in this manner would acquire the ability to adapt quickly to the harsh Soviet climate.[97] Despite Lysenko's later reputation as emblematic of the errors of Soviet science, vernalization generated broad acceptance in both the Soviet Union and the United States. Vavilov himself endorsed the practice as a method to conduct breeding experiments on new plants, and USDA's Brandes explicitly instructed his colleagues in the kok-sagyz project to employ the method.[98]

Thus American rubber scientists soon joined both Vavilov and Lysenko in one of the most significant disputes in the history of twentieth-century science, albeit on the fringes. Since the mid-1930s, Lysenko, the preeminent spokesman for Soviet scientific ideology, had viciously challenged Vavilov's genetic theories on the grounds that they represented the "bourgeois," "undialectical," and "idealist" values of the capitalist West. In the words of one historian, Lysenko charged that Vavilov's system represented the "bourgeois penchant for stability" rather than the revolutionary potential of Soviet science. With the support of Stalin, Lysenko rose to the top of the Soviet scientific community; Vavilov and hundreds of his political and scientific rivals suddenly were at risk. Imprisoned in 1940 and later sentenced to death—despite the pleas of defenders who stressed his expertise on strategically important plants—Vavilov died of malnutrition in a prison camp in 1943. To this day, Vavilov is well known as one of the most significant victims of the Soviet system and among the most famous martyrs in the history of science.[99]

Meanwhile, news of the kok-sagyz possibility sparked considerable enthusiasm and optimism among industrialists, scientists, and engineers who hoped to prove their value to the war effort. Henry Ford had an employee in spring 1942 use a portable vacuum cleaner to collect seeds from dandelions that grew in a Dearborn road median, just in case the native variety was as valuable as the Russian.[100] At Cornell University, scientists hoped to get their hands on one thousand pounds of seed, enough to plant one million acres in 1943 and yield one hundred thousand tons of rubber for the war effort.[101] In Washington State, officials predicted that kok-sagyz could soon become a staple crop and end problems with the overproduction of wheat.[102] Individual engineers peppered government officials with requests for the precious seed on the grounds that they could make unique contributions to the rubber emergency.[103] As research translated from Russian experts trickled in, dozens of questions about the plant's anatomy, cytology, physiology, pathology, irrigation needs, and rubber extraction technologies remained unanswered. Some scientists, like G. Ledyard Stebbins of the University of California, found opportunities to experiment with the plant's genetics especially attractive. Unlike guayule, which had been domesticated and carefully improved since the first decade of the twentieth century, kok-sagyz had just recently been removed from the wild; therefore, he considered it likely that American geneticists could bring rapid improvements through crossbreeding and polyploidy techniques.

The kok-sagyz portion of the ERP proved far more decentralized than the guayule work, but hardly less comprehensive. Federal officials distributed kok-sagyz seeds "as though they were diamonds" to dozens of different experiment stations, land-grant universities, and rubber-company research plots.[104] Eventually, about two hundred American and Canadian plant physiologists, pathologists, agronomists, entomologists, soil specialists, and others worked on the kok-sagyz project. In the summer of 1942, scientists across North America (including Alaska) tested the Russian dandelion in different soil types and with various fertilizing practices; compared the "Lysenko soak" with other seed preparation methods; recorded optimal seeding rates and depths; and speculated on ways to mechanize cultivation, harvesting, and rubber extraction techniques. According to USDA experts, "in breadth of conception, both as regard to geographic scope and intensity of the program of research and testing conducted under the pressure of time limitations," the kok-sagyz work was "unexampled in the annals of plant introduction."[105]

Other motivations contributed to kok-sagyz enthusiasm around the world. In Canada, experts predicted that the plant could solve the decades-long problem of underemployment and soil infertility of the northern "cutover" districts.[106] In Chile, American agricultural scientists hoped the crop could play a useful role in the Good Neighbor policy.[107] And in Germany, the Nazi regime made kok-sagyz a centerpiece in their efforts to master a natural-rubber crop. After capturing many of the rubber research facilities and kok-sagyz germplasm that had been under Soviet control, Albert Speer made plans to plant up to 350,000 hectares in Nazi-occupied Europe with rubber plants. Hermann Göring, Heinrich Himmler, and other SS officials directed a team of German botanists, agronomists, and chemists to conduct an extensive array of kok-sagyz research, much of it conducted by prisoners at the Auschwitz concentration camp complex.[108] Although the Nazi efforts failed to yield any rubber, the case of kok-sagyz offers yet another illustration of the tremendous national commitment to domestic and agricultural rubber crops in 1943. These hopes, however, soon were dashed.

The Cryptostegia Boom

The wartime crisis also revived interest in cryptostegia, the vine that Edison, Ford, and others had embraced in the 1920s before shifting their focus to goldenrod. Indeed, for nearly three decades the American-born chemist Charles S. Dolley had tried to convince anyone who would listen that cryptostegia offered an easy and inexpensive source of rubber that could grow readily in the Western Hemisphere. A chemist trained at the University of Pennsylvania, Dolley began his rubber crop work in 1906, when he worked through Bernard Baruch and an IRC subsidiary on guayule projects in Mexico. By 1912, Dolley ended up in the Bahamas, where he spent most of his career in service to the British Crown at the Botanical Gardens in Nassau. During World War I, he had several thousand cryptostegia plants growing on a Bahamas plantation.[109] Dolley's enthusiasm for the vine brought him into contact with Edison, Ford, and the USDA scientists who studied rubber crops in the 1920s, although they all soon determined that cryptostegia did not offer a real alternative to the other natural-rubber possibilities.[110]

Researchers from the private sector were the first to mobilize after the rubber crisis began. In May 1942, H. L. Trumball, a rubber chemist with the B. F. Goodrich Company, offered a confidential report to government officials touting cryptostegia as perhaps the most promising of the domestic rubber crop possibilities.[111] Then the United States Rubber Company announced plans to begin experimental plantings near Yuma, Arizona, and to try X-ray technology as a means to determine the rubber value of the plants.[112] Browns-ville, Texas, was another center of cryptostegia research. Peter Heinz, a former USDA scientist once stationed in Mexico, brought a native variety of the plant to his home in Texas. Heinz began his privately funded research in 1940, long before the rubber crisis hit.[113]

Convinced that results from experiments of the mid-1920s had not been promising, many USDA scientists had not paid much attention to the vine. Thus, as the rubber crisis reached a peak in the summer of 1942, the cryptoste-gia issue offered yet another opportunity for critics to attack the government's apparent failure to develop domestic rubber resources. In June, Drew Pearson and Robert S. Allen, authors of the nationally syndicated "Washington Merry-Go-Round" column, wrote that Heinz had been "bombarding" Jesse Jones and other rubber bureaucrats with "stacks" of correspondence that touted the vine. Indeed, Heinz reportedly offered to give seed to farmers for free if they would commit to grow ten acres, and he promised to pay growers fifty-seven dollars per acre upon harvest for a crop that would be ready in ninety days. The *Chicago Tribune* and the *New York Daily News* announced they would lead their own rubber research projects in a friendly competition to compare two pos-sibilities. To that end, the New York paper launched a twenty-acre cryptoste-gia project in Brownsville, Texas, while the Chicago counterpart studied a method to obtain rubber from the wastes of the newsprint industry. Neverthe-less, Jones seemed still committed to the expensive synthetic-rubber program based on petroleum, while Vice President Wallace urged a cautious approach and to wait for more experimental results.[114]

The journalist Charles Morrow Wilson offered a particularly scathing at-tack, claiming that British and American officials had ignored the work of Dolley, Heinz, and others because they were beholden to the Dutch and Brit-ish merchants who controlled the international rubber trade.[115] According to Wilson, cryptostegia seemed ideal: it flourished in almost every climate; it could be propagated easily from cuttings; its seeds were "exceptionally virile"; it had few insect enemies; it required no fertilizer; and it could produce more high-quality rubber than any other annual plant.[116] *Dallas Morning News* col-umnist Victor Schoffelmeyer asked similar questions: "One is forced to wonder why . . . when the future of the nation is at stake, carte blanche orders are not given for all recognized rubber processes . . . whether [from] oil, corn, wheat or sweet potatoes, guayule, rabbitbrush or cryptostegia?"[117]

Cryptostegia enthusiasm rose rapidly. Rubber company scientists John McGavack and H. L. Trumball called for the immediate creation of a one-hundred-thousand-dollar federally funded project that would test cryptostegia in at least five southern states.[118] Trumball also delivered a talk before the American Chemical Society, claiming that cryptostegia's rubber was of much higher quality than guayule's and could be harvested thirty times a year through regular trimming of the vines.[119] Meanwhile, Dolley lobbied the heir to the British throne, the duke of Windsor, to invest in cryptostegia as a meaningful way for colonists in the Bahamas to contribute to the Allied effort.[120]

The Baruch committee took note of the sudden excitement for cryptostegia. Their September 1942 report included hopeful comments about the vine's potential, as there was "little to be lost and much to be gained by pursuing this program vigorously."[121] Within weeks, national leaders mobilized in the search for rubber from cryptostegia. Vice President Wallace now converted to the cryptostegia cause and worked aggressively to reach deals with the United Fruit Company and others who might be able to help expand production. As he explained to President Roosevelt, "We are going ahead vigorously . . . Mr. Baruch thinks it might be our 'white hope' on the natural rubber front within two years."[122] In the end, the Board of Economic Warfare, which Wallace directed, did not settle for a one-hundred-thousand-dollar project as the rubber-industry leaders had proposed; instead, it spent millions. The government and the United States Rubber Company made "cost-plus" arrangements with SHADA (Société Haïtiano-Américaine de Développement Agricole), an entity wholly owned by the Haitian government but wholly funded by the United States, to make cryptostegia grown in Haiti a centerpiece in efforts to combat the rubber emergency.[123] Under the leadership of Thomas A. Fennell, a USDA employee with close ties to Wallace and David Fairchild, SHADA rapidly changed from a project intended to uplift the Haitian export economy in general to one almost wholly committed to cryptostegia.[124]

SHADA mobilized quickly and soon transformed the Haitian rural economy. With the backing of Haitian president Élie Lescot, who was eager to showcase his nation's commitment to the war effort, the organization quickly appropriated tens of thousands of acres of privately held lands. Some Haitians, however, angrily resisted these appropriations. Fennell feared for his life and requested permission to carry a .38-caliber revolver.[125] But SHADA offered relatively high wages, perhaps thirty cents per day (twenty cents for women), and many Haitians flocked to the project. Peasants collected hundreds of pounds seeds per day—seventeen tons in all—from wild vines to help the project get started. Meanwhile, the United Fruit Company provided seeds from wild cryptostegia growing near its banana properties in Central America, and others engaged scientists at the Boyce Thompson Institute to induce rapid propagation of cryptostegia seedlings through innovative hormonal treatments. The

Figure 5.5. Clearing land in Haiti for SHADA's cryptostegia rubber project, 1943. Photograph by Thomas A. Fennell. Courtesy of Thomas Dudley Fennell.

USDA now brought Dolley to the cause and sent him on expeditions across the western United States and western Mexico in search of different strains of the plant.[126]

Initial reports were promising. Cryptostegia grew easily and rapidly in virtually any climate, its rubber quality tested favorably against *Hevea* rubber, and its by-products showed potential as a source of fiber, drugs, and other useable wastes. Although questions regarding effective harvest and extraction technologies remained, hope remained strong that the plant offered promise as an emergency rubber supply. A report by M. D. Knapp, an American consultant who toured SHADA facilities in March 1943, downplayed any hints of trouble with land procurement, housing, disease, or labor control (because the "Haitian peasant seems quite amenable to discipline").[127] John McGavack of the United States Rubber Company toured the Haitian operations in May 1943 and concluded that production of two to three hundred pounds of rubber per acre was possible within twelve months.[128] Dolley concurred, writing in August 1943 that cryptostegia could produce more pounds per acre at less cost than other alternatives: "My persistent efforts to bring cryptostegia rubber into production on a commercial scale have finally met with a gratifying reward."[129]

The plant also promised to serve a political purpose by strengthening the regime of an American ally. In a series of speeches before nearly fifty thousand workers involved with the project in July 1943, Haitian president Lescot predicted that the project's promise of success ensured that Axis "barbarians" would fail in their goal to bring slavery back to the nation.[130] SHADA's official publication also embraced the metaphor, asking every Haitian peasant to become a "soldier in the battle for United Nations rubber production." By the summer of 1943, some five hundred million nursery seedlings were ready for transplanting. According to Fennell, the American-led project in Haiti could produce 10 percent of the nation's natural-rubber needs by 1944. Once again, scientists seemed on the verge of an agricultural solution to the nation's rubber crisis.[131]

Revival of Goldenrod Work during World War II

Goldenrod research at the USDA plant introduction station in Savannah had continued through the 1930s, but only on a modest and poorly funded scale.[132] Savannah's investigators whittled the number of possible varieties of goldenrod from four hundred down to four, all strains of *Solidago leavenworthii*. By 1939, government scientists were claiming that they could obtain 6 to 8 percent rubber yields from their goldenrod varieties on a regular basis. In the next year, Harlan Hill of the Savannah station declared that goldenrod had developed to the point of yielding up to 250 pounds of rubber per acre, well above the levels achieved in Fort Myers. Still, with only a "few grams" of goldenrod rubber on hand in the 1930s, surely no one imagined that the plant would again become a national priority within just a few years.[133]

In the aftermath of Pearl Harbor, David Bisset of the Savannah station received orders to drastically intensify goldenrod production. Bisset soon hired a crew of twelve women to expedite the harvesting of seeds and cuttings from the plants that remained in the field. Because Savannah's facility lacked sufficient acreage for much expansion, he suggested that the government bring Harry Ukkelberg back as an adviser, as well as contract with Henry Ford's nearby plantation to plant goldenrod on some of its acreage.[134]

In any case, the project expanded in 1942. The budget for goldenrod research exploded from $8,000 in fiscal year 1942 to $301,000 in 1943. To address this expansion, Savannah's staff began searching for landowners willing to lease their land and sought to secure enough equipment to handle the suddenly larger scale of production. They also hired an assistant, Dorothy Druggers, to process 140,000 goldenrod cuttings as the basis for the expanded acreage. In all, the Savannah station had planted about 10 acres of its own land in goldenrod, plus about 140 acres of land leased from local farmers. Drought conditions caused an exceptionally weak crop in 1942, however, and researchers only bothered to harvest 74 acres of poor-yielding plants.[135]

Those results did not slow hopes for a dramatic expansion in the goldenrod project for the 1943 crop year. For one thing, tests on the quality of goldenrod rubber were promising: according to the USDA's Southern Regional Research Laboratory, goldenrod rubber achieved tensile strength 2,160 pounds per square inch and elongation of 700 to 800 percent (comparable figures for *Hevea* rubber are 3,800 and 900, respectively). At least one rubber industry expert urged an immediate expansion of the goldenrod project. Then, on the heels of the Baruch report's call for an urgent response to the rubber crisis, several USDA scientists met in Savannah in September 1942. They hoped for nothing less than to make goldenrod rubber a permanent and sustainable crop, one that could make a lasting impact on the nation's demand for rubber and overturn the long tradition of cotton monoculture in the South.[136]

By January 1943, a team of USDA scientists put onto the table a proposal that far exceeded the scale and scope of any previous efforts. In brief, the goldenrod experts proposed to increase the program from the seventy-four acres harvested in 1942, to eight to twelve thousand acres in 1943, and then to *one million* acres in 1944. The 1943 crop could potentially yield 165 tons of rubber, but they proposed instead that the 1943 crop be used as the vegetative basis for a 1944 crop that could yield about 21,000 tons of rubber (about 3.5 percent of the nation's peacetime consumption). To accomplish this would require a vast commitment of time, energy, and money. First, several new technologies were needed: modified celery transplanters to get the crop into field; a new combine capable of mowing the crop and stripping the leaves; a huge number of leaf-drying machines to place in the fields at harvest time; and more. Almost two thousand cars, trucks, and pieces of farm equipment would be needed in 1943 and over seven thousand such items in 1944, which would in turn consume significant amounts of critical war materials like iron, steel, copper, and, of course, rubber. Meanwhile, rubber chemists had to perfect the technologies needed for industrial applications, and they hoped engineers could convert existing naval stores and turpentining operations into rubber-extraction factories. Perhaps most significantly, the government planners would need to find thousands of landowners to participate and as many as twenty-five thousand field workers for the 1944 project. A team of six hundred "shock troops" of experts would be sent into the field in 1943 to "intelligently present the program" to rural southerners. If funded, the proposed expanded project would cost eighty-eight million dollars by December 1944. As vast as this project seems, the team was confident that goldenrod could grow on a wide range of southern soils, that its expansion would not negatively affect the harvest of cotton and other crops, that plenty of "unskilled" African American and female labor was at hand, and that the cost of the rubber obtained would decline rapidly. In all, the political emergency demanded an approach to the problem that virtually ignored local tradition and market reality. As will be seen in chapter 7, however, the project's reality fell short of these goals.[137]

The Search for Rubber Crops in Full Gear

By the end of 1942, the American struggle for a domestic rubber crop was in full gear. Agricultural science and the war effort had converged to an unprecedented degree. Within weeks of Pearl Harbor, an important consumer commodity suddenly had become a vital war commodity, and rubber research shifted dramatically from the private to the public sphere. Through the ERP, the government mobilized land, labor, and capital on a vast scale in order to develop agricultural crops to serve the American war effort. Even without any assurance that these projects would yield results immediately beneficial to the war economy, or help turn the tide on the battlefields, many presumed that domestic agricultural sources of rubber loomed on the horizon. Tens of thousands of people understood the significance of rubber crop investigations: political opponents of government intervention into the private sector faded away; diplomats worked to improve the nation's relationships with the Soviet Union, Haiti, and other unlikely partners that could help the rubber research; journalists aggressively exposed the government's failures to mobilize more quickly; Japanese Americans toiled behind barbed wire to probe questions that remained unanswered in the vast ERP; and ordinary citizens committed themselves to transplanting guayule seedlings, weeding rows of kok-sagyz, and offering President Roosevelt and other leaders their own suggestions for a suitable rubber crop.

The crisis highlighted the role of botanists and other agricultural scientists as crucial participants in the war effort. About a thousand agricultural scientists worked in laboratories of the USDA, state agricultural experiment stations, agricultural colleges and universities, and similar institutions in an intense and well-funded study of the physiology, genetics, agronomy, and economic potential of agricultural rubber crops. At the "Big Four" rubber companies and at the USDA's regional research laboratories, polymer chemists, rubber engineers, and other technicians investigated how to turn the latex derived from these plants into something that could be useful on an industrial scale. As a whole, the intense search for rubber from guayule, kok-sagyz, goldenrod, and cryptostegia had shifted from underfunded and hypothetical projects of a few rubber enthusiasts to a national priority supported by millions.

This commitment did not last long. By the end of 1942, against signs that the synthetic-rubber program was finally getting off the ground, it began to appear that synthetic rather than agricultural rubber offered the better solution to the nation's needs. Hopes for the alternative rubber crops waned as physicists and industrial chemists replaced botanical and agricultural scientists as war heroes. Guayule, kok-sagyz, cryptostegia, and goldenrod each presented agricultural, scientific, technical, economic, and geopolitical hurdles that even the vast and multipronged commitment of the ERP could not overcome. Nevertheless, the intense mobilization for rubber crops that took place in the twelve months after Pearl Harbor indicated that in the search for a solution the rubber crisis, the soil was the first place to look.

Chapter 6

Sustainable Rubber from Grain

The Gillette Committee and the Battles over Synthetic Rubber

THE CRISIS INTENSIFIED as it became clear how vital rubber was in modern warfare. Each Sherman tank—and the United States eventually produced 50,000 of them—required about a half a ton of rubber. Each of the nation's 30,000 heavy bombers needed about a ton. Each battleship contained more than 20,000 rubber parts, totaling about 160,000 pounds on each ship. Americans produced 1.4 million rubber airplane tires in 1944 alone. American soldiers wore 45 million pairs of rubber boots, 77 million pairs of shoes with rubber soles, and 104 millions pairs of shoes with rubber heels. Every industrial facility contained rubber conveyor belts and wheels; every hospital contained miles of rubber tubing and other rubber equipment. Anyone familiar with the logistics of modern warfare understood that access to rubber and other tropical resources, and protection of shipping lanes that connected the tropics with the United States, was an essential factor in the nation's preparation.[1]

In view of the extreme demands for rubber, even the thousands of Americans committed to the ERP did not expect agricultural rubber to do more than make a modest contribution to the nation's immediate wartime needs. Only synthetic rubber offered any realistic hope. Yet unlike Nazi Germany and other nations that had invested in synthetic rubber in the 1930s, the United States had to start essentially from scratch in the months after Pearl Harbor. Synthetic rubber is a complex substance that can be produced in many ways and with various raw materials as the starting point. By the summer of 1942, though, this difficult problem of organic chemistry had been simplified into a public, emotional, and politically charged battle between those who favored alcohol and those who favored petroleum as the initial raw material. In simple terms, in the first case, alcohol is derived from grain, molasses, or similar products of the soil, and converted first into butadiene and then into rubber. In the second case, petroleum is converted into butadiene, and the butadiene into rubber. As a further complication, most of the nation's existing industrial alcohol plants faced shortages of their typical raw material, molasses, and were operating below capacity due to wartime transportation complications. Yet at the same time, midwestern storage bins were filled with grain, and prospects for a bumper crop looked increasingly likely as the debate unfolded in 1942.[2]

These circumstances generated public discussions and congressional investigations oriented as much at assigning blame as at finding solutions to the crisis. Historical narratives of the "rubber mess," as some called it, generally emphasize the timid leadership, personality disputes, and bureaucratic infighting that delayed a solution, or stress the eventual triumph of the synthetic-rubber industry as a key element in the American war economy.[3] The American search for a domestic rubber supply entered this debate as well, because a domestic, renewable, and sustainable source for America's synthetic rubber seemed on the verge of becoming a reality. In the spring and summer of 1942, this topic appeared on the front pages of every major American newspaper. Led by Iowa senator Guy Gillette and a subcommittee he chaired, a wide range of politicians, bureaucrats, media personalities, and common citizens wrangled over the possibility of synthetic rubber from American grains. Yet when industry leaders, Washington, DC, bureaucrats, and even President Roosevelt himself seemed to continually brush off this intriguing possibility, mounting voices of protest responded with direct challenges to the nation's wartime planners and bureaucrats. The debate over the political economy of rubber thus became one of the war's most important domestic issues, one that only slowly found resolution through a national commitment to consumer austerity and government investments in the synthetic-rubber industry. The episode also highlights early debates about the development of renewable versus nonrenewable resources, the value of a nationalist versus internationalist approach to economic development, and agriculture's position in the nexus of industry, science, and geopolitics.

Origins of the Debate

The notion that American grain could solve the rubber crisis arose rather suddenly. After a decade of depressed prices for imported rubber and complex efforts to manipulate the price and supply of American grain, few people contemplated that a solution to the American dependence on imported rubber might lie in midwestern grain fields. Brief reports from Poland, Italy, Germany, and the Soviet Union indicated that farm crops could indeed be the feedstock for synthetic rubber, but these generated very little attention among government and industry officials who anticipated no threat to the supply of cheap imported natural rubber.[4] Even the prominent chemurgist Billy Hale—whose 1936 text *Prosperity Beckons* passionately portrayed industrial alcohol as the "king of chemicals" and the "industry of industries"—virtually ignored the potential production of rubber from grain.[5] In 1938, Albert F. Swanson campaigned for an Iowa congressional seat on a platform that highlighted the conversion of corn into synthetic rubber.[6] USDA officials had little interest in such schemes, however, because government aid for small distilleries might adversely affect the U.S. industrial alcohol businesses—an industry already linked to large corporations on the eastern seaboard that acquired their raw materials from the by-products of sugar distilleries.[7]

Also in 1938, the USDA quietly acquiesced to some of the chemurgists' basic arguments and established a network of four regional research laboratories.[8] Under authority of the Bureau of Agricultural Chemistry and Engineering (BACE), these facilities soon began pursuing dozens of research lines devoted specifically to the industrial utilization of agricultural resources. Although planners at the BACE considered studying the twin issues of domestic natural rubber and synthetic rubber from natural sources, rubber did not prove significant in their initial research agendas.[9] Henry Granger Knight, chief of the BACE, met with USDA secretary Henry A. Wallace and rubber-industry experts. After hearing assurances that the petroleum industry could promptly supply synthetic rubber should the need arise, the group decided to make rubber from domestic renewable sources only a secondary priority.[10] One exception to this pattern emerged in July 1940, when William Sparks of the Northern Regional Research Laboratory (NRRL) in Peoria, Illinois, urged a new program of investigating rubber extenders from vegetable oils and synthetic rubber from alcohol. Sparks's proposal questioned the administration's hopes for projects around the globe when the solution could be found in midwestern crops, and he articulated a vision of sustainability in his hope that synthetic rubber could be derived from "continuously producible (agricultural) supplies of raw materials."[11]

In 1941, a few experts warned government authorities about impending shortages in industrial alcohol, which they argued could be alleviated by stepping up the utilization of midwestern grain crops. In February 1941, R. E. Buchanan, director of the agricultural experiment station at Iowa State College, met with the economist John Kenneth Galbraith, then of the National Defense

Advisory Commission (NDAC), the group most responsible for preparing the nation for a war economy. Buchanan convinced Galbraith of alcohol's importance for rubber production; by April 1941, plans for stockpiling alcohol were in place.[12] A few months later, the grain-alcohol expert Leo Christensen, then at the University of Idaho, approached Edward Ray Weidlein, the well-known leader of the Mellon Institute of Industrial Research in Pittsburgh, in search of current insights on synthetic-rubber technology. Weidlein—whose benefactor, the Mellon family, also dominated the Gulf Oil Company—had concluded, however, that petroleum offered the most economical and likely feedstock for future synthetic-rubber projects, a judgment that became ingrained into government planning and policy for months to come.[13]

Beyond Washington, DC, even fewer Americans took the threat of a rubber crisis very seriously, despite the world war raging around them. Press reports implied that the government's rubber stockpile program was a success, that voluntary efforts to limit rubber consumption were having an effect, that rubber trade with the East Indies remained uninterrupted, and that other solutions to any potential rubber shortage were just around the corner. Other reports and industry advertisements promised that synthetic rubber was imminent and downplayed the possibility of rubber shortages. Perhaps inevitably, the media highlighted progress and new developments in synthetic-rubber research; rarely or barely did it mention the limitations, pitfalls, or time needed to make a real impact on the nation's rubber demand.[14] In 1940, the Standard Oil Company of New Jersey (SONJ) promised that it could move synthetic-rubber production to a "very large scale" as soon as the government asked it to do so.[15] J. J. Newman, a vice president at the B. F. Goodrich Company, praised the foresight of government stockpiling programs and predicted that a synthetic-rubber industry could quickly move in to replenish the stockpiles.[16] In March 1941, A. L. Viles of the Rubber Manufacturers' Association forecast that expansion of synthetic-rubber production would bring about lower prices for both natural and synthetic types, to the benefit of all.[17] In September 1941, the United States Rubber Company boasted that its new synthetic-rubber plant in Rhode Island offered an important safeguard against emergency.[18] The Goodyear Tire and Rubber Company tried to reassure the American public with a series of newspaper advertisements that highlighted progress made in conservation and in the quality of new synthetic-rubber discoveries.[19]

Meanwhile, only a few dared to challenge the assumption the Americans could quickly produce synthetic rubber should the need arise. The case of Ernst Hauser, rubber expert from MIT, offers a telling illustration. In August 1940, in a commencement address delivered at the University of Wyoming, Hauser cautioned of America's vulnerability to losing access to vital raw materials such as rubber. Hauser dismissed the government's preparations as ineffective, but he praised his fellow scientists for their response. Should the emergency arise, Hauser reported, chemists had found the methods to produce "all the rubber we need" from domestic coal, oil, and natural gas.[20] Yet Hauser also supported the guayule alternative and by July 1941 he was warning against media reports that synthetic rubber was at hand. Hauser dismissed such news as petroleum-company propaganda and stated that the current quality of synthetic rubber was poor. "Even if viewed in the very brightest light," Hauser cautioned, synthetic rubber "cannot be considered a satisfactory answer to our rubber problem."[21]

Pearl Harbor and the Beginning of the Rubber Mess

It soon became obvious that the nation did not have four years to solve the problem. Within weeks of the Japanese attack, the rubber crisis became a reality. Debates over rubber policy moved from trade journals like *India Rubber World* and *Rubber Age* to the front pages of American newspapers.[22] The Office of Production Management (OPM) suspended tire production by 10 December, then froze automobile production and forbade sale of rubber tires to the general public. By February 1942, the government controlled the sale of virtually all rubber goods. The Federal Bureau of Investigation issued warnings about the likelihood of tire thieves, and various officials debated the possibility of speed limits, gasoline rationing, and which interest groups might be exempted from such restrictions.[23]

In this new context, war-planning bureaucrats demonstrated an eagerness to build synthetic-rubber factories; further, they had secured an apparently bottomless source of funds to pay for these facilities. By spring 1942, the Reconstruction Finance Corporation (RFC) had budgeted $650 million to expand or develop thirty-one synthetic-rubber plants, almost all of which would follow the petroleum-based processes developed by the SONJ.[24] Despite emerging enthusiasm for guayule, Latin American rubber, and synthetic rubber from petroleum feedstocks, only a few experts recalled the possibility of rubber from grain. In January 1942, for instance, Vice President Henry A. Wallace remarked that rubber specialists were missing an easy and inexpensive source of rubber from American grain.[25] Meanwhile, Charles Friley, president of Iowa State College, essentially ordered his chemists to revive the research on the manufacture of rubber from corn by-products that Leo Christensen had left unfinished during his stint at the college in the early 1930s.[26]

The multiple and rushed responses did not convey confidence that the American government had any real solution to the rubber mess. The agencies involved—such as the WPB, the Rubber Reserve Company (RRC), the RFC, the USDA, and others—seemed to work at cross-purposes, and President Roosevelt did not speak out in a way that indicated a preferred solution. Thus it became increasingly acceptable for Americans to criticize government policy makers over the rubber issue, even in a wartime atmosphere when most criticism of military and foreign policies was taboo.

Senate Republicans spoke out against the government's failure to prepare for the crisis, and the House Agricultural Committee held hearings on the guayule shortage, the members of which quizzed chief Henry Knight on the bureau's failures to seriously pursue research on synthetic rubber.[27] Senator Harry Truman's Special Committee to Investigate the National Defense Program exposed several failures in the American preparedness program and brought the Missourian into national prominence. The committee focused particularly on the government's failure to prepare adequately for the rubber shortages. In March, Assistant Attorney General Thurman Arnold revealed the degree to which SONJ had allowed agreements with the German chemical cartel IG Farben to interfere with war preparedness.

In brief, Standard Oil had agreed, beginning in the early 1930s, not to develop processes for the manufacture of synthetic rubber in exchange for the German firm's agreement not to compete in the American petroleum market. Despite multiple signs that IG Farben and the Nazi war effort were intertwined, SONJ engaged in what one scholar has called a "conscious policy of deception" and avoided steps that could have aided American rubber manufacturers to develop a domestic industry. Remarkably, the companies reaffirmed their partnership in September 1939 at a secret meeting held in the Netherlands, even after the war in Europe had begun. The companies adjusted their arrangements to meet war circumstances, with IG Farben agreeing to transfer hundreds of patents and access to markets to the American company, including those for the German cartel's patented *buna* rubber. IG Farben remained unwilling, however, to transfer the real know-how necessary to complete the process. Moreover, SONJ turned over to the German firm, and not companies in the United States, the process it had developed to make butyl rubber. These agreements further stipulated that the companies would continue their partnership even if the United States and Germany went to war and that the American and German companies would plan to redistribute wartime profits when hostilities ceased. As news of these agreements reached the Truman committee and the press in March 1942, senators and journalists expressed shock at SONJ's seemingly unpatriotic behavior. Nevertheless, the company escaped harsh punishment because it had finally agreed to push vigorously for a synthetic-rubber program that used petroleum as the principle feedstock.[28]

Introducing the Gillette Committee

Also in March, the Senate authorized Iowa senator Guy Gillette to form a subcommittee of the Committee on Agriculture and Forestry to study the production of industrial alcohol, synthetic alcohol, and synthetic rubber. The subcommittee immediately focused exclusively on the production of rubber from surpluses of midwestern grain. For the next several months, the Gillette committee gained national attention almost daily for its efforts to expose the government's reluctance to adopt a renewable and sustainable solution to the wartime crisis.[29]

The personal and political traits of the five members of the Gillette committee contributed to the volatility of the rubber issue. Each member hailed from west of the Mississippi, could claim a heritage of progressivism and independence from the political establishment, and had a background that suggested a preference for an agricultural and sustainable solution to the rubber crisis. Chair Guy Gillette, a Democrat from Iowa, had made his reputation in 1937 when he led the challenge from within the party to President Roosevelt's proposal to "pack" the Supreme Court with sympathetic nominees. Roosevelt had responded with an active but unsuccessful effort to defeat Gillette in Iowa's 1938 Democratic primary. Thereafter, Gillette had opposed the president's plans to run for an unprecedented third term in 1940 and would do so again in 1944. Burton Wheeler, Democrat of Montana, first gained national prominence in 1924 as the nominee for vice president on Robert La Follette's Progressive Party ticket. By 1940, Wheeler had become the Senate's most vocal opponent to American intervention in the world war; indeed, he had established himself as a real alternative to Franklin D. Roosevelt and might have won the Democratic presidential nomination had Roosevelt declined to seek a third term. Elmer Thomas, a progressive Democrat from Oklahoma, campaigned tirelessly for the remonetization of silver as a strategy to expand the money supply and thus aid indebted farmers. Pertinent to the rubber issue, Thomas had once used the filibuster to challenge the large oil companies, which he accused of working against the small, independent oil producers of his state. The ranking Republican member, Charles McNary of Oregon, served as minority leader of the Senate, although he often veered away from the party line. In 1940, he ran for vice president on the Republican ticket, placed there to balance the eastern establishment credentials of presidential nominee Wendell Willkie. After the defeat, Roosevelt offered him a position in the cabinet, but McNary declined it in order to remain a representative of western progressive interests. Perhaps the best-known member was George Norris of Nebraska, whose many accomplishments included the push for the

Tennessee Valley Authority (TVA). After a long career as a Republican, Norris declared himself an Independent Progressive and became a curmudgeonly thorn in the side of both parties.[30]

The Gillette committee opened its proceedings by hearing the case of George E. Johnson. An ally of Senator Norris, Johnson was an engineer who worked with the "Little TVA" project that established government-funded irrigation systems in central Nebraska. At this time, Nebraska leaders remained committed to the industrial utilization of agricultural products, even as the chemurgy movement had faded on the national level. In 1941, the Nebraska legislature funded a "chemurgy project," one that lured alcohol scientist Leo Christensen from Idaho in the fall of 1941 to serve as head of a new Department of Chemurgy at the University of Nebraska. Later that year, with Christensen as technical ally, Johnson had organized a plan to build small plants in rural Nebraska to supply the grain alcohol that could eventually be processed into synthetic rubber. On 6 December 1941, Christensen brought his appeal for industrial alcohol and synthetic rubber to the secretary of the navy, Colonel Frank Knox, a former Roosevelt critic who had been among the founders of the chemurgy movement back in 1935. At this meeting, held the day before Pearl Harbor, Knox (the successor to Charles Edison as naval secretary) brushed aside warnings of an impending rubber crisis.[31]

Washington, DC, bureaucrats continued to block Johnson's proposals to produce rubber from renewable and domestic sources even after Pearl Harbor and the Japanese advances into the rubber-producing areas of Southeast Asia. In January 1942, with the support of Senator Norris's office, Johnson and his colleagues came to Washington, DC, to obtain the necessary permits to build their alcohol plants. Donald Nelson, head of the WPB, repeatedly dismissed Johnson's effort as unnecessary in view of many other plants in the nation that could produce industrial alcohol. Johnson's testimony focused especially on the obstinacy of Frazer Moffett, a bureaucrat in charge of distribution of strategic materials for the WPB. Because Moffett also was president of the United States Industrial Chemical Company, a subsidiary of Standard Oil that controlled 16 percent of the industrial alcohol market, he was subject to criticism that conflicts of interest made him unwilling to support a new, decentralized system of industrial alcohol production. Yet Johnson and his group intended to risk their own money, without any support from the government; they appealed for permission to obtain only 150 tons of steel and other tightly controlled strategic materials necessary for new construction projects. Nevertheless, the WPB repeatedly denied Johnson's requests.[32]

The "Polish Process"

The Gillette committee then took up the cause of Waclaw Szukiewicz, a Polish refugee scientist who had developed an alcohol-based method to manufacture rubber products from potatoes in 1926. His experiments had reached the pilot plant stage by 1935, and he had a viable production process in place in 1939 when the war broke out in his home country. Szukiewicz's movements over the next few years contained elements of a spy thriller, or what Office of Price Administration (OPA) official Leon Henderson called "somewhat of a governmental romance." In 1939, he dramatically destroyed his factory in advance of the Nazi invasion and fled to France. The remnants of the Polish government, now in exile, cooperated with Vichy France to send him to Italy, where he became caught up in one diplomat's hopes that if the Poles shared the alcohol rubber technology with the Italians, the Mussolini government might be convinced to call for peace with Poland rather than join the Nazi war effort. Representatives of the Publicker Commercial Alcohol Company of Philadelphia, the largest producer of grain alcohol in the country, also tried to contact him soon after hostilities in Europe began but without success.[33] The saga also reached the desk of economist John Kenneth Galbraith, who as deputy to the defense commissioner recognized the process's potential to reduce the perennial crop surpluses that diminished farm income. Galbraith encouraged the USDA's Knight to pursue the possibility. Meanwhile, Szukiewicz was shuttled to Brazil and Argentina, and American leaders (and possibly the British Security Coordination spy ring) also tried to track him down. By June 1941, an official from the Polish Foreign Office who then lived in exile in New York approached William Lacy of the OPA about the chemist's availability. Lacy helped to arrange a visa and related paperwork necessary for Szukiewicz to immigrate to the United States. Finally, in November 1941, the elusive rubber chemist landed in New York City.[34]

Lacy immediately brought Szukiewicz to Washington, DC, but he proved reluctant to divulge the details of his process to American officials without patent protection. Next, Szukiewicz and his Polish colleague M. M. Rosten met with the USDA's Henry Knight. Szukiewicz and Rosten shared prospective blueprints and other clues about his alcohol rubber methodology but declined to reveal the secret catalyst that facilitated the conversion of alcohol into butadiene. Knight bluntly rejected the Poles' terms in their offer to help; according to his diary entry of 14 November 1941, "I informed them as far as the Department of Agriculture was concerned we were not interested in any secret process of making rubber."[35]

In January 1942, Szukiewicz and Rosten traveled to the NRRL in Peoria, where the effort to produce rubber from alcohol had intensified considerably. Again, the Poles refused to reveal the catalyst without assurances of patent protection, for which they applied that month. Also around this time, Szukiewicz received a telegram from an adviser in the Office of Production Management (OPM), who saw a potential conflict in awarding a patent to a foreign national amid U.S. war efforts. The OPM issued a formal "secrecy order," which instructed the Poles and anyone else who knew about the plan to "keep quiet" about their methods. The OPM told Szukiewicz and Rosten to leave Peoria at once to return to Washington, DC, for meetings with bureaucrats there. Upon consultation with Weidlein, chemical adviser to the RRC, the government lifted the secrecy ban on 11 February, on the grounds that the rubber crisis would be solved through petroleum rather than agricultural feedstocks. Thus, after considerable intrigue and effort to bring the Polish chemists into the war effort, American officials suddenly abandoned an opportunity to develop a potential solution to the rubber crisis based on renewable and sustainable domestic agricultural crops.[36]

The Gillette committee found that this was not the end of the story. Left to fend for themselves in the private sector, Szukiewicz and his colleagues agreed to cooperate with the Publicker Company in Philadelphia. By April, Publicker's officials were convinced that the Polish process was so likely to succeed that there was little need to put it through the pilot plant stage; they expected the process to yield one ton of butadiene per day by 15 May. On 30 April 1942, Publicker's president Dr. Lewis Marks testified before the Gillette committee that the process could promptly supply about 70 percent of the nation's needs. The next day, chemical engineering consultant John W. Weiss validated Publicker's claims that the Polish alcohol process deserved immediate priority over the "Jersey" petroleum process linked to SONJ. The Weiss report circulated widely among the war-planning agencies, congressional committees, and newspaper columnists, which fueled a public battle over rubber policy and brought a steady stream of embarrassment to the Roosevelt administration. Even after the government's decision had been made, Senator Gillette fanned the fire when he bluntly charged that the Washington, DC, agencies that wanted to prevent "agriculture from getting into the picture" had told Szukiewicz to "keep his mouth shut."[37]

Rubber and Zionism

Meanwhile, Zionist leader and biochemist Chaim Weizmann came to the United States from Great Britain to pressure Americans to speed up the development of synthetic rubber. Famous as a dedicated scientist who helped secure the British victory in World War I, Weizmann had predicted the seriousness of the impending rubber shortages long before World War II erupted. As seen in chapter 1, Weizmann had developed his own tentative method for producing synthetic rubber from alcohol even before World War I. In World War II, the Ministry of Supply granted him a small London laboratory to further research his process. Early in 1942, the American ambassador to Britain, John C. Winant, asked Weizmann to consult with President Roosevelt and other Americans on the rubber crisis. To do so, Winant pointed out, would offer Weizmann the chance to subtly advance the long-term cause of Zionism while focusing on the immediate wartime issues.[38]

Weizmann arrived in the United States in March 1942 and immediately gained access to the leaders of American rubber policy. Although State Department officials expressed little interest, both of Weizmann's initiatives—the development of rubber from agricultural crops and the establishment of a postwar Jewish homeland in Palestine—intrigued Vice President Wallace. Wallace presumed that the opportunity for the grain alcohol rubber process had not yet run out, so he arranged for Weizmann to meet Knight and other USDA officials several times in April and May 1942. Weizmann repeatedly argued that the use of his alcohol process could produce rubber on a large scale in just a few months. But Weizmann soon became "disgusted" with the political nature of the rubber issue in the United States and complained that the oil interests seemed determined to control the synthetic-rubber industry, despite the likelihood that high-octane fuels and other war needs would place pressure on petroleum resources. He chided the government for relying on the analysis of the Mellon Institute chemist Weidlein, whom he denounced as "the mouthpiece of the petroleum concerns" and a person he did not "trust any distance at all." Convinced that it would be "extremely dangerous" to rely exclusively on the petroleum process, Weizmann appealed directly to Wallace to undercut Nelson and Weidlein's plans to develop the petroleum process.[39] According to Wallace's key adviser, Earl Bressman, discussions over the rubber options took place virtually every day on street corners, in exclusive clubs, and at the vice president's home.[40]

The Rubber Mess Intensifies

When WPB head Nelson suggested that synthetic rubber would supply all of America's future rubber needs, Wallace and Bressman sprang to action. They agreed that Nelson's comments were "ridiculous" from the standpoint of rubber technology, damaging from the standpoint of efforts to develop rubber-producing allies in Latin America, and delays in the alcohol rubber possibility "tragic from the standpoint of the war effort."[41] USDA officials arranged for special meetings with OPM bureaucrat George Ball and others to secure information that would challenge Nelson, Weidlein, Office of Scientific Research and Development (OSRD) head Vannevar Bush, and others who favored the petroleum process.[42] For Carl Hamilton, the young Iowan who served as assistant to Claude Wickard, the new secretary of agriculture, the episode proved that the "dollar-a-year men" who worked for the WPB did not deserve their patriotic reputations. Their real priority, he concluded, was to control the hundreds of millions of dollars to be spent on synthetic rubber. Despite the dire rubber situation and the multiple demands on petroleum resources, however, Hamilton found that WPB bureaucrats repeatedly insisted that only petroleum feedstocks offered a solution. Bressman heard rumors that officials had been told "to stay away from this man Weizman [sic]" and that Weidlein's ties to the Mellon family and Gulf Oil meant he had "no intention" of allowing any alcohol process to get off the ground. Only when forced to testify before the public scrutiny of the Gillette committee, Hamilton observed, did WPB officials admit that any of the alcohol processes were feasible. In the end, Hamilton became "absolutely convinced" of a conclusion that he called a "shocker": that these men selfishly put their own interests ahead of the war effort, even at the cost of soldiers' lives.[43]

Members of Congress heard other evidence of the government's resistance to rubber production schemes. Not surprisingly, some of the most volatile testimony came from the flamboyant chemurgist Billy Hale, then president of the National Agrol Company. On Capitol Hill, Hale proclaimed that "this war cannot be won without full chemurgic enterprise," labeling efforts by "pseudo-economists" to block the industrial utilization of farm products the "working of the Anti-Christ." In particular, Hale charged that government and petroleum interests intended to crush all alcohol fuel and alcohol rubber initiatives. Hale insisted that alcohol rubber would soon sell for five cents per pound, less than one half of a low price for natural rubber in a peacetime economy. He mercilessly mocked the work of the RFC's "impartial committee" made up of representatives of at least five major oil companies, which had allocated to petroleum interests virtually the entire $650 million government rubber project.[44] In subsequent testimony to the House Mining Committee, Hale suggested the establishment of one hundred new alcohol plants, which purportedly could produce enough butadiene to meet the nation's rubber needs within one year at one-sixth the amount budgeted to petroleum-based plants. Hale dismissed the official line that new alcohol plants would require significant amounts of strategic materials and claimed, "We can build the plants from stovepipe or brickbats." Only the oil company "Hitler cohorts" who did not want to win the war, Hale charged, were blocking the distribution of the strategic materials needed to implement his proposal. The only hope, Hale believed, was to get President Roosevelt himself to see the value of the alcohol rubber process.[45]

The Gillette committee also heard from others who testified to the obstacles that they had encountered in the pursuit of a natural and renewable rubber supply. One was Clarence Bitting, president of the United States Sugar Corporation (USSC) in Florida, who claimed to be able to produce up to one hundred thousand tons of synthetic rubber from Florida Everglades waste products but had been told by RRC officials that petroleum firms had such contracts all "in the bag."[46] In another case, Senator Thomas described the case of investors who wanted to set up a small rubber plant at Seminole, Oklahoma, this time with local petroleum as the feedstock. The case was similar to what Johnson had experienced with his Nebraska project: even though the investors offered to use their own capital and had lined up most of the necessary equipment and strategic materials, they could not secure the governmental approval necessary to complete the project.[47]

The Gillette committee then uncovered yet another obstacle to production of rubber from grain alcohol. Alcohol firms typically depended on molasses from Cuba and other Caribbean locales as their source of feedstock, but wartime needs brought shipping cargo space under tight control.[48] Gillette committee members assumed that grain surpluses offered an obvious alternative as a source of carbohydrate feedstocks. Other witnesses testified that small alcohol plants could in fact minimize the use of strategic materials and reduce stress on the nation's rail system, because the plants would be closer to the grain supply. In addition, the waste products of grain-based alcohol plants could easily be returned to the farm as livestock feed. Moreover, Thurman Arnold, the assistant attorney general who had exposed the cartel agreements between IG Farben and SONJ, now investigated another chapter in the "old and monotonous story" of monopoly, this time among large industrial alcohol firms. Arnold complained that a "Big Five" of large alcohol plants based mainly on the eastern seaboard had virtual control over the market and thus had little desire to sanction upstart competitors from Nebraska and elsewhere in the Midwest. Arnold also uncovered a new form of collusion between the petroleum industry and the major industrial alcohol manufacturers, because the petroleum firms controlled the patents that permitted the production of alcohol from the hydrocarbons of petroleum.[49] As a result, farmers were not simply losing a struggle to obtain a new market for farm surpluses and farm wastes, but they also were on the verge of losing an existing market for their farm by-products to a nonrenewable alternative.

As the Truman and Gillette committees revealed a variety of blunders and oversights in the government's rubber policy, public rhetoric on the rubber issue became increasingly heated, politically damaging, and even violent. Various politicians and journalists charged that certain bureaucrats protected special interests, especially oil companies, even when it hindered the war effort. Editorialists praised the Gillette committee's willingness to investigate the mess, political cartoonists poked fun at the administration's rubber policy, and investigative reporters like I. F. Stone and Drew Pearson demanded explanations for the nation's failure to prepare for the emergency.[50] Republican Party leader Thomas Dewey attacked the Roosevelt administration for bungling policies and poor leadership on the rubber issue.[51] Senator Ellison Smith of South Carolina, chair of the Senate Agriculture Committee, quipped that WPB really stands for "we pass the buck."[52] These complaints, however, also brought troubles for Gillette, who noted that his committee was "meeting with opposition in some very powerful and highly placed circles."[53]

Jesse Jones, the secretary of commerce and in charge of the rubber program, became especially embittered over the blame that had been directed at him. In April, Jones got into a shoving match and actually came to blows with *Washington Post* editor Eugene Meyer at the prestigious Alfalfa Club in Washington, DC. Fed up with Meyer's sharp editorials and verbal taunts, Jones grabbed Meyer by the coat and smashed his eyeglasses. Meyer tried to punch the secretary, but friends restrained him; Jones escaped the hotel unharmed. Significantly, President Roosevelt's response to the episode—he told the reporters he "hoped there would not be a second round"—again suggested a nonchalance in coming to grips with the rubber crisis.[54]

Yet signs that the rubber mess remained unsolved hinted at serious political repercussions. Henry Knight of the USDA urged the head of the U.S. Chamber of Commerce to dampen the public's enthusiasm for grain alcohol rubber.[55] Vice President Wallace heard reports that new synthetic-rubber factories were nearing completion, yet they lacked the petroleum, or any other raw material, needed to actually produce any rubber.[56] On 9 May, Ferdinand Eberstadt of the Army-Navy Munitions Board warned that if synthetic rubber were not available by July 1943, the country might "have no alternative but to call the whole thing [World War II] off."[57] In late May, presidential adviser Isador Lubin warned WPB head Donald Nelson of an "unhealthy state of public morale" if the speedier alcohol process were not funded; according to Lubin, dollars would "scarcely be a valuable yardstick if we lose the war." Lubin's report explicitly asked Nelson to expand the alcohol rubber program from sixty thousand to three hundred thousand tons, to admit that strategic materials were no excuse to delay construction of alcohol rubber plants, and to form a committee, perhaps led by Harvard University president James Conant, to bring about some solution to the rubber crisis.[58] As some Americans expressed urgency, President Roosevelt's casual attitude seemed off the mark. In a press conference held on 27 May 1942, the president confidently predicted that a new synthetic-rubber tire would be available before a real crisis arrived.[59]

By late spring 1942, the five members of the Gillette committee were unanimous in their frustration. Senator Thomas concluded, "There is somebody some place . . . which I will call the rubber monopoly" blocking any alternative synthetic-rubber program. "What it is I don't know, but I can feel it," Thomas added, implying that petroleum interests had access to wartime bureaucrats that even U.S. senators could not identify.[60] Senator Norris charged that representatives of "monopolistic control" were opposing rubber production schemes that permitted control and ownership at the local level. Senator Wheeler surely agreed, for he mixed metaphors with racism in his charge that the WPB had unnecessarily denied attempts to establish grain alcohol plants in Nebraska: "I don't know where the 'nigger in the woodpile' is, but . . . I say there is something rotten in Denmark."[61] Senator Gillette suggested that petroleum interests opposed renewable rubber through "either carelessness or a definite desire to control the program." According to Senator McNary, the time for listening to testimony had reached an end. "Even if we lose it all," Norris asserted, the time had arrived to draft legislation to fund the grain alcohol rubber program.[62] Thus by the end of June, the Gillette committee's bill, which called for a new rubber agency devoted to the production of synthetic rubber from renewable resources, had made it through the Senate Agricultural Committee and was headed to the floor.[63]

Pressure from the Gillette committee failed to force the government to commit to the alcohol process. On June 5, the bureaucrats most involved met with Roosevelt to urge a "fireside chat" on rubber and find a way to settle the surrounding controversies. The president surprised the group, however, with a call for further studies. "Personally," he admitted, "I am not worried about the rubber situation."[64] Hesitant to commit the nation to the unpopular notion of gasoline rationing, Roosevelt instead announced a massive scrap-rubber drive. Backed by a huge media campaign, the labor of thousands of civic organizations and their members, and the public relations efforts of the Petroleum Industry War Council (whose members hoped to delay gasoline rationing), the rubber drive of June and July 1942 signaled one of the first visible symbols of the nation's commitment to the war effort. At the end, over five hundred thousand tons of rubber were collected, enough to supply rubber reclaiming mills for months to come. As a result, officials had again succeeded in deflecting the popular impact of the rubber crisis, and the petroleum indus-try continued to sell un-rationed gasoline to motorists.[65]

But the scrap-rubber drive did not halt the broadening and deepening criticism against the petroleum process. Alf Landon, the Republican presi-dential nominee in 1936, and also an independent oil producer, blasted Jones and especially Roosevelt for the roadblocks that delayed the alcohol process.[66] Francis Townsend, head of the controversial but influential Townsend Old Age Pension movement, urged farmers to demand synthetic rubber from grain.[67] Howard Doane, head of a firm that managed large farm enterprises, called rub-ber from farm crops the "most significant opportunity that US farmers have ever had," particularly because it could lead to the elimination of subsidies for agriculture. National leaders faced a simple choice: to help farmers or to help oil companies.[68] In an open letter published in the *Fort Dodge (Iowa) Messen-ger* on 10 July, Edward Breen, an Iowa radio station executive and candidate for Congress, expressed concern that Vice President Wallace might not realize that common people perceived the rubber issue as a scandal. Breen praised Senator Gillette's exposure of "almost incredible stupidity, incompetence so gross as to verge on treason."[69]

Thus the political implications of the rubber crisis intensified. No longer just a question of whether alcohol or petroleum was the better feedstock to produce butadiene, the rubber mess had turned into a political embar-rassment for the government and a potential threat to the war economy. In late June—after a long interview with Wallace—nationally syndicated radio commentator Fulton Lewis Jr. addressed the issue on six of his daily broadcasts. Lewis, known for an aggressive style that repeatedly hounded the Roosevelt administration, essentially endorsed the Gillette committee's find-ings and dismissed the government's delays and alibis as "99 percent bun-combe." "The actual production of synthetic rubber is as simple as boiling potatoes on the back of the stove," Lewis reported, implying that hypocriti-cal government officials were guilty of a scandalous degree of favoritism to the petroleum interests. Nearly three million adults heard Lewis's exposé, which the broadcaster claimed generated more interest than anything he had produced in five years on the air.[70] Conspiracy theories abounded. "Ding" Darling, political cartoonist for the *Des Moines Register*, told a friend of an "underground report" that implied that President Roosevelt had "signed his name to a document which is proving very embarrassing."[71] George Johnson, promoter of the Nebraska scheme for new alcohol plants, charged that Stan-dard Oil "would rather see this country lose the war than lose control of the new business they are developing in the manufacture of rubber."[72] According to historian and journalist Bruce Catton, "The emotional temperature in

Washington by the end of June was beyond measurement by any thermometer known to man."[73]

Beyond the hearing rooms of the Senate Office Building, the heated controversy also sparked a profound interest in the domestic rubber solution on the grassroots level. Lewis's broadcasts prompted an active Wisconsin prohibitionist, George H. Halbert, to call for the conversion of "whiskey trust" distilleries into more useful rubber plants.[74] Rex Price, president of the Watherville (Washington) Commercial Club, complained to USDA secretary Wickard, "Obviously the liquor and oil interests are suppressing our cause."[75] J. Jacobs of Sioux City, Iowa, urged Senator Gillette's committee to distribute small alcohol distillation stills among small farmers as a means to alleviate the crisis and generate natural livestock feeds.[76] M. I. Browne of Emmett, Kansas, complained that efforts to discourage the alcohol process seemed rooted in a cynical government plan to keep food prices low and control inflation.[77] J. C. Cranmer and T. W. Reddington of Campbell County, Virginia, suggested that the government's refusal to support the alcohol process indicated that "someone in authority is guilty of treason in time of war."[78] A newspaper editor from Pendleton, Oregon, organized a meeting of local farmers and ranchers to lobby for a local alcohol plant.[79] Labor activists in Iowa and Nebraska organized committees that fanned out across their states to find suitable sites for new grain alcohol plants.[80] Numerous local chapters of the National Grange issued resolutions that implied that big oil companies had conspired to retard the alcohol rubber program.[81] The chambers of commerce of Emporia, Kansas, and Fremont, Nebraska, issued resolutions that echoed the Gillette committee's findings and called for farmers and small oil producers to work together for the solution that would stave off gasoline rationing, utilize local grains, and benefit both parties.[82] Three hundred Nebraskans came to Lincoln in July to create a new organization to coordinate alcohol rubber projects in that state.[83] Grain dealers in Oklahoma lobbied to get their "fair share" of any funds appropriated for synthetic-rubber projects.[84] In all, the rubber from alcohol issue offered an opportunity for many Americans to express their frustration over a tangible war issue.

The crisis had reached a decisive moment and calls for action intensified. Chaim Weizmann met with Roosevelt on 7 July and urged him to create an independent and unbiased commission that would settle the rubber issue once and for all; once again, Weizmann linked the Zionist cause to an appeal for the grain process of rubber production.[85] Meanwhile, pressure on oil industry interests continued to mount. Press reports indicated that the thirty-one plants expected to employ the "Jersey" petroleum process had not progressed beyond the blueprint stage.[86] Yet as the threat of serious rubber shortages came closer, mystifying details concerning the complexities of synthetic processes emerged. Clearly, synthetic-rubber production involved chemical complexities beyond the simple question of grain versus petroleum sources of feedstock. In July, officials from the NRRL in Peoria proclaimed that the rubber made from corn oil and soybean oil possessed an encouraging degree of tensile strength and elasticity, and officials from Joseph Seagram and Sons distilleries announced that their process had successfully passed through the pilot plant stage.[87] Also in July, the Houdry Process Corporation of Wilmington, Delaware, purchased a series of full-page advertisements in major newspapers claiming that their process could produce synthetic rubber faster and more cheaply than other petroleum-based processes.[88] In contrast, Per K. Frolich of SONJ touted his firm's progress but in vague terms and platitudes that generated further condemnation and embarrassment. In all, the competing claims threatened to spark another round of tests before any decisive action could be taken.

Showdown over the Gillette Bill

Activities of the Gillette committee came to a head in July 1942. Rubber expert John Weiss testified that the Publicker process was ready to scale up to war-level production. Eleven members of Congress and other government officials made a well-publicized journey to the Publicker plant in Philadelphia, which left most of them convinced that the pilot plant project could be scaled up within just a few months.[89] But Vice President Wallace intervened to speak out against the Gillette committee's proposed legislation, arguing that a "new isolationism" in rubber policy could be dangerous to other American concerns, cut American producers off from the global marketplace, and exacerbate international tensions.[90] Meanwhile, WPB chief Nelson announced that he had taken personal responsibility for solving the rubber mess. Nelson steadfastly resisted the Gillette committee's recommendations, named a new committee of experts to study the issues, and held a series of closed-door meetings with his advisers in the hope of coming to some kind of conclusion.[91] In his next round of testimony, Nelson described the WPB's first priority as bringing the currently planned factories to full capacity, and thus he tried to dissuade the committee from its enthusiasm for new facilities for the Nebraska and "Polish" processes. Nelson made clear that contracts were in place to produce some two hundred thousand tons of synthetic rubber with alcohol as the feedstock for butadiene, albeit in large eastern chemical factories rather than the small plants that midwestern farmers preferred. More importantly, Nelson conceded in retrospect that if more time were available, he would have devoted a much larger portion of the synthetic-rubber program to the alcohol processes.

Figure 6.1. Des Moines Register cartoonist "Ding" Darling's portrayal of political controversies surrounding the Gillette committee. Published 27 July 1942. Reproduced courtesy of the "Ding" Darling Wildlife Society.

Senator Gillette thanked Nelson for the "death-bed repentance," but this did not stop his assault.[92]

On 21 July 1942, Gillette used a nationally syndicated radio broadcast to announce his committee's findings. Not one witness, he insisted, could dispute that rubber from alcohol could be produced quickly, readily, and cheaply. Anticipating his opponents' response, Gillette denied "with all the vehemence of which I am capable" that he endorsed the alcohol solution simply because it could benefit his Iowa constituents.[93] Three days later, Gillette brought his bill, S 2600, to the Senate floor where it passed by voice vote with only nine senators in the chamber. The House passed its version the following day. The Gillette bill called for a dramatic reduction in the powers of existing rubber administrators and created instead a new rubber czar with a mandate to develop synthetic rubber, but only from agricultural or forestry products. The bill stated further that no alcohol producer could control more than 10 percent of the total, and it explicitly called for the rubber director to enter into contracts with nonprofit corporations and farm cooperatives. Supporting documents asserted in blunt language that the Gillette committee had uncovered "undisputed" evidence that the alcohol process would yield more rubber, more quickly, and use fewer strategic materials that the petroleum process. Moreover, the alcohol process had scored successes on the pilot plant scale, while many of the petroleum processes remained untried. The "only plausible explanation," Gillette frankly asserted, was an unpatriotic attitude that placed market strategies and postwar plans ahead of military and civilian needs.[94]

At this moment, the United States stood on the verge of creating a sustainable rubber industry based on the use of natural and renewable raw materials. But the Gillette bill also represented a remarkably direct attack on the Roosevelt administration; not surprisingly, the president promptly vetoed S 2600. In his veto message, Roosevelt raised the logical objections that creating a second rubber czar would be inefficient, that grain surpluses could not be assured year after year, that limiting the search to farm and forestry products created an artificial roadblock when the real priority was to obtain rubber from any source, and that Gillette's bill amounted to an attempt to intervene into wartime strategic issues "by legislative fiat."[95] The veto signaled at last the administration's readiness to mobilize on the rubber issue. Most editorial writers agreed that the Gillette bill had to be vetoed and that the president needed to do something else in response.[96]

The Baruch Committee

The Baruch committee seemed to offer one last hope for opponents of the petroleum process. Baruch had supported the guayule industry since 1906 and knew already that the petroleum processes were far from ready. Moreover, he and Compton, a former member of the board of directors of the National Farm Chemurgic Council, both had supported the expanded industrial utilization of agricultural raw materials in other contexts. On the other hand, Baruch and his colleagues understood that many Americans believed that the Gillette bill reeked of farm-bloc politics, and they seemed eager to find a tactful and non-partisan answer to the crisis. The committee heard conflicting testimony about various alcohol processes, including the debates between those who supported, and those who opposed, Weizmann's isoprene process and the "Polish" process linked to Szukiewicz and the Publicker Company.[97] Supporters of alcohol as the feedstock for synthetic rubber, such as Nebraskans George Johnson and Leo Christensen and the powerful Congressman H. P. Fulmer of South Carolina, lobbied Baruch with charges that oil company greed was hindering the war effort.[98] As Compton wrote in his diary of 20 August, "Sound reasons must be advanced against the [alcohol] plan or else it should be recommended."[99]

Yet the Baruch committee's report, issued with much fanfare in September 1942, stopped short of a blanket endorsement of any of the alcohol processes. The Baruch report admitted that alcohol processes could have been promoted much earlier, but it evidently also concurred with reports that clearly indicated alcohol rubber could not remain price competitive.[100] The report urged an expanded push for alcohol rubber, supported by new grain alcohol plants to be situated in the Midwest, but did not call for any reduction in the allocation of resources for the petroleum processes. Indeed, one of the Baruch report's useful contributions was to clarify for the general public that several processes could create synthetic rubber, some based on petroleum, some based on alcohol, and some based on either. The report made clear that debates over petroleum and alcohol processes had become overly simplified and politicized, and thus they had obscured the complicated science involved.[101]

The Baruch committee's high esteem and image as a neutral and scientific authority rendered its findings the final word on the matter.[102] Gillette, Weizmann, and others expressed disappointment that synthetic rubber from alcohol did not warrant a higher profile, and columnist I. F. Stone complained that the report still reeked of the "Standard-Mellon-du Pont interests."[103] RRC head Jesse Jones complained loudly that the report was unfair to him because Baruch had taken credit for a program little different from one that Jones had been working on for months.[104] Most Americans, however, appreciated the Baruch report's clear articulation of the importance of rubber conservation, its call to speedily expand synthetic-rubber programs at any cost, and its support for a "rubber czar" to administer it all. Moreover, most were tired of the bickering among government agencies and were pleased that the call for a new rubber czar might bring such conflicts to an end.[105] For his part, Baruch reacted strongly to complaints that the report did not provide enough support for the alcohol processes. As Baruch asserted in a letter to Senator Sheridan Downey, "All the twaddle about interests being able to affect me runs like water off a duck. Let those men who heard my warnings in 1937, 1938, and 1939, and my entreaties since then, search their conscience for gross neglect. . . . Their neglect has costs the lives of hundreds of thousands and men, women, and children around the world."[106]

Rubber from Grain in 1943

In the aftermath, the heated and public battles between petroleum and agricultural interests faded into the background. Some final skirmishes, however, remained to be fought. The Gillette committee resumed hearings in 1943 and secured from WPB chief Nelson an admission that synthetic rubber from grain stood on surer methodologies than the still untested petroleum-based methods. Pressures from farm leaders continued, such as complaints that civilian morale suffered when contracts for alcohol plants went exclusively to "certain vested interests" (the distillery industry giants).[107] Others were outraged at the news that the United States had exported over one hundred thousand tons of rubber to lend-lease partners and that a delegation of experts sent to study the synthetic-rubber situation in the Soviet Union did not include anyone who supported agricultural feedstocks.[108] Various obstacles still delayed construction of grain alcohol plants in the Midwest. "We have been given somewhat of a run-around, Senator," testified the farm leader Ezra Taft Benson. Senator Gillette quickly concurred: "I move to strike out the 'somewhat' from the record." M. M. Rosten, the Polish refugee engineer who had supported Szukiewicz in the 1942 hearings, angrily testified that Nazi Germany and the Soviet Union were responding effectively to their war emergencies, in contrast with the "economic psychosis" that caused some Americans to think more of wartime profits and postwar markets than the current crisis. In all, Gillette and his committee continued to push for an agricultural solution for the rubber problem and managed to keep the issue on the fringes of the public limelight.[109]

Meanwhile, midwestern politicians continued to lobby on behalf of rubber from grain. In January 1943, Minnesota congressman August Andresen, a longtime supporter of chemurgic projects, introduced a bill similar to Gillette's in most ways except that it allowed for Roosevelt's rubber czar to remain in power.[110] In February, Iowa's state legislature issued a resolution that called for new factories to be located in the state so that "our state, nation, and humanity" could benefit and to bring the war "to a successful and more rapid conclusion."[111] That same month, legislators in Minnesota passed a resolution that cited the success of the "the Great Russian War Machine, as well as that of our arch enemy Hitler" as justification for extensive research in Minnesota on rubber from agricultural sources. Nebraskans, unsatisfied with the promise of an industrial alcohol distillery in Omaha, pushed for similar plants in other parts of the state. Congressman Karl Stefan, fearing a revival of the Dutch and British rubber monopoly, lobbied to keep synthetic-rubber factories open after the crisis had passed.[112] In general, however, efforts to establish an alcohol rubber industry in the Midwest fizzled because widespread perceptions that the farm state politicians pushed the idea for local political reasons, and because the grain surpluses of 1942 transformed into shortages in 1943 and 1944.[113]

As synthetic-rubber plants came online and the rubber crisis faded, critics tended to express their frustrations behind closed doors rather than in the public arena. For instance, OPA official William Lacy wrote to Roosevelt adviser Isador Lubin, "It is a crying shame that the Polish process remains unused, tucked away in the plant of the Publicker Commercial Alcohol Company in Philadelphia."[114] Weizmann continued his dual crusades and complained in a meeting with Wallace that the emphasis on unproved petroleum methods over proved alcohol methods cost the war effort "at least six months and probably more."[115] In his diary, Wallace was blunter: "It is evident that the oil people, interested in building up an industry which will be profitable to them, have sacrificed the national welfare to their own cupidity or ignorance."[116] For their part, representatives of the petroleum industry made overtures to chemurgists and agricultural leaders, suggesting on one occasion that oil companies would relinquish their lead in the synthetic-rubber industry should the alcohol process prove more economical under peacetime conditions.[117]

A final irony in these lengthy battles was that the alcohol processes contributed much more rubber to the war effort than the petroleum processes. Just as its supporters had predicted, the new alcohol rubber factories became operational more quickly and proved more efficient than those that followed the original petroleum processes. In January 1943, Reichhold Chemicals Inc.—a company with close ties to the chemurgy movement—announced the availability of a "chemurgic rubber" dubbed "Agripol" for commercial applications. In full-page advertisements depicting a combination of soybean plants and rubber products, Reichhold boasted that American dependence on foreign rubber sources was near its end. Touching on another common theme, Reichhold suggested that its collaboration with USDA scientists in Peoria justified the long-standing faith in "American scientific skill and manufacturing ingenuity."[118] By the end of that year, alcohol processes yielded 130,000 tons of synthetic rubber, or 77 percent of the nation's total production. In 1944 and 1945, the respective figures were 362,000 tons and 63 percent, and 233,000 tons and 39 percent. As some historians have concluded, "the availability of alcohol butadiene was the salvation" of the government's rubber program.[119]

In contrast, petroleum-based plants faced numerable delays both in construction of the factories and in scaling up the process. Even late in 1943,

Figure 6.2. Advertisement depicting chemurgic synthetic rubber derived from American farm crops. Appeared in *Chemical Industries*, February 1943.

rubber czar William Jeffers reported, "By and large, butadiene from petroleum is a pretty sick picture."[120] Butadiene from petroleum plants operated at a fraction of their rated capacity with "woefully inadequate" output. In contrast, the Carbide and Carbon butadiene plant in Institute, West Virginia, which first used grain as a feedstock, operated at 213 percent of rated capacity during parts of 1944. All alcohol plants that produced butadiene operated at 164 percent of capacity for the whole of that year.[121] Further, the relatively small consumption of petroleum in the synthetic-rubber program meant that significant petroleum supplies could be diverted to other crucial wartime needs, such as the production of aviation fuel.

These circumstances suggested a future in which renewable grains, rather than petroleum, would supply America's synthetic-rubber program. In remarks delivered at the 1943 opening of a new alcohol plant in Ontario, Senator Wheeler asserted that Americans and Canadians should base their postwar synthetic-rubber industries on farm surpluses and forestry wastes, no matter how well the petroleum process may function. Such an approach, Wheeler stated, would protect jobs, prevent "the paralyzing effects of another rubber shortage" and keep the new synthetic-rubber industry off the international bargaining table.[122] Moreover, Senator Gillette continued his investigation of utilization of farm crops into 1944. As before, Gillette argued that rubber from grain offered advantages in terms of sustainability, kept farmers gainfully employed, and possibly lowered net costs. Above all, grain alcohol rubber contributed to national security by freeing up the irreplaceable petroleum resources necessary for modern warfare.[123]

Yet the petroleum processes had undeniable advantages. Petroleum feedstocks remained relatively inexpensive, consistently available throughout the year, and had fewer impurities than alcohol feedstocks derived from grain. As many had predicted, the vast grain surpluses of 1942 did not last; indeed, they largely evaporated when spring floods threatened the 1943 harvest. As prices rose, farmers had little incentive to supply industrial alcohol facilities with grain, and many turned to the black market to bring scarce grain products to eager livestock producers.[124] More to the point, the costs to produce rubber from grain, already greater than producing it from petroleum, became even more out of balance. Thus fundamental economic issues favored the petroleum process, and by an increasing degree. Indeed, although their products helped win the war, several grain alcohol rubber plants closed very soon after V-J Day.

Despite such postwar setbacks, the Gillette committee had established with relative certainty that the alcohol process could produce synthetic rubber more quickly and with fewer difficulties than rubber derived from petroleum. Moreover, the hearings generated a surprisingly bold rhetoric rooted in agrarian and environmental themes, suggesting that America's dependence on imported petroleum for rubber and other chemicals might have taken a different turn. Iowa State College agricultural expert R. E. Buchanan, for instance, argued that "with petroleum, we are drawing constantly upon our capital and our reserves, with agricultural products we need only use the interest upon our capital stock of land and climate, yet maintain our reserves."[125] Buchanan's colleague, the chemical engineer Orland Sweeney, testified that the Midwest was "unquestionably America's greatest asset," for its soil ensured the continuity of "biologic functions" necessary for a people's well-being.[126] Senator Wheeler expressed amazement at the "stupidity" of those who were not interested in helping farmers, the "backbone of this democratic republic."[127] Florida sugar executive Clarence Bitting urged the senators to "save for future generations those irreplaceable natural resources of coal, petroleum, and minerals."[128]

Others recognized that the development of connections between any sector of the American economy with foreign and nonrenewable resources might well lead to future shortages. Earl Bressman, the USDA bureaucrat at Wallace's right hand, reminded lawmakers not to lose sight of the fact that "our resources of petroleum are not unlimited" and could run out in fifteen to forty years.[129] Ralph K. Davies, the deputy petroleum coordinator, cautioned that current oil reserves might not last twenty years; he strongly favored the use of renewable grains wherever possible. In his 1944 report, Gillette contrasted grains, which he characterized as "replaceable and in constant surplus," with petroleum reserves that were "irreplaceable and being constantly depleted."[130] For a number of reasons, prominent Americans ignored these warnings and the nation became increasingly dependent on the petroleum-producing nations of the world. Among other implications of the alcohol rubber crisis, the issue contributed to the greatly expanded role of the United States in the postwar politics of the Middle East: both through its early support for the Zionist cause and its increasing dependence on the oil resources of the Arab world.

The "rubber mess" became a metaphor for the government's unpreparedness for war. For those on the Left, the situation reeked of a scandalous collusion of rubber companies, petroleum interests, and the dollar-a-year men in the government. For those on the Right, it offered ample evidence of naiveté, mismanagement, and poor planning among administration officials that seemed to lead up to Roosevelt himself. Any citizen with access to the media would have been aware that synthetic rubber could be obtained from an entirely domestic and virtually unlimited source: the crops of American farms. Thanks to the Gillette committee hearings and associated activities, the public gained increased consciousness on the issue, learned the flaws of the existing approaches, and heard widespread complaints against the administration's rubber policies. The committee and its allies of farmers, chemurgists, and midwestern politicians successfully demonstrated that grain alcohol offered a renewable domestic solution that could quickly and effectively answer America's rubber crisis. Gillette's farm rubber bill of July 1942 directly induced President Roosevelt to create the Baruch committee and bring final resolution to the crisis. The ensuing Baruch report created a national commitment to rubber austerity, gasoline rationing, and investing millions of dollars in factories that produced synthetic rubber from both grain and petroleum feedstocks. In the final analysis, the Gillette committee, perhaps more than any other factor, provided the impetus that spurred the Roosevelt administration into finally addressing the rubber crisis of 1942.

In the long run, however, an internationalist interpretation of the postwar world prevailed over the nationalist schemes that Gillette committee members favored. The prewar notion that placed agriculture at the nexus of industry, science, and geopolitics faded from view. Internationalists argued that the

American leadership in the global economy offered an opportunity to exploit any economically viable resource, and they brushed aside the search for renewable, domestic, or even hemispheric resources. They also believed that the Gillette committee and the chemurgists, with their demands for a focus on domestic farm and forest products, placed an unnecessary limitation on the potential power and wealth of a global superpower. Moreover, supporters of the internationalist solution anticipated the restoration of access to Southeast Asian rubber and Caribbean molasses. Thus the only viable solution utilized large industrial sites and secured a continuous flow of raw materials independent from seasonal rhythms and environmental variations.

All of this remained unclear in the spring and summer of 1942. Conceivably, the vision of agrarian nationalists might well have prevailed over the internationalists had events taken a different turn. The production of rubber from renewable and sustainable domestic resources proved effective during the wartime crisis of 1943 and 1944; and in view of the fact that petroleum is not an inexhaustible resource, circumstances may favor a renewable process someday in the future. It seems less likely, though, that those circumstances will embrace the agrarian reformers' vision of small projects managed by local investors, part of a comprehensive agricultural policy that considers rubber production in terms of by-products used for feeds and fertilizers. Nonetheless, the search for synthetic rubber stands as one of the most remarkable success stories of World War II, and it contributed significantly to the Allied victory in 1945.

Chapter 7

Resistance to Domestic Rubber Crops and the Decline of the Emergency Rubber Project

As the rubber mess moved off the front pages in late 1942, political tensions began to subside. The assertive tone of the Baruch report, the manageable hardships of gasoline rationing, and the promise of synthetic-rubber successes assured most Americans that the rubber crisis would soon pass. Costly but steady successes on the battlefields offered further hope that the United States had begun to marshal the combination of scientific, industrial, and military might needed to ultimately succeed. In this milieu, the notion that agricultural productivity offered a key to geopolitical and industrial power faded from view, and few agricultural scientists and botanists became central players in an emerging scientific establishment dominated by physicists, engineers, and industrial chemists. In particular, the notion that obscure rubber plants were essential for military victory or economic security disappeared from the public arena.

Nevertheless, the ERP remained a massive undertaking, and thousands of scientists, farmworkers, and policy makers remained committed to the search for a domestic rubber crop. That search faced a continual series of ups and downs, and prospects for these plants ebbed and flowed with shifts in both agricultural and geopolitical circumstances. In general, however, proponents of domestic rubber crops found themselves facing almost insurmountable opposition on multiple fronts. Much of this resistance originated in the plants themselves. Guayule, kok-sagyz, cryptostegia, and goldenrod all proved difficult to master, both in the agricultural problems of breeding, growing, and harvesting; and in the engineering problems of extracting and processing rubber compounds. In contrast to other agricultural crops, which for years had been bred to conform to the rhythms and machines of industrialized agriculture, domestic rubber plants could not suddenly be manipulated to meet similar demands under the time pressures of global war.[1] Perhaps more importantly, domestic rubber plants continued to encounter serious political resistance. Like other ambitious wartime programs, proposals that seemed ideal in principle disintegrated in practice. Under pressure from various members of Congress, landowners, political interest groups, and an emergent cultural embrace of synthetic products, the broad political support for domestic rubber crops evident in 1942 faded. By 1945, only a small circle of rubber crop enthusiasts still warned of the nation's dependence on faraway strategic resources, challenged the assumption that synthetic substitutes could be devised for virtually any useful agricultural product, or questioned the rhetoric that highlighted industrial chemists, not agricultural scientists, as the heroes who solved the nation's rubber crisis.

New Directions at Manzanar

Political tensions were especially acute at Manzanar, although the situation calmed for a while. Ralph P. Merritt, who took over as administrator of the camp in November 1942, proved quite willing to support the guayule research of Caltech's Robert Emerson and his team of Japanese American researchers who worked there, especially because evidence indicated that the Japanese American scientists had made specific scientific successes. In their thirty years of trials with guayule propagation techniques, William McCallum and the IRC staff had abandoned attempts to grow guayule from vegetative cuttings. Yet the team of interned scientists who worked behind the barbed wire succeeded with this approach within weeks of beginning their research. By applying plant-rooting hormones to the cuttings and heat to the seedbeds, they managed to get tens of thousands of cuttings discarded from the ERP project in Salinas to grow at Manzanar.

Similarly, the interned scientists had considerable success with guayule hybridization. According to one reporter, the yellow flowers received the same care as the "rarest orchid," as they worked to hybridize the ERP's strains with wild guayule from Texas. Frank Akira Kageyama, an experienced gardener who had been offered a university scholarship to study horticulture, reported that his shirtless work earned him the nickname "Black Boy" as he tested and transplanted hundreds and even thousands of plants per day. In the end, Manzanar breeders developed hybrids that could survive the tough high desert winter and were better suited to the unirrigated and marginal environment of the Owens Valley. All varieties bred from the ERP's Salinas seed had perished. This opened the possibility of establishing a long-lasting guayule program in the American Southwest even after the wartime crisis subsided, a significant contrast to the ERP's focus on the irrigated and valuable farmland of the San Joaquin and Salinas valleys of California.[2]

The group also made progress with guayule extraction, first by using a small milkshake blender that yielded latex one pint at a time. Led by Homer Kimura, Manzanar researchers also developed an innovative, effective, and energy-efficient method of milling the guayule that yielded a low-fiber rubber extract. One observer suggested that expanded to an industrial scale, the process could "revolutionize and modernize" the extraction process. In December 1942 the Kirkhill Rubber Company of Los Angeles proved that samples vulcanized from Manzanar guayule were more pure and more elastic than any other guayule rubber produced by IRC or ERP experts.[3]

Such evidence mobilized defenders of the Manzanar research, who could argue that the internees deserved support, as Millikan put it, "quite apart form the humanitarian and sociological angles."[4] Arizona governor Sidney Osborn still hoped that guayule might become a new crop suitable for the arid lands of his state and eagerly endorsed the use of Japanese American internees to help with the project.[5] In January and early February 1943, Osborn, Millikan, Emerson, and others lobbied several Washington, DC, officials in an effort to secure greater government cooperation with the Manzanar researchers. Members of Senator Truman's committee that monitored the defense industry similarly urged rubber czar Jeffers to devote more attention to the progress at Manzanar.[6] Meanwhile, Emerson pointed to the project's potential for improving Japanese Americans' "future relations with their Caucasian fellow citizens." He also explained that a committee of Caltech professors had concluded that the production of synthetic rubber could not be sustained in the long run, either from petroleum or from grain feedstocks. Emerson thus stressed that the interned scientists' methods could save the nation thousands of tons of steel that might otherwise go into the proposed infrastructure of new synthetic-rubber factories.[7]

These efforts proved successful. In February 1943, government officials agreed to allow research at Manzanar and Poston to continue so long as Caltech remained officially in charge. Merritt restored the researchers' access to water and suggested that their work might help guayule eventually become a suitable crop for small-scale farmers, even if it could not immediately help the war effort. After carefully scrutinizing the Japanese scientists' qualifications, research successes, and loyalty to the United States, Merritt could find no reason to deny recognition and funding.[8]

Guayule research expanded at Manzanar in spring 1943. Caltech's Millikan secured grant money and other help from several industrial and philanthropic groups, and he urged authorities to "release the energy and skill" of the Japanese Americans.[9] Milton Eisenhower, the general's brother and a leader in the military's publicity arm, the Office of War Information, also urged full support for the Japanese Americans' research.[10] Over the next few months, Takashi Furuya and Homer Kimura designed larger and more effective mills from parts salvaged from washing machines and automobiles. Camp director Merritt witnessed tests of this contraption in the barracks' laundry room and enthusiastically described the "globs of rubber" that floated to the surface.[11] As production continued, the Emerson home became filled with guayule "by the bale" and dozens of prototype products derived from extracted guayule rubber.[12] Using his 1928 Studebaker—and Greyhound buses when gas coupons ran out—Emerson made countless trips among Pasadena, Salinas, and the internment camps in California and Arizona. Other prominent scientists began to assist the Manzanar researchers, such as G. Ledyard Stebbins and Ernest B. Babcock of the University of California, Berkeley; Reed Rollins of Stanford University; Frederick Addicott of the University of California, Santa Barbara; Katherine Esau of the University of California College of Agriculture; and Arthur Galston of the University of Illinois. Several significant scientific publications resulted from this collaboration, particularly on the plant's complex genetics.[13]

As reports of research success continued to come in, the ERP's new director Paul Roberts argued that it would be "indefensible" if the government failed to help fund a portion of the Manzanar research. With skilled but interned scientists working for salaries of sixteen dollars per month, Roberts guessed that the "possibility is remote that we not get full value from the investment." Despite concerns about his "bombastic" style, Roberts also defended Emerson, who had demonstrated sincere, able, and enthusiastic leadership of the project. As a result, government officials offered a few thousand dollars to support research at Manzanar and brought Emerson back onto the payroll as a consultant for three hundred dollars per month (nearly twenty times the interned scientists' pay).[14] Roberts's supervisor, Gordon Salmond, agreed that working with Emerson and funding the Japanese American scientists was worth the political difficulties of this "delicate subject." "If we can get something out of these Japanese of value," Salmond wrote, "we should bend every effort to do so."[15]

Sympathy for the Japanese American internees surfaced in another form after the photographer Ansel Adams visited Manzanar in late fall 1943. Like Dorothea Lange, Adams was drawn toward the rubber research as symbolic of the internees' efforts to build a normal community despite their predicament. Unlike Lange's, however, Adams's photographs were published in *Born Free and Equal*, a 1944 work that suggested the government's willingness to atone for the internment embarrassment. Under the simple caption "Here Is a Rubber Chemist," Adams captured an image of a proud and determined Frank Hirosawa, who by then had demonstrated his loyalty and had made important scientific contributions to the war effort.[16]

Making public the interned scientists' contributions to the war effort did little, however, to lessen widespread public hostility. Guayule expert David Spence conceded that the Japanese Americans had developed "exceedingly interesting" methods, but he wanted no part of any effort that would "help the cause of these Japs."[17] When Emerson put on a demonstration of the Manzanar scientists' improved guayule extraction method before a group of California politicians, he was explicitly told to not mention that the idea came from the interned scientists.[18] Soon after the war ended, Emerson's Caltech colleagues pressured him to leave, in part because he had made the Manzanar project such a priority. By then, Emerson had accumulated a debt amounting to about three years' salary, but he continued to provide money, odd jobs, and rooms in his home to his Manzanar colleagues.[19] And despite its scientific and historical importance, the official history of the Emergency Rubber Project—a work of over 240 pages—devoted less than two sentences to Japanese Americans' work. All other official histories and media treatments of the rubber project ignored the Manzanar work altogether. As much as any other episode, this blatant and official erasure of the Japanese American and Caltech discoveries lays bare the controversies and contentions that surrounded the American search for a domestic rubber crop.[20]

Opposition to the ERP in California

Although the Manzanar issue was a distraction, real trouble came to the ERP and its leadership in the Central Valley. The problems centered in Kern County, California, the fertile agricultural district that surrounds Bakersfield. Leaders in the Kern County Chamber of Commerce, the Kern County Farm Bureau, the local newspaper, and many of the leading farmers and agribusinesses orchestrated opposition to the government's heavy-handed intrusion into area farm practices. "Food first, then rubber" became a local battle cry, as residents insisted that potato production would do more to help the war effort than work with experimental rubber crops.[21] By the end of December 1942, however, a "truce" of sorts had been negotiated in the "guayule wars," and nearly fifty Kern County farmers and agribusinesses had signed on to the ERP.[22]

The peace did not last. Many regretted signing contracts to lease their lands through the ERP rather than holding out for more lucrative terms, because wartime demand for traditional food and fiber crops brought attractive prices and pushed up land values. Some residents claimed that government negotiators had coerced them into signing by threatening to condemn their lands and expose them as uncommitted to the war effort.[23] Indeed, land leases were inherently unfair to local farmers and business owners, at least in the sense that they bound the lessor to the government for ten years but allowed government to terminate the arrangement on a year-by-year basis. The guayule project also put pressure on the region's irrigation networks, farm machinery supplies, and, above all, its labor supply. Wartime demand made farm labor scarce and expensive, and California agribusinesses became convinced that the ERP exacerbated the problem. Local growers also criticized worker camps as too luxurious (complaining especially about the size of the sinks) but wages were clearly the more important issue. So desperate for help that it hired recruits at skilled workers' wages, ERP officials would move guayule workers into unskilled fieldwork without reducing their wages. As one might expect, residents complained that the government agricultural experts had a poor understanding of local farm circumstances. Many charged that ERP officials used labor inefficiently, employed farm practices that would harm the long-term fertility of local soils, and were responsible for new infestations of birds, insects, and nematodes. Despite nationwide calls for wartime sacrifice, some threatened to sue the federal government. Perhaps at the center of the dispute, many California farmers had been accustomed to receiving price guarantees from the Agricultural Adjustment Administration (AAA) for growing cotton and food crops, and the rubber project threatened to cut into that source of income.[24]

Farmers and business owners also grumbled that ERP officials took many of the best local lands for the desert shrub, acreage that could have been used to produce potatoes, beets, beans, cabbage, and other crops, as well as by-products that could be used to support twenty thousand beef cattle. Moreover, some leading Kern County farmers recently had invested three hundred thousand dollars in a dehydration plant designed to process cabbage, carrots, and potatoes, crops that they did not want taken out of production. The *Wall Street Journal* picked up the story and warned that the project might be "the entering wedge for collectivized farming in the US" on a Soviet model.[25] The Associated Farmers of California (a group of powerful landowners best known from John Steinbeck's portrayal of their vigilante violence against migrant farmworkers in *The Grapes of Wrath*) condemned the project as amounting to an "enormous wastage of labor" and a "superfluous . . . inefficient . . . [and] bungling Socialistic experiment." In a similar vein, local landowner and cotton gin operator Frank Jeppi, Sicilian by birth, complained that the project

Figure 7.1. Guayule harvest in the Coachella Valley. Photograph by Walter Fiss Photography. Courtesy of Margaret Crowl MacArthur.

"seemed un-American in principle" and "smacked of fascism."[26] In March 1943, the Kern County Farm Bureau passed a resolution declaring the guayule program "doomed to failure."[27] In short, many landowners of the Central Valley were in open revolt.

A similar rebellion surfaced in the Santa Maria Valley near Santa Barbara. Much of the land there had been planted in lettuce, carrots, cauliflower, and other high-priced crops. Fearful that a shift to guayule would cut into profits and affect investments in newly built packing houses, a local organization called the California Lettuce Growers, which had gained control of many land leases previously held by Japanese American farmers, openly refused to surrender these properties to the ERP. On 19 February 1943, local farmers met and criticized the entire project. Soon thereafter, they placed a large display advertisement in the local paper that declared raising food more patriotic than growing rubber. This especially upset farmers who had been motivated by patriotism to offer their land to the guayule project.[28] Other Californians joined the fray. One resident called for a federal investigation of how "Uncle Sam [was] so easily led astray" by the guayule "fad," while another warned that farmers themselves would suffer from malnutrition because so much food was being taken out of production.[29]

Few dared to defend the project. One letter writer complained that the farmers' smear of President Roosevelt and the guayule project played "into the hands of the Axis" and "border[ed] on treason." Another newspaper editorial writer dismissed the Central Valley farmers' newfound passion for growing food as nothing more than "face-saving hooey," because the fifty-five thousand acres proposed for guayule in California was but a fraction of the thirty million acres planted then in crops. In fact, much of the disputed land involved actually had been planted in cotton and alfalfa, not food crops. Some recognized the grumbling about high wages for guayule field workers (who received forty cents per hour for a sixty-hour week) as merely a ploy to redirect workers to fruit and vegetable operations at lower wages. If tires remained scarce in 1944, one writer noted, Californians would have no one to blame but the Associated Farmers and their "blah-blah [about] socialism."[30]

Even as their program was imploding, some ERP officials speculated that guayule might make a comeback in Kern County once the war ended if synthetic rubber proved unviable and a "wave of isolationism" returned.[31] Yet the large landowners' arguments prevailed, at least in the short term. Without consulting ERP officials, rubber czar William Jeffers informed Secretary of Agriculture Claude Wickard that it would be a "grave injustice" to place the 1946 and 1947 guayule crop ahead of the food needs of 1943 and 1944.[32] On 30 March, the same day that he posed for photographers with the first all-synthetic U.S. rubber tire, Jeffers proclaimed that the nation could no longer wait for results from alternative rubber crops.[33] Jeffers and Wickard publicly announced that the project had been placed on "stand-by" to minimize interference with California's food production. All unfinished plantings were halted, and the ERP officials who had turned Salinas, Bakersfield, and other places into rubber-growing boomtowns in 1942 now were charged with quickly returning lands leased for guayule production back to their owners.[34] Barely one year after its launch, the ERP came to a "sudden and jarring stop."[35] Moreover, the photograph of Jeffers with the synthetic tire imprinted a lasting image onto memories of the World War II home front by suggesting that synthetic rubber, not guayule, had facilitated the American military victory.

The Decline of Kok-Sagyz

Despite the tremendous push for kok-sagyz in 1942, that crop also lost favor in subsequent years. In preparation for the 1943 crop year, kok-sagyz experts determined that a full-size factory might require an investment of about four million dollars, involve hundreds of farmers, and demand hundreds of thousands of acres of cropland. In an important January 1943 judgment, Jeffers ruled it was premature to make a large-scale commitment. Instead, he permitted experimental work with kok-sagyz to continue in dozens of states, and the government funded kok-sagyz processing on the pilot plant scale through the USDA's Eastern Regional Research Laboratory (ERRL) near Philadelphia.[36]

Shaky research results caused interest in kok-sagyz to decline as rapidly as it had ascended. At Cornell, the first of three attempts to plant kok-sagyz in 1942 failed completely, the second showed little promise, and the last had very disappointing yields. Scientists also found seed quality unpredictable, which caused germination rates and rubber yield to vary tremendously.[37] In some cases, scientists unfamiliar with the plant's growth habits plowed it under before young seedlings had the chance to emerge through the topsoil. The extensive research projects spread over several states in the western United States proved difficult to coordinate and keep on schedule. Kok-sagyz was highly susceptible to weeds and disease, and hordes of ants, aphids, beetles, grubs, worms, slugs, gophers, and rabbits seemed to feast on the plant. Kok-sagyz also required continual hand cultivation, but wartime labor shortages made it difficult to find workers for the government's kok-sagyz plots in many parts of the country. In Plainfield, Wisconsin, for instance, growers used migrant laborers from Jamaica, but many refused to work on cold and rainy autumn days. Kok-sagyz experts also fretted over the government's long delays in meeting payroll. According to one geneticist, "We are in pretty bad shape here. Uncle Sam has forgotten that we are working for him." Most importantly, kok-sagyz failed to produce much rubber. In contrast to the fantastic predictions of about two hundred pounds per acre, the actual result usually was between thirty and sixty.[38] In all, the crop seemed unviable for American agricultural circumstances. Despite all the hoopla that surrounded kok-sagyz in 1942, it did not take long for many American experts to conclude that Soviet scientists had simply exaggerated the plant's potential for propaganda purposes.[39]

In September 1943, Secretary of Agriculture Wickard asked the new rubber czar, Bradley Dewey, to more than double funds for the kok-sagyz project in the fiscal 1944 budget, to about seven hundred thousand dollars. Another year of research would be very important, Wickard asserted, and he assured the new czar that the project would have no impact on American food production.[40] C. E. Steinbauer, head of much of the kok-sagyz research, sent an appeal to Washington, DC, that outlined the extensive and sophisticated research that his team planned for 1944 in rubber crop genetics, biochemistry, and agronomy.[41] These appeals, however, failed. In their extensive discussion over the future of the ERP, political and USDA officials decided to abandon the

Figure 7.2. Weeding kok-sagyz, the Russian dandelion, at Cass Lake, Minnesota, May 1943. Photograph by Paul H. Zehngraff. National Archives.

kok-sagyz project altogether. In state after state, agricultural scientists received letters instructing all operations to cease by 30 June 1944, even if it meant plowing perfectly healthy rubber plants into the soil.[42] Only a few American scientists continued to pay much attention to kok-sagyz after 1944.[43]

The Embarrassing Collapse of Cryptostegia

The collapse of cryptostegia proved an especially embarrassing blow in the struggle for a new rubber crop. It also offered clear evidence of the increasingly extensive reach of American agricultural science and the implications for international affairs. Pressure to replace lost export revenue transformed SHADA's modest aims to uplift the Haitian rural economy into a massive effort to address the Allied rubber crisis. Its budget of some $6.5 million was roughly equivalent to the Haitian government's entire annual revenue. SHADA eventually took control of sixty-three thousand acres, often lands that had already been in agricultural production, and assigned workers to plant by hand up to three million seedlings each day. In all, some 150,000 workers—about 5 percent of the nation's entire population—participated in the project.

Amid the wartime emergency, SHADA swelled to become a huge project long before several important questions had been answered. Should cryptostegia be planted from seeds or seedlings? On irrigated or unirrigated plots? Planted closely together or spaced for cultivation? Could diseases and pests be controlled? Most problematic, could engineers find a way to efficiently harvest the drops of latex? Answers to such questions proved difficult to obtain in a tropical nation that virtually lacked electricity, telephones, paved roads, and other basic infrastructure. Despite these fundamental uncertainties, SHADA pushed ahead, before experimental plot and pilot plant work had been completed, and before President Élie Lescot could persuade Haitian citizens that research on the exotic rubber vine served the public good.[44]

Researchers soon found that unlike the rubber plants that were too difficult to grow, cryptostegia was too easy. If not cut back on a regular basis, the plant grew out of control, rapidly turning into impenetrable jungles of vines that defied every model of modern agriculture. With a density of 10,000 plants per acre in Haiti, compared with 150 trees per acre on a *Hevea* rubber plantation, the frequent manual harvest of billions of rubber plants seemed impossible.[45] One report estimated that obtaining each pound of latex required three "man-days" of labor (or six thousand "man-days" per ton). Because wartime conditions negated utilization of most chemical methods, the whole project would be in vain unless engineers could develop a method to extract rubber mechanically.[46]

But extracting rubber from these plants proved virtually impossible. On the *New York Daily News*'s plots near Brownsville, Texas, one engineer devised a contraption "resembling a miniature crossbow" that encouraged the plant to drip its latex into small cups, but reports indicated that "more blood than rubber was collected."[47] In Haiti, SHADA officials developed a rather primitive device to collect rubber that involved cutting a bundle of vines by hand to allow latex to drip down a slanted bamboo frame into a small bottle. And although SHADA workshops produced hundreds of these tapping devices by "assembly line methods," the process remained a long way from industrial-scale production. Meanwhile, Haitian natives—described as "urchins" in press reports—developed a method to coagulate the latex collected inside coconut shells. The next stage recalled preindustrial modes of production in that workers processed coagulated sheets of cryptostegia rubber through a hand-operated printing press. Each operation produced a sun-dried sheet of rubber no larger than a piece of paper.[48]

Cryptostegia failed for a variety of political reasons as well. Although American officials denied the charges, many Haitians claimed that SHADA routinely cut down and burned healthy breadfruit and mango trees, bulldozed hundreds of peasant homes, and ignored the promise of preserving small plots around homes so that peasants could grow food crops.[49] SHADA's influx of

Figure 7.3. The "Van den Brughe device" used for collecting latex from cryptostegia, part of the SHADA project in Haiti. Photograph by Thomas A. Fennell. Courtesy of Thomas Dudley Fennell.

cash also brought widespread price inflation to Haiti. SHADA's defenders presented this as something that permitted Haitians to buy shoes, better clothing, and more food, but the project undoubtedly jolted the Haitian economy in several ways. Labor turnover became chronic and strikes were common; overseers dismissed and hired hundreds of workers each day. Many peasants traveled great distances to gain jobs on the cryptostegia lands, which disrupted labor markets and left fields untended in other areas.[50]

Problems intensified in late 1943, as some half a billion seedlings were ready for transplanting. Many cryptostegia plantations employed a *Garde nationale* force of fifty or more members to maintain the peace. Rumors that Haitian police were "escorting, recruiting, and retaining" workers led to fears that the project might be seen as using forced labor. As the regime became increasingly intertwined with the controversial project, anti-American protests and pressures on President Lescot intensified.[51]

Expanding the project into the Artibonite Valley of central Haiti proved especially contentious. Peasants there threw eggs and vegetables at American officials, and military advisers warned that locals might be plotting to cut a dike and flood the valley.[52] Malaria outbreaks compelled SHADA officials to begin emergency medicine distributions; one witness reported that workers were "dying like flies."[53] Then came an insurmountable outbreak of June bug larvae—thirty thousand bugs per acre devoured the rubber plants. In fall 1943, SHADA abruptly abandoned over seven thousand acres of recently planted vines in the Artibonite Valley, resulting in sudden unemployment for over eight thousand men.[54]

These circumstances impelled officials in Washington, DC, to question the SHADA project. Some started looking for ways to pull out, even if it meant, in the words of one State Department official, that some Haitians would consider the United States "not only crazy but malevolently so."[55] In response, both President Lescot and SHADA's director Thomas A. Fennell traveled to Washington, DC, to request more time to deal with various problems and the "considerable amount of anti-American propaganda."[56] More negative publicity came in November, when Senator Hugh Butler of Nebraska issued a lengthy report that decried billions of dollars "wasted" on various aspects of the Good Neighbor policy, including unneeded "boondoggles" such as the cryptostegia project.[57] Several American scientists saw the end coming and chose to abandon the project.[58]

Matters soon came to a head. In the United States, the New York Daily News abandoned research on the vine in Texas, complaining of the "beautiful but cantankerous and disappointing cryptostegia."[59] A USDA scientist in California concurred; he hoped to find a "man-sized man" who would stop pouring money "into a rat hole" of cryptostegia research.[60] In Haiti, Fennell continued to defend the effort as one that could offer the nation long-term economic opportunities, but Haitian politicians complained that SHADA had abandoned its original mission of economic development.[61] Lescot pleaded with American officials to continue the program so that he would not lose prestige in his country.[62] A delegation of American diplomats returned from Haiti with the unanimous recommendation, however, that the project cease immediately. Within weeks, the RDC severed its ties to SHADA and the project. Although war raged around the world, and rubber remained in short supply, officials ordered Fennell to oversee the burning of tens of thousands of acres of healthy rubber vines.

The program's failures fueled bitterness on all sides. For Fennell and other cryptostegia enthusiasts, the episode marked a lost opportunity to improve the Haitian economy over the long run, as well as reduce American dependence on faraway rubber. Indeed, Fennell suspected that chemical and petroleum companies connected to the synthetic-rubber industry had caused the program's sudden collapse.[63] For supporters of synthetic rubber, the greater scandal was that the Haitian project cost at least $6 million yet yielded only about five tons of rubber—rubber that cost about $546 per pound.[64] Despite some public relations claims to the contrary, the government did little to help landowners and peasants remove cryptostegia and resume normal cultivation of their lands. Few American officials made it a priority to relocate property lines blurred in the denuded landscape, return the lands to their former owners, or help restore them for food production.[65] The project also cost President Lescot a great deal of political capital. Some scholars suggest that the episode doomed Lescot's career and that American influence in Haiti has never recovered from the debacle.[66] In any analysis, cryptostegia research in Haiti proved an embarrassing waste of time, effort, and money, a quintessential example of a well-intentioned but failed government program.

Waning Interest in Goldenrod

The ERP's 1942 proposal to plant one million acres in goldenrod, to employ twenty-five thousand people in its production, and to establish the plant as a long-term addition to southern agriculture also fell far short of its original goals. Undersecretary of Agriculture Paul Appleby presented a version of the proposal to Jeffers in January 1943, but the rubber czar ruled that goldenrod needed another year of testing as a field crop and feasible source of rubber before he could contemplate such an ambitious program. As a result, ERP officials in 1943 planted only 570 acres in Burke County, Georgia, and forty acres in Chatham County near Savannah. Additional test plots that year were located in Alabama, Mississippi, and Louisiana.[67] Four hundred sixty-eight workers, one-fiftieth the proposed numbers, were involved in the project. The budget for goldenrod in fiscal year 1944 was $426,000, in contrast to $54 million in the larger proposal.[68]

The goldenrod project also ran into difficulties with extraction technologies. Beginning in June 1942, the ERP assigned this problem to the USDA's Southern Regional Research Laboratory (SRRL) in New Orleans. A team of twenty investigators explicitly built on Edison's research to design a scheme for rubber production on a pilot plant level. SRRL researchers systematically studied the morphology and physiology of goldenrod's rubber-bearing cells, developed solvents to permit the separation of rubber and resin globules, and designed appropriate recovery and compounding techniques. As Edison had suspected, the overall yield and quality of goldenrod rubber was less than that of *Hevea* due to its lower molecular weight and smaller concentration of isoprene molecules. The SRRL sent goldenrod rubber samples to various rubber manufacturing companies for fabrication tests. Manufacturers found that goldenrod rubber had a higher quality than recycled rubber and would be ideal for commercial products such as shoe heels and water bottles. In addition, the companies produced about forty goldenrod rubber bicycle tires. Although the

Figure 7.4. Stripping goldenrod leaves by hand, Burke County, Georgia, 1943. National Archives.

project by no means offered a commercially competitive means for an industrial scale of production, the results once and for all justified Edison's faith in the potential of goldenrod as an emergency source of rubber.[69]

But the goldenrod project of World War II ended almost as suddenly as it began. Project directors had to be vigilant not to upset local labor markets, yet rules that controlled wartime publicity prevented saying much in local newspapers about the project. Agricultural scientists complained that the only housing available in Waynesboro resembled "slum dwellings." The 1943 harvest was again disappointing, the entire production totaling less than seven hundred *pounds* of rubber. In light of the nation's annual consumption of about six hundred thousand *tons*, the goldenrod yield was not remotely sufficient. Moreover, pilot plant work at the SRRL met with numerous difficulties because technologies for screening, precipitating, and purifying goldenrod rubber simply were not yet in place.[70] Perhaps unsurprisingly, the hasty experiments in the field also brought "inevitable difficulties" and "insurmountable problems."[71] Crop experts found that certain soil types were ill-suited for goldenrod, fertilization was more expensive and more complex than anticipated, leaf-stripping technologies were difficult to design, and drying the crop in converted lumber kilns was costly, time consuming, and unsuccessful. By April 1944, the proposed agricultural appropriation bill for the new fiscal year offered no funds for goldenrod research. As with kok-sagyz, the ERP's work with goldenrod officially ended on 1 July 1944; only a few lightly funded projects continued. As the project came to end, USDA officials could only report that their effort had fallen short: "Knowledge concerning goldenrod has not progressed to the point where a large scale program can be undertaken. Conjectures along such lines would be purely a mental fabrication."[72] Although the best funded of the USDA's rubber crop efforts in the 1930s, the goldenrod project fizzled out long before the war came to a close. By early 1944, even though rubber remained in short supply for both military and civilian projects, the government began to hire tractors and mule teams to plow up stands of goldenrod and bring the Georgia farmlands back into cotton production.[73]

Yet Another Guayule Revival

Despite the many blows to the struggle for a domestic rubber crop, the guayule program came back to life in the summer of 1943. With rubber still in short supply, natural rubber still necessary to conduct the war, the military still facing a determined enemy in both theaters of war, and an emergent military-industrial complex on the horizon, it is not surprising that calls for government action with rubber crops continued. In this context, the WPB sent its rubber expert John Caswell on a monthlong survey of the guayule alternative. In his July 1943 report, Caswell admitted that the ERP had made errors in California, particularly in its relationship with the farmers who controlled valuable and irrigated acreage in the heart of the Central Valley. But Caswell consulted with several guayule experts, such as Francis Lloyd, David Spence, Robert Emerson, and Robert Millikan, and became convinced that the shrub still could make a difference in the war effort and beyond.[74]

Caswell proposed an immediate resumption of the project. So that it would not compete with agricultural crops, he called for a new focus on inexpensive and nonirrigated lands in California, Arizona, New Mexico, and Texas. He found Texans especially eager to jumpstart the guayule industry and believed that low labor and land costs in that state could make guayule competitive with imported rubber on a long-term and sustainable basis. William O'Neil of General Tire also asserted that the guayule program should shift from the irrigated farmland of California to dry lands in Texas and Mexico.[75] The ERP plan to expand guayule "in a big way" breathed life back into Arizona governor Osborn's long interest in developing the arid regions of his state.[76] Thus Jeffers resumed the program in August 1943, with a new goal of

Figure 7.5. Pulling and loading wild guayule at the 02 Ranch near Alpine, Texas, October 1943. Photograph by W. Baxter. National Archives.

producing twenty thousand tons of guayule rubber annually from unirrigated lands.[77] This policy reversal caused headaches for the ERP, which yet again had to leap into "high-speed action." After spending the spring and summer trying to sever its lease arrangements with farmers, ERP officials now shifted course and began seeking cooperation from enough farmers to cultivate the shrub on 107,000 acres in four southwestern states.[78]

Meanwhile, national shortages of natural rubber induced the ERP to harvest the last wild guayule stands in southwestern Texas. Overharvesting of the shrub had caused the guayule factory in Marathon, Texas, to close in both 1909 and 1926, but some regrowth of wild stock had occurred by the 1940s.[79] The project took another bizarre turn when the ERP contracted the General Air Conditioning Company of San Francisco (which was unable to continue its main business during the war) to conduct the harvest of Texas guayule. This proved to be an imprudent choice, as its pipe fitters, plumbers, and air conditioning salespeople were ill prepared to harvest rubber plants from the remote corners of western Texas. Bringing more embarrassment to the project, the press revealed that the company had employed illegal Mexican immigrants secreted across the Rio Grande. The company faced several lawsuits for damages from Texas ranchers who considered guayule a good forage crop for sheep and thus objected to its harvest. According to one government official connected with the Texas project, only the USDA's direct intervention "saved us from going insane," while the press reported that the government abandoned the wasteful and expensive effort "without producing so much as a nipple for a baby's bottle."[80] The impression that the ERP was in chaos was starting to spread.

Work on the guayule project also extended beyond American borders. As indicated by the internationalist message of President Roosevelt's veto of the original guayule legislation, Mexico became an important collaborator in the effort. The Mexican government lifted the restrictions that had hampered IRC operations for decades, while the U.S. government agreed to purchase all Mexican guayule production at a guaranteed price of thirty-one cents per pound for the duration of the war. The policy benefited both sides: the Mexican government received nearly a million dollars in tax revenue, while the IRC earned a handsome profit; further, the IRC built a new factory in Mexico and received government funds and gasoline rations that helped it aggressively harvest and process the remaining stands of wild guayule. The IRC's Mexican production exceeded 6,500 tons in both 1943 and 1944. Inspired by substantial profits, the IRC hinted that it might rehabilitate operations in Arizona, which had closed in 1923. Significantly, though, controversy over the IRC's war profiteering returned, because many of seeds and seedlings planted in Mexico had been obtained from the U.S. government's purchase of the IRC's American assets back in 1942.[81] Despite the harsh desert conditions, shortage of pack animals, impenetrable roads, and labor strife, IRC operators (assisted by poachers who served a black market) had eliminated virtually all the world's wild guayule by 1945.

Guayule as a Permanent Part of Southwestern Agriculture

As the federal government's guayule program became increasingly chaotic and controversial, new authorities and institutions tried to salvage guayule as a domestic rubber crop. In contrast to the ERP, which by definition focused on the wartime emergency, new proposals centered on the idea that guayule could become a permanent addition to the agricultural economy of the American Southwest. As hopes for the future of the ERP reached a low point in spring 1943, longtime guayule promoters lobbied for California governor Earl Warren to prevent both "Wallaceites" and petroleum interest groups from killing progress on the nation's most promising domestic rubber crop. Notably, the lobbyists also hinted that leadership on the guayule issue might help Warren position himself as an alternative to internationalists like Wallace in the 1944 presidential campaign. In response, Warren created a five-member State

Guayule Committee in April 1943.[82] The committee soon heard from powerful interest groups that divided sharply over the question of guayule's future. It also heard warnings of dire consequences should the project be abandoned: Texas politicians might beat Californians and take control of the project; a Dutch company might buy up the ERP's seed stock and plant it in Mexico; and the world's nonrenewable petroleum supplies might dry up within just a few decades. Even if waste or inefficiency could be attributed to the ERP, such costs were minuscule compared to other wartime programs, like synthetic rubber.[83]

By late 1943, California's State Guayule Committee came up with a concrete list of recommendations: (a) the government should continue to lead research and experimental work with guayule and similar plants; (b) the government should help transfer the guayule program to private enterprise, especially through a farmers' cooperatives and privately owned guayule processing facilities; (c) leaseholders should be allowed to renegotiate their contracts at the end of the harvest; and (d) the government should establish a five-hundred-thousand-dollar program to provide a minimum of fifteen fifty-acre plots for each milling site and offer American farmers who commit to the plant guaranteed price supports at a level no less than what the United States paid Mexican producers.[84]

The committee struggled to make headway with this program. Opposition came from an ironically named Kern County Guayule Committee, led by large landowners who opposed any program that smelled of government support for the plant. According to the Kern County group, the rubber emergency had passed; the fact that Edison, Firestone, and the IRC all eventually had given up on domestic guayule meant that California's government should do the same. The group's vocal expression of their patriotic commitment to grow vital food crops left unstated that other crops would bring them more money.[85]

Meanwhile, Congressman John Anderson of California called for a thorough congressional investigation of the guayule program to help restore its viability on the national level.[86] The House Committee of Agriculture put Texas congressman Bob Poage in charge of a subcommittee to study the guayule situation. A quintessential Texas Democrat from Waco, Poage soon became committed to the struggle for American rubber crops; over the next few years, he became a national expert on guayule and fought its opponents in both the political and agricultural arenas. In April 1944, Poage and four congressional colleagues held hearings in the heart of California's guayule country. In Salinas, most of the witnesses who spoke publicly presented a favorable view of the ERP and the progress that had been made. Salinas officials treated the

Poage subcommittee to a luxurious banquet, and the local newspaper declared it "ridiculous" to even consider halting the rubber project.[87]

The tone changed when the members of Congress reached Bakersfield. There they found a vocal cadre of landowners eager to reiterate their complaints of 1942 and 1943. The Bakersfield group dismissed the ERP as a prime example of wasteful government spending, with inflated wages for inefficient workers, unwelcome advice, and unreasonable requirements for farming techniques. As before, critics considered guayule completely disruptive to Kern County agricultural practices. Guayule's opponents complained that each ton of guayule rubber cost far more than any other rubber and that millions of sacks of potatoes and other food crops could have been produced from the same land. Yet Poage tersely questioned the Kern Countians' expertise on rubber issues, and others testified that the ERP had been reasonably successful considering the pressures of wartime circumstances. In the end, Poage and his colleagues became convinced that guayule offered a potential long-term solution for American rubber needs, preferably through private enterprise.[88]

Poage's hearings had an immediate impact. In response, Congressman Clarence Cannon of Missouri resumed his campaign against the extravagant costs of the search for a domestic rubber crop. As head of the House Appropriations Committee, Cannon pushed through a proposal that reduced the ERP's funding and called for its complete liquidation in fiscal year 1945.[89] The Senate intervened, however, and held hearings at which guayule's defenders from the House presented their case. Poage and members of his subcommittee were unanimous in their defense of the shrub, mainly as a plant that had the potential to contribute to southwestern agriculture over the long term. The rhetoric over guayule took another turn when Poage invoked Dwight Eisenhower's confidential 1930 report that called for guayule as a sensible insurance policy for American strategic interests.[90] In June 1944, as Eisenhower's armies moved forward from the Normandy beaches, debates over guayule reached the floor of the House. Cannon insisted that the project had been "ill advised" from the start, that "every dollar" put into domestic rubber crops had been wasted, and that the concept had been approved only because of fears of a ten- to thirty-year war. Yet the Poage committee's study, in concert with repeated references to General Eisenhower's warning from 1930, carried the day. In an unusual move, House members rejected the recommendations of Cannon's Appropriations Committee and agreed to the Senate's terms, which called for additional funds and a continuation of the ERP.[91]

Remarkably enough, the struggle for domestic rubber crops staged yet another revival in early 1945. While some stressed that military victory in Europe seemed inevitable and that synthetic-rubber production was approaching levels adequate for a full-bore war economy, others saw victory in Asia as still uncertain and considered natural rubber still essential for various war goods such as airplane tires. The new rubber czar, John Collyer, called for a quick survey to determine if guayule could help. A committee of four rubber-industry executives recommended the immediate harvest of whatever guayule remained in the ERP's California fields. According to rubber expert John Caswell, "every pound—not ton, but pound—of natural rubber" must be harvested. Although the plants in the ground were at least two years short of ideal maturity, ERP officials planned to strip guayule fields bare. Officials called for the immediate construction of four new processing mills in the Central Valley, hoping to garner about thirteen thousand long tons—enough to make a small difference in the war economy.[92]

Meanwhile, Congressman Poage continued his efforts to keep the remaining guayule project alive. He went directly after Phil Ohanneson, a prominent landowner and chair of the Kern County Guayule Committee—the group that had tried to block Warren's State Guayule Committee and had called for the quick destruction of the region's fifteen thousand acres of guayule. Poage countered that natural-rubber stockpiles remained ominously low and urged Kern County landowners "to forgo some degree of war profit in the interest of our common war effort." In a sharp rebuke to Ohanneson, Poage asked if Kern County farmers really wanted inferior synthetic rubber in their tires, or if they just wanted such products for everyone else.[93] The rubber chemist and longtime guayule supporter David Spence also challenged the Bakersfield group, urging Ohanneson not to allow tens of millions of dollars worth of guayule in the field to go to waste.[94]

In early 1945, the House finally began to consider HR 2347, the by-product of Congressman Poage's thorough study of the rubber situation the previous year. Poage still argued, as he had done for years, that Americans desperately needed natural rubber, that government investments in guayule research were about to pay off, and that the program was in position to be transferred from public project to private enterprise. In a nutshell, Poage's bill called for: (a) a continuation of government investment in scientific and technical research on rubber crops; (b) continued maintenance of all guayule growing in the field until their maturity, even if the same lands could be profitably grown in food crops; (c) discontinuation of government role in the harvest and processing of agricultural rubber; and (d) incentives for small farmers to invest in guayule as a long-term project by establishing a guaranteed price of twenty-eight cents per pound for up to four hundred thousand acres, the magic number that Eisenhower had proposed in 1930. In another challenge to the norms of California agribusiness, Poage's bill offered the guaranteed guayule price subsidies only to small farmers. The bill passed the House Agricultural Committee in March 1945 but got no further.[95] Debates over the proposal dragged on throughout the summer. Although Poage's call for a long-term national commitment to domestic rubber crops almost became a reality, in the end it did not. In all, the combination of political controversies, technological barriers, and agricultural constraints proved too much even for healthy guayule plants to bear. As the war came to a close, the circle of guayule enthusiasts grew ever smaller.

1945: Final End of the Rubber Crisis?

Not surprisingly, the American military victory in World War II helped bring an end to the struggle for domestic rubber crops. It soon became clear that the Japanese had not used a "scorched earth" policy to destroy the *Hevea* rubber plantations in Southeast Asia, as had been feared. The urgency of the natural-rubber problem abated. Congress officially liquidated the ERP on 25 August 1945 and ordered that the thirty-seven-million-dollar investment in California guayule end as quickly as feasible. Also, just after the cement foundations had been poured, a telegram from Washington, DC, halted construction of the four additional mills intended to quickly process the harvest of 1945. The mill in Salinas, which had opened in 1930 and operated continuously from late 1944 to late 1945, went on the auction block in January 1946. Officials also quickly sold off the ERP's nurseries, farm equipment, worker housing, and other assets.[96]

Next came the destruction of twenty-three thousand acres of healthy American guayule, enough to yield millions of pounds of rubber. Farmers eager to plant conventional agricultural crops demanded the quick suspension of their government leases for guayule. Because those leases called for lands to return to their original condition, local officials hastily organized crews that plowed the plants into the ground. In the Salinas Valley, German prisoners of war were still burning living guayule plants in February 1946, nine months after the Nazi surrender. With rubber plants smoldering, harvest equipment rusting in the fields, and crops like lettuce and lima beans deemed more important than a domestic rubber source, Salinas officials were embarrassed that their project had become a "national joke." In perhaps the most ironic illustration of the new thinking, 230,000 pounds of guayule seeds, ten times the amount that had been so precious when the ERP began, were sold to a Bakersfield farm supply company for resale as a part of protein-enriched cattle feed.[97]

In the words of one ERP veteran, hardly any other program had been "attacked more viciously than guayule."[98] For Henry A. Wallace, Fred McCargar, Hugh Anderson, and other natural-rubber enthusiasts, the triumph

of synthetic rubber and the defeat of the ERP could be explained by a conspiracy of petroleum and rubber-industry officials.[99] Yet no evidence of such a
conspiracy has been uncovered, and there were other factors to be considered.
Indeed, the drama of the American struggle for a domestic rubber crop reveals
the extent to which decisions regarding science and technology were based on
far more than rational judgments and scientific facts; they were also are shaped
by political, social, cultural, and even psychological factors.

The end of the ERP also fits in with broader trends in American history,
particularly a new understanding about the relationships among government,
agriculture, and technology. Circumstances had changed. Older arguments,
such as Congressman Anderson's cries that it was "utter folly" to allow the
United States to "become the international sucker for the notorious British-
Dutch rubber monopoly," no longer carried much weight in an era when internationalism replaced isolationism.[100] Social reformers like Hugh Anderson
failed to recognize that most Americans were not very interested in using rubber crops as the core of social experiments.[101] Scientists who had devoted the
war years to mastering the genetics, physiology, morphology, and other traits
of the guayule plant soon moved on to other projects. And in the Salinas Valley, local growers finally convinced Fred McCargar that his tireless promotion
of guayule had caused him to ignore crops that had a real future in "America's
Salad Bowl."[102]

Tellingly, after 1944, claims of newly discovered domestic sources of rubber did not attract much attention. The climate of rubber crop politics stood
in contrast to the 1920s, when Edison, Ford, Hoover, and other national
leaders had championed the cause of domestic rubber crops. Indeed, many of
the rubber crop enthusiasts of the mid-1940s had little political or economic
clout, and a few were crackpots.[103] Meanwhile, other natural-rubber crop promoters like Baruch and General Eisenhower had moved on to other issues of
concern. Along the same lines, the chemurgists' pursuit of agricultural sources
for industrial products became increasingly irrelevant and poorly funded.
Those who promoted synthetic rubber derived from renewable resources in
American corn and wheat fields could not compete with those who favored
synthetic rubber that used Cuban molasses or Middle Eastern petroleum as
the feedstock. Threats of political instability in these parts of the world faded
from the conversation, replaced by a somewhat arrogant assumption that
American leaders could handle such interruptions if and when they occurred.
Similarly, those outraged by government price supports for Mexican guayule
greater than those for American-grown guayule also missed the point that the
nation's commitment to the Good Neighbor policy was more intense than its
interest in domestic rubber crops.

Beyond these explanations for the collapse of rubber crop enthusiasm, a few more fundamental changes also deserve attention. As an editorial in the *Wall Street Journal* explained, proposals for government support for guayule stood in opposition to the nation's postwar goals as articulated in the Four Freedoms, the Atlantic Charter, Bretton Woods, and Dumbarton Oaks. This attitude had become evident by 1943, when at the height of enthusiasm for cryptostegia, President Roosevelt toasted President Lescot of Haiti with the promise that the United States would not use tariffs to protect its infant synthetic-rubber industry; instead, it would allow natural rubber to compete freely.[104] Several legislators were reluctant to support Poage's proposal for a lasting commitment to agricultural rubber in view of potential disruptions to postwar trade relationships. Most government advisers embraced the rhetoric of "free trade" as a mantra for the postwar world; thus they strenuously opposed government subsidies for the rubber industry, whether that meant help for the emerging synthetic-rubber industry, the possible revival of a rubber industry in Latin America, or domestic rubber crops.[105]

The emergent cold war fit in as well. In general, postwar Americans became convinced that peace and prosperity depended in part on the well-being of citizens in other nations, and many linked global prosperity with international security. American diplomats sought to build strategic partnerships with a large number of non-Western nations, including many potential rubber producers. The fact that the United States required more imported rubber and petroleum than ever would provide suppliers of these raw materials with the funds needed to purchase American grains and manufactured products. Meanwhile, the cooperation that had developed between American and Soviet rubber scientists and policy makers died out even before the war with Germany ended, as had the last fields of the Russian dandelion. In 1944, American scientists alleged that their Soviet counterparts were not delivering promised technical information on kok-sagyz growing techniques, while Soviets complained that Americans were not delivering promised guayule harvesting equipment.[106]

The Narrative of Synthetic Rubber

As shown in chapter 2, the prominent science writer Edward Slosson had argued back in the 1920s that World War I offered lessons concerning the American dependence on imported chemical raw materials. With a somewhat dire tone, Slosson had urged Americans to invest in their chemical industries, to develop nationalistic measures to protect against their vulnerability to war, and to search for chemical raw materials in American-grown agricultural products.[107] The science writer Williams Haynes expanded Slosson's message in three books published in 1942 and 1943, but he projected a much more confident and hopeful message. Amid a war that caused incredible suffering and millions of deaths around the world, Haynes presented the case that American chemists were war heroes, for they had developed scores of artificial materials that would become the basis for a postwar "economy of abundance." According to Haynes, synthetic products were more consistent in quality, more likely to fall in price, and more readily altered to fit custom needs than natural products. Haynes also described the geopolitical implications of these discoveries. International tensions would decline, he predicted, because access to synthetic materials was less subject to war, diplomatic disputes, currency fluctuations, and the accidents of ecological advantage than agricultural materials that came from the soil. Synthetic chemicals had "made wastepaper of all the old tenets of international relations." Haynes concluded that it would be a "grievous mistake" to allow the search for raw materials to bring back an era of "unreasoning isolationism."[108] MIT rubber chemist Ernst Hauser preached a similar message, explaining that despite all of the hype over guayule and similar crops, the bleak situation of 1942 led to the conclusion that "our only hope" was synthetic rubber. Synthetic rubber had to be a national priority, and informed citizens needed to learn the basics of rubber chemistry. Even if natural rubber returned to dominate the marketplace, he had no doubt that some of the rubber chemists' discoveries had come to stay.[109]

In this milieu, rubber industry officials could challenge hopes of the guayule enthusiasts on a fundamental level. Chemical products, they argued, rather than agricultural products, stood at the foundation of postwar prosperity. In summer 1945, the Goodyear Tire and Rubber Company released a widely distributed pamphlet that argued that it was "high time" for candid comment on guayule. Paul W. Litchfield, Goodyear's chair of the board, presented a long list of reasons to move away from the guayule alternative, deeming it inconceivable that American national defense would ever again be as lax as in 1941.[110] Litchfield was bullish on Latin American and Caribbean *Hevea* sources, explaining that guayule's quality remained inferior and that the war had proven that synthetic rubber, not natural, was the ideal insurance policy. In addition, he believed that the synthetic option would inevitably place pressure on British and Dutch producers to remain price competitive.[111]

Meanwhile, rubber-industry and government officials began to construct a new narrative connecting the triumph of synthetic rubber with the emergence of the United States as a truly modern nation. Positive publicity about the synthetic-rubber program kept well ahead of reality, but it has remained embedded in the popular understanding of the World War II home front ever since. Construction of the synthetic-rubber story was already under way by 1943. Rubber director William Jeffers quickly gained the reputation as a hero whose investment in synthetic rubber had cleaned up the rubber mess of 1942; some floated his name as a candidate for the presidency in 1944.[112] Upon his resignation in September 1943, Jeffers declared a rather premature victory over the rubber problem. The seven hundred million dollars had been well spent, he said, and the emphasis on synthetic rubber rather than domestic rubber crops had been justified.[113]

The Allied victory allowed rubber and petroleum companies to claim that they had responded to the rubber crisis with patriotic and ingenious efforts to develop synthetic rubber. In July 1945, for instance, the Rubber Manufacturers' Association issued a press release that called synthetic rubber "one of the epochal achievements of the war."[114] Another telling example appeared in a Firestone Company brochure released in 1946: its back cover featured nothing more than the word "Rubber" and an imposing picture of modern chemical factory. The brochure barely mentioned that rubber still could come from living plants, and its description of synthetic rubber plainly ignored that alcohol from grain could be used as a feedstock.[115] The B. F. Goodrich Company released a series of small booklets for its everyday customers that touted the company's foresight on the synthetic rubber issue, one of which called for the return to absolute free trade conditions for both natural and synthetic rubber by the end of 1946.[116] Several rubber and petroleum producers published corporate histories during the war and postwar years, all of which placed considerable emphasis on the companies' patriotic commitment to solving the rubber crisis through synthetic rubber and other sacrifices.[117] Amid the nascent cold war, Frank Howard, a vice president of SONJ, highlighted the history of synthetic rubber as emblematic of American scientific genius and the American ability to adapt to a world filled with competing societies.[118] John Collyer of B. F. Goodrich stressed the role of entrepreneurship and private capital—not the government's investment of seven hundred million dollars—while insisting that "no achievement . . . has been of greater value in the preservation of free nations" than the triumph of "American-made" synthetic rubber.[119] In the public and political consciousness, the rubber crisis of 1942 had passed, and most Americans accepted the common narrative that credited oil and chemical companies for the heroic and timely development of synthetic rubber from petroleum. Few people recognized that much of the synthetic rubber developed in World War II derived from grain, and perhaps now even fewer realize that far more natural rubber is consumed today than before synthetic rubber existed.[120]

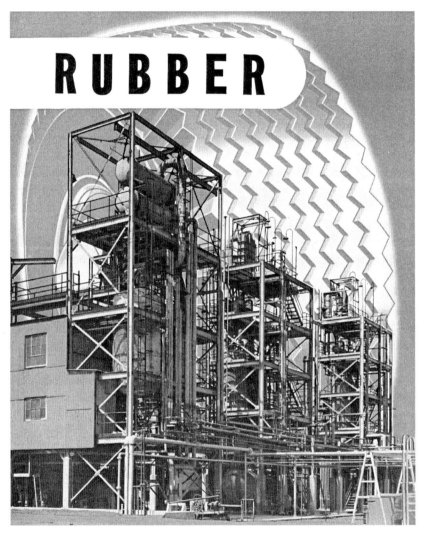

Figure 7.6. Firestone Tire and Rubber Company pamphlet, 1946, depicting rubber as a product of chemical engineering.

Such developments illustrate how the decline of the ERP and rise of synthetic rubber offers insights into an important shift in the history of Americans' understanding of relationships among agriculture, science, technology, and geopolitics. As part of a "progress ideology" that emerged in the middle of the twentieth century, American consumers and industrialists learned to favor inorganic over organic materials, even if technical reality might have allowed for alternative outcomes. Whereas early synthetic fibers, like rayon, and early pesticides, like pyrethrum, had been derived from agricultural and renewable resources, these were soon replaced by synthetic products derived from petroleum in the years surrounding World War II. Projects that the chemurgists had touted in the 1930s—ethanol from corn, plastics from milk casein, paints from tung oil—lost their luster as technologists developed cheaper and logistically simpler alternatives derived from petroleum and other hydrocarbons. As in many other cases, rubber from petroleum fit expectations for uniform quality, steady supply chains, and professional chemists' and engineers' mastery of the production process. The rubber crop enthusiasts' efforts, so much of which focused on crops that most regarded as little more than bothersome weeds, fared poorly in comparison. The notion that agriculture could provide solutions to the nation's resource needs faded from view, and the late twentieth century marked the heyday for synthetic materials in the American economy.[121]

Chapter 8

From Domestic Rubber Crops to Biotechnology

In 1934, ALVIN HANSEN, a leading New Deal economist, asked famed USDA plant explorer David Fairchild about the prospects of producing rubber in the United States. Fairchild's response was probably not the direct answer that Hansen was seeking: "What do you mean by rubber? What is your idea of possibilities? *When* are you talking about? Tomorrow? Next week? Fifteen years hence?" He continued:

> If you could tell me what the price of rubber will be fifteen years from now; or if rubber and not something else will be used to cushion the tires of automobiles; or if flying will have so developed that there will be a million or so miles of abandoned thoroughfares; or whether there will be a revolution in the Dutch East Indies and Malaya and the culture of rubber there abandoned; then perhaps I could predict with more certainty that somehow or other Americans would work out the cultivation of one or other of the various rubber producing plants which we know can be grown in this country.[1]

Fairchild's answer reveals a few of the complications that accompanied the long struggle for a domestic rubber crop. Had American rubber companies and political leaders chosen to develop a domestic rubber strategy at the end of World War II, they certainly could have. The hundreds of scientific and technical experts who worked on the ERP (and under the guard towers of Manzanar) had made considerable progress on increasing the yield and quality of American-grown rubber products. Investigations of blight-free strains of *Hevea*, intended to restart a rubber industry in the Western Hemisphere, were also on the verge of breakthrough discoveries. Within just a few years, the United States had gone from the world's number-one rubber-importing nation to its premier rubber exporter. The United States also controlled over 90 percent of the world's capacity to produce synthetic rubber and the potential to control know-how in that field for decades to come.[2] Yet American political and industrial leaders abandoned the search for rubber self-sufficiency, persuaded as much by social and cultural circumstances as by the agricultural, scientific, and technical problems associated with the possibility.

Thus the struggle for domestic rubber crops collapsed in the postwar world. In view of the economic and geopolitical complexities involved, rubber crop enthusiasts' visions of flipping on the switch of a new bio-based industry proved illusory. New cultural attitudes also played a role, for many Americans felt supremely confident of the nation's economic, political, scientific, and technological might in the aftermath of their victories over Germany and Japan. Along with penicillin, plastics, DDT, and, of course, the atomic bomb, synthetic rubber contributed to the illusion that American scientists and engineers had developed mastery over the natural world. World War II's crash projects created the impression that if enough capital and infrastructure were allocated into a problem, American experts could extract a solution. These projects also suggested that if science helped win this war, then the next war would be won in laboratories, especially those that focused on physics and chemistry. As articulated in Vannevar Bush's blueprint for postwar research, *Science: The Endless Frontier*, the time had come for a greatly expanded federal role for research in the basic sciences. This report—released in mid-1945, just as officials were liquidating the ERP—made clear that policy makers now conceived of frontiers far different from the nineteenth-century agriculturally centered expansion. Indeed, the resultant National Science Foundation (NSF), established in 1950, distributed 95 percent of its monies to physicists, chemists, and engineers; only 5 percent of its funds were left for all of the life and social sciences. Farmers and botanists seemed of minor importance in comparison with the contributions of chemists, physicists, and others who came to dominate the scientific and industrial economy of the late twentieth century. Postwar agricultural research shifted to relatively unheralded programs that enhanced foods and other products for the convenience of the American consumer. Its geopolitical implications were also relatively minor, focusing on the search for markets for continued surpluses of basic commodities and the transfer of American agribusiness methods to allied nations in the developing world.[3] If the beginning of this history confirms historian Richard Drayton's observations about the centrality of agricultural products in the rise of the Western industrial and imperial powers, as well as Philip Pauly's emphasis on the role that plants have played in American history, the postwar history suggests the converse. By the end of World War II, global struggles over geopolitical influence and strategic resources rarely centered on the products of agriculture.[4]

Indeed, the whole notion that the United States should pursue nationalist strategies of economic development virtually disappeared. Economic nationalist groups like the Chemurgic Council found that their message had little resonance, and institutions like the World Bank, the International Monetary Fund, and the International Trade Organization ensured that the United States became deeply involved in global economic development. As President Truman stated in a message to Congress in 1947 concerning rubber policy, the nation's priority was to restore trade networks rather than debate details concerning raw materials. Despite his explicit recollection of the rubber crisis of 1942, and although he admitted that the supply of natural rubber was "still inadequate to meet world needs," Truman called for no change in rubber policy. In brief, President Truman seemed little worried that natural rubber might become an important geopolitical issue in the postwar world.[5]

The old IRC concurred. After forty years in the business, the firm began efforts to liquidate its entire operations. Financial losses mounted in the late 1940s, and hopes to develop alternative revenue streams proved fruitless. The Korean War brought a last-gasp effort to harvest native guayule near Torreón, a project that decimated the remaining stands of wild guayule. Negotiations with the U.S. government to plant guayule as a living strategic reserve for emergency purposes proved unsuccessful. Finally, in October 1953, the IRC sold its assets to an upstart company interested in acquiring the IRC's valuable listing on the New York Stock Exchange. That firm, Texas Instruments, proved a better fit than the IRC for the late twentieth-century American economy.[6]

A Few More Guayule Revivals

Interest in domestic rubber crops has resurfaced from time to time in the sixty years since the end of World War II. Postwar predictions that synthetic would replace Hevea rubber proved to be quite incorrect. The emergence of radial tires in the 1970s was a special boon to natural-rubber producers, for they require elasticity, strength, and "tack" in the sidewalls that only natural rubber can provide. Demand for natural Hevea rubber doubled quickly, and it continues to increase. But natural rubber still derives from a small corner of the world. Three nations—Indonesia, Thailand, and Malaysia—now produce 67 percent of the world's natural-rubber supply.[7] Fortunately for the industrialized world, those nations have been relatively peaceful and stable since the 1950s, and there has been no repeat of the rubber crisis of World War II.[8] Yet the threat of a rubber emergency remained, and guayule's enthusiasts continued to find allies in their ongoing quest to secure an American rubber source. The intermittent episodes of guayule research generally have coincided with periods of diplomatic tension, such as the peak of the cold war, the oil crisis of the 1970s, and the emergency that followed the terrorist events of 11 September 2001.

For instance, guayule emerged again as an issue in the early 1950s. In the words of Senator Lyndon Johnson, "If the Eastern nations crumble and communism chokes off our supply of natural rubber, the industrial might of this nation will have its muscles slashed." Thus in 1951, Johnson's Preparedness Subcommittee of the Senate Armed Services Committee asserted that "it is important to reach a stage where we can, if necessary, go into large scale domestic natural rubber production" and called for major investments in guayule seeds and nursery plantings to shave off time needed to restart domestic rubber production.[9]

Other guayule proponents sought funds to plant a million acres in guayule, a vast project that potentially could supply nearly half the nation's natural-rubber needs on a permanent basis. Some dabbled with aerial seeding of guayule on military bases, sought permission to plant guayule on the protected lands of Big Bend National Park, and proposed efforts to keep guayule on western lands as a "living reserve" that could be maintained with minimal effort and be available for tapping in the case of natural rubber emergency.[10] All of these ideas fell short of success. In all, the investment of five hundred thousand dollars yielded a mere 9,600 pounds of new seed, and some critics saw the effort as another domestic-rubber boondoggle. History was repeated in the early 1950s when hundreds of acres of healthy guayule plants were destroyed before any rubber could be retrieved.[11] In 1953, government officials flatly terminated programs to develop blight-free strains of natural *Hevea* rubber reserve in the Western Hemisphere. By the end of the 1950s, Congress cut funding for the Natural Rubber Research Station in Salinas, California, and transferred the remaining genetic stock of ERP-era guayule to the National Seed Storage Laboratory at Fort Collins, Colorado. Congress also used progress in synthetic-rubber technology to justify the liquidation of the nation's stockpile of nearly half a million tons of natural rubber. Efforts for a domestic rubber crop had reached another dead end.[12]

Yet another guayule revival occurred in the 1970s. The oil crisis of 1973 provided the main impetus. The case of Caltech scientist James Bonner, who had been involved with guayule since the early 1940s, is illustrative. In April 1973, before the oil crisis, Bonner warned guayule supporter Ed Flynn, "I am now absolutely and totally convinced [guayule] has no future." Within a few years, however, Bonner told Flynn to "count me in on any guayule development."[13] Other veterans from war-era projects reemerged. Guayule enthusiast

Hugh Anderson lobbied incoming Carter administration officials even before inauguration day in 1977, while Thomas A. Fennell, leader of SHADA's cryptostegia project during World War II, asked James Schlesinger of the Federal Energy Commission to reconsider the rubber vine in any program designed to reduce the nation's dependence on imported petroleum.[14]

A new cadre of guayule supporters also joined the cause. Some explicitly compared the explosion in petroleum prices with the Stevenson Plan of the 1920s, while others feared that the Association of Natural Rubber Producing Countries (ANRPC) could imitate the Organization of the Petroleum Exporting Countries (OPEC) cartel and leave American rubber consumers in a similar bind.[15] Thus in 1975 the Office of Arid Land Studies at the University of Arizona organized a comprehensive conference and the National Academy of Sciences sponsored a thorough investigation, both of which asserted that modern technology might make guayule a "commercial crop once again."[16] Guayule's supporters insisted that World War II's ERP had been a "most magnificent program," mistakenly and shortsightedly cut amid the emerging enthusiasm for synthetics. Significantly, these new studies made an effort to acknowledge contributions by the Japanese American researchers interned at Manzanar. Scientists promised that advances in genetic research techniques could lead to rapid improvements in guayule purity, yield, and performance as an alternative to Hevea rubber. Thanks to the plant's quality of apomixis, it was a "plant geneticist's dream," for genetic improvements could be locked into seed clones without risk of cross-pollination.[17]

Many other circumstances contributed to a more favorable political climate for the guayule alternative. Research from the World Bank and other institutions indicated that demand for natural rubber was stretching the production capacity of Southeast Asian plantations.[18] Dire predictions suggested that demand would outstrip supply by about 1990, coinciding with the time when some forecasted a peak of global petroleum production, which would in turn hamper the synthetic-rubber industry. In short order, political leaders and the mainstream news media caught another case of guayule fever.[19] In Congress, New Mexico senator Pete Domenici and California congressman George Brown Jr. led the campaign to provide federal funding. Tellingly, congressional hearings began on the very day that the Senate approved the treaty that would relinquish American control of the Panama Canal, another reminder of the nation's vulnerability to international instability. Experts on national defense testified that the Soviet Union had increased its consumption of natural rubber, while war-preparedness experts admitted that the nation had fallen far short of its goals in stockpiling rubber and other strategic materials. Humanitarian arguments also entered into the debate, for it seemed that both Native Americans on poverty-stricken reservations and Africans in the drought-stricken Sahel Belt might benefit from a new crop suited for arid lands. Others warned that the Ogallala Aquifer could become exhausted, cause a new Dust Bowl, and force thousands of farmers to find alternatives to wheat and other traditional crops. In the case of widespread political upheaval in the rubber-producing nations, it seemed feasible that guayule might produce 60 percent of the nation's natural-rubber needs by 1990.[20] Noel Vietmeyer of the National Academy of Sciences was especially blunt, testifying that never before had the nation faced such a "compelling need for *renewable* resources crucial for the nation's very survival."[21]

As a result, Congress passed the Native Latex Commercialization Act of 1978, a bill that offered up to thirty million dollars to the departments of Commerce and Agriculture to coordinate domestic rubber crop research. Plans called for extensive genetic research, field-testing, database creation, and experiments with harvesting equipment and extraction techniques, including those using solar power.[22] Additional funding from the NSF went toward seed-collection projects, technology assessments, and analyses of market and production costs.[23] Political leaders such as Governor Jerry Brown of California and Texas's highly visible commissioner of agriculture Jim Hightower found funds to support guayule research at the state level.[24] Private industry also jumped on the guayule bandwagon, as Goodyear, Firestone, Weyerhauser, and even Pennzoil invested in experimental guayule plantings in several southwestern states.[25]

Yet this flurry of enthusiasm proved another disappointment to guayule's supporters. Nearly twenty-seven million dollars of the monies made available in the Native Latex Commercialization Act were never appropriated, because various bureaucrats could not agree on research priorities. Generally speaking, agricultural officials wanted to wait until market conditions for the new crop seemed promising before asking farmers to expand guayule acreage. Commerce Department officials assumed that experiments with processing technologies and market research would be fruitless until sufficient quantities of high-yielding strains became available. Proposals to break this deadlock in the name of national defense went unheeded. By the early 1980s, Reagan administration officials targeted guayule as a case of an unwarranted government subsidy and thus made it part of their budget-cutting agenda. In the congressional hearings of 1984 intended to revitalize the project, officials from both the departments of Agriculture and Commerce testified against continued funding for these projects. To their chagrin, Congressman Brown, Hugh Anderson, and other guayule enthusiasts witnessed yet another guayule collapse.[26] Hundreds of new scientific papers resulted from guayule's boomlet of the late 1970s and early 1980s, but guayule experts once again found themselves on the fringes of the scientific community.[27]

Figure 8.1. In the twenty-first century, guayule research has focused on the development of drought- and salt-tolerant lines that are appropriate for a changing climate, on agronomic techniques that can complement cotton production, and on methods to utilize the plant's biomass for energy production. Photograph courtesy of the Yulex Corporation.

Into the Twenty-first Century

Remarkably, yet another guayule revival is currently under way. In the words of one guayule researcher, the plant has had "more resurrections than Madonna."[28] Once again, a combination of economic, geopolitical, social, and cultural circumstances has made the plant a possible rubber alternative. The AIDS crisis helped spark this resurgence, for significantly more Americans used rubber gloves in its wake. Many complained of allergic reactions to latex, which is caused by certain proteins found in *Hevea* rubber. USDA scientists discovered in 1994 that guayule contains none of these proteins. As in the 1920s, when the IRC promoted rubber crops when few others would listen, one company has taken the lead. The Yulex Corporation, founded in 1997 and now based in Chandler, Arizona, acquired rights to the USDA patents, hired technical experts, and has aggressively preached the message of allergen-friendly domestic rubber ever since. Yulex has had contracts to sell its liquid guayule latex to manufacturers of surgical gloves, catheters, and other medical devices. More recently it has found other specialized markets for guayule, including athletic clothing, footwear, and mattresses..[29] Another Arizona firm, PanAridus, has used its genetic and agronomic know-how to develop guayule as a potential raw material for energy derived from biomass, and for the largest market of all—tire manufacturing.[30] Kok-sagyz, the Russian dandelion, also has made a comeback in the early twenty-first century, with domestic research underway in Ohio and Oregon, as are extensive projects in Canada, China, Kazakhstan, and several European nations.[31]

Such developments may bring into question the list of "The Ten Most Overblown, Covered-to-Death Stories in Rubber History," published in 2004 in the trade journal Rubber and Plastics News. Editor Edward Noga could think of only six, and guayule topped the list. Like David Fairchild's response of seventy years earlier, Noga concluded that the "cute little desert shrub" was still a long way from making any kind of meaningful impact.[32] Yet even if the future of domestic rubber crops remains uncertain, there can be little doubt that some of the issues that sparked interest in domestic rubber crops remain pertinent today.

Like a century ago, the interrelationships among war, raw materials, and the environment are becoming clear once again. The centrality of petroleum in past, present, and future geopolitical conflicts is an obvious example, but it also seems that both international and internal "resource wars" are bound to become increasingly common in the face of twenty-first-century demographic and environmental pressures.[33] Further, experts have rather suddenly turned their attention to bio-security; in this new conception of war preparedness, bio-defense spending is one of the fastest growing sectors in the federal budget.

The threat of agricultural bioterrorism is particularly acute, for a terrorist would need relatively little technical expertise to cause mass devastation with lasting economic and social repercussions.[34]

The relationship between war and the environment also may affect the world's rubber supply. In the 1980s, Tamil insurgents in Sri Lanka purportedly attempted to sabotage rubber plantations with unspecified biological agents.[35] More recently, ethnic tensions and violence against rubber industry properties have emerged as part of a Muslim-led separatist movement in southern Thailand.[36] Most ominously, there may be no other plant that is more susceptible to an act of biological terrorism than *Hevea brasiliensis*. According to one expert, a bioterrorist could conceal enough fungal spores inside a shoe to decimate the Asian rubber plantations—and perhaps eventually wipe out the natural-rubber industry altogether—because the genetic material of virtually all natural-rubber plantations is cloned from the handful of seeds that British botanists first propagated in Singapore over a century ago.[37]

Such factors may bring agricultural and botanical experts back into the forefront of geopolitical issues. It is now commonplace for both advocates and critics to label the twenty-first as the "biotech century."[38] Likewise—and in contrast to the NSF's message of the 1950s—the National Research Council recently has declared that "the biological sciences are likely to make the same impact on the formation of new industries" in the twenty-first century as the physical and chemical sciences played in the late twentieth century.[39] As industrial nations search for ways to replace depleted fossil fuels, reduce carbon dioxide emissions, and deal with persistent overproduction of commodity crops, plants may again become a focal point in the relationship between science and power. Modern agricultural scientists (now often known as biotechnologists) can do more than simply increase crop yields—they can engineer entirely new products. With the backing of well-capitalized and often transnational corporations (as well as national governments concerned about food security), life scientists are becoming more adept at manipulating the natural world. Politicians, scientists, industrialists, and investors also have jumped on the biomass and renewable resources bandwagon. A widespread search for agricultural raw materials, new crops, and new uses for crop residues and wastes is under way.[40]

These developments may also signal a significant reversal in cultural attitudes about the natural world. After a half century of a world dominated by plastics, synthetic materials, and a long list of products derived from fossil fuels, people are becoming more aware that agricultural products can provide sustainable solutions to the world's resource needs. Many now question the postwar notion that synthetic material frees consumers from nature's variability, and it seems clear that applied organic chemistry has not delivered on its promise of universal access to nature's resources. Citizens are increasingly cognizant of the interconnections between the natural world and virtually every aspect of human society, and the narrative that once highlighted synthetic materials and petrochemicals may fade from view.[41]

Thus, despite its countless ups and downs, the struggle for domestic rubber crops may have a lasting legacy and significance. There is little doubt that global industrialization, automobility, and worldwide improvements in the standard of living will increase pressures on both natural and synthetic-rubber producers. As consumption in China and other developing nations soars, rubber-industry officials are again warning of rising costs and impending shortages, while concerns about climate change and carbon emissions are pushing more of the rubber market away from synthetic and toward natural and sustainable options.[42] In the event that a rubber crisis materializes again, the long struggle for domestic rubber crops will prove significant because much of the basic research in rubber plant genetics, physiology, and biochemistry has already been completed.

Early in the twentieth century, when Americans first searched for a rubber alternative, they initially looked at domestic botanic and agricultural resources. By the end of World War II, American citizens had turned away from products of the soil and were seeking chemical and "high-tech" solutions to social and economic problems. American policy makers of the postwar era were confident that nation's geopolitical power and scientific resources would be enough to obscure its continued demand for vast amounts of imported rubber, petroleum, and other strategic materials. In the twenty-first century, however, issues at the intersections of geopolitics, science, technology, and the environment suggest that the efforts of Thomas Edison, Henry Ford, Dwight Eisenhower, Bernard Baruch, Guy Gillette, Barukh Jonas, Chaim Weizmann, Shimpe Nishimura, Hugh Anderson, and the thousands of others who sought American-grown rubber may become important again. The struggle for domestic rubber crops may not be finished.

Notes

SHADA   Société Haïtiane-Americaine de Développement Agricole
          (Haitian-American Partnership for Agricultural Development)
SMN   *Savannah Morning News*
SMP   *Salinas Morning Post*
SRRL   Southern Regional Research Laboratory
TAE   Thomas A. Edison
TAF   Thomas A. Fennell
USDA   U.S. Department of Agriculture
WPB   War Production Board

Introduction

1. *NYT*, 2 July 1942. The juxtaposition of these stories is discussed in the video-cassette *Genocide*, directed by Arnold Schwartzman (Los Angeles, CA: Simon Wiesenthal Center 1982).
2. Harold Ickes to BMB, 29 July 1942, unit 6, vol. 56, BMB Papers, Public Policy Papers, Department of Rare Books and Special Collections, Princeton University Library, Princeton, New Jersey (hereafter cited as Baruch Papers); and William M. Tuttle Jr., "The Birth of an Industry: The Synthetic Rubber 'Mess' in World War II," *Technology and Culture* 22 (January 1981): 35–67.
3. Jason Clay, *World Agriculture and the Environment: A Commodity-by-Commodity Guide to Impacts and Practices* (Washington, DC: Island, 2004), 333–345.
4. See Howard Wolf and Ralph Wolf, *Rubber: A Story of Glory and Greed* (New York: Covici-Friede, 1936); Lucile H. Brockway, *Science and Colonial Expansion: The Role of the British Royal Botanic Gardens* (1979; repr., New Haven, CT: Yale University Press, 2002), 141–150; Barbara Weinstein, *The Amazon Rubber Boom, 1850–1920* (Stanford, CA: Stanford University Press, 1983); and Wade Davis, *One River: Explorations and Discoveries in the Amazon Rain Forest* (New York: Touchstone, 1996), 233–239, 302–309.
5. See William Schell Jr., "American Investment in Tropical Mexico: Rubber Plantations Fraud and Dollar Diplomacy, 1897–1913," *Business History Review* 64 (Spring 1990): 217–254; Michael Edward Stanfield, *Red Rubber, Bleeding Trees: Violence, Slavery, and Empire in Northwest Amazonia, 1850–1933* (Albuquerque: University of New Mexico Press, 1998); Warren Dean, *Brazil and the Struggle for Rubber: A Study in Environmental History* (Cambridge: Cambridge University Press, 1987); Peter V. N. Henderson, "Modernization and Change in Mexico: La Zacualpa Rubber Plantation, 1890–1920," *Hispanic American Historical Review* 73 (May 1993): 235–260; J. Forbes Munro, "British Rubber Companies in East Africa Before the First World War," *Journal of African Studies* 24 (1983): 369–379; Adam Hochschild, *King Leopold's Ghost: A Story of Greed, Terror, and Heroism in Central Africa* (Boston: Mariner, 1998), 243–249; and Henry Hobhouse, *Seeds of Wealth: Five Plants That Made Men Rich* (Emeryville, CA: Shoemaker and Hoard, 2003), 125–185. For additional case studies of the rush to find alternate sources of tropical rubber, see Cuthbert Christy, *The African Rubber Industry and Funtumia Elastica ("Kickxia")* (London: Bale, 1911); Robert Hall, *Report of the First Inspector to the Stockholders of the Wisconsin Rubber Company* (Madison: Cantwell, [1905?]); and "Rubber Planting Interests," *IRW* 30 (1 February 1904): 166.

6. Henry Ridley quoted in O. F. Cook to David Fairchild, 17 August 1928, Bertram Zuckerman Garden Archives, Fairchild Tropical Garden Research Center, Miami, Florida (hereafter cited as Fairchild Papers).

7. Brockway, *Science and Colonial Expansion*, 156–165; James C. Scott, *Seeing Like a State: How Certain Schemes to Improve the Human Condition Have Failed* (New Haven, CT: Yale University Press, 1998), 262–269; and Richard P. Tucker, *Insatiable Appetite: The United States and the Ecological Degradation of the Tropical World* (Berkeley and Los Angeles: University of California Press, 2000), 236–243. For a reevaluation of Wickham, see Donald Kennedy and Marjorie Lucks, "Rubber, Blight, and Mosquitoes: Biogeography Meets Global Economy," *Environmental History* 4 (July 1999): 369–383.

8. Jonathan Marshall, *To Have and Have Not: Southeast Asian Raw Materials and the Origins of the Pacific War* (Berkeley and Los Angeles: University of California Press, 1995), ix–x; Richard P. Tucker, "The Impact of Warfare on the Natural World: A Historical Survey," in *Natural Enemy, Natural Ally: Toward an Environmental History of Warfare*, ed. Richard P. Tucker and Edmund Russell (Corvallis: Oregon State University Press, 2004), 15–42; Edmund Russell, *War and Nature: Fighting Humans and Insects with Chemicals from World War I to Silent Spring* (Cambridge: Cambridge University Press, 2001); Jeffrey K. Stine and Joel A. Tarr, "At the Intersection of Histories: Technology and the Environment," *Technology and Culture* 39 (July 1998): 601–640; and Paul A. C. Koistinen, *Arsenal of World War II: The Political Economy of American Warfare, 1940–1945* (Lawrence: University Press of Kansas, 2004).

9. See Roy MacLeod, "Introduction: Science, Technology, and the War in the Pacific," in *Science and the Pacific War: Science and Survival in the Pacific, 1939–1945*, ed. Roy M. MacLeod (Dordrecht: Kluwer, 2000), 1–9; and MacLeod, "Introduction," in *Nature and Empire: Science and the Colonial Enterprise*, ed. Roy MacLeod, *Osiris* 15 (2000): 10–11.

10. See John Krige and Kai-Henrik Barth, "Introduction: Science, Technology, and International Affairs," in *Global Power Knowledge: Science and Technology in International Affairs*, ed. John Krige and Kai-Henrik Barth, *Osiris* 21 (2006): 1–21; Alex Roland, "Science and War," in *Historical Writing on American Science*, ed. Sally Gregory Kohlstedt and Margaret W. Rossiter, *Osiris* 2 (1985): 247–272; and Richard Howard, "The Role of Botanists During World War II in the Pacific Theatre," *Botanical Review* 60 (April–June 1994): 197–257.

11. On the scale and scope of World War II science, see Peter Galison, "Introduction: The Many Faces of Big Science," in *Big Science: The Growth of Large-scale Research*, ed. Peter Galison and Bruce Hevly (Stanford, CA: Stanford University Press, 1992), 1–17; Robert Seidel, "The Origins of the Lawrence Berkeley Laboratory," in Galison and Bruce Hevly, *Big Science*, 21–45; James H. Capshew and Karen A. Rader, "Big Science: Price to the Present," in *Science After '40*, ed. Arnold Thackray, *Osiris* 7 (1992): 3–25; and Peter Bacon Hales, *Atomic Spaces: Living on the Manhattan Project* (Urbana: University of Illinois Press, 1997).

12. See Alfred W. Crosby, *The Columbian Exchange: Biological and Cultural Consequences of 1492* (Westport, CT: Greenwood, 1972); Alfred W. Crosby, *Ecological Imperialism: The Biological Expansion of Europe, 900–1900* (Cambridge: Cambridge University Press, 1986); and David Landes, *The Wealth and Poverty of Nations: Why Some Are So Rich and Some So Poor* (New York: Norton, 1999).

13. A. J. Lustig, "Essay Review: Botanists Sow, Historians Reap," *Journal of the History of Biology* 34 (Winter 2001): 581–591; and Londa Schiebinger, *Plants and Empire: Colonial Bioprospecting in the Atlantic World* (Cambridge, MA: Harvard University Press, 2004).

14. See Stuart McCook, *States of Nature: Science, Agriculture, and Environment in the Spanish Caribbean, 1760–1940* (Austin: University of Texas Press, 1992); Brockway, *Science and Colonial Expansion*; and Antonio Lafuente and Nuria Valverde, "Linnaean Botany and Spanish Imperial Biopolitics," in *Colonial Botany: Science, Commerce, and Politics in the Early Modern World*, ed. Londa Schiebinger and Claudia Swan (Philadelphia: University of Pennsylvania Press, 2005), 134–147.

15. See Richard Drayton, *Nature's Government: Science, Imperial Britain, and the "Improvement" of the World* (New Haven, CT: Yale University Press, 2000); and Daniel R. Headrick, "Botany, Chemistry, and Tropical Development," *Journal of World History* 7 (Spring 1996): 1–20.

16. Alan I Marcus, *Agricultural Science and the Quest for Legitimacy: Farmers, Agricultural Colleges, and Experiment Stations, 1870–1920* (Ames: Iowa State University Press, 1985); and Deborah Fitzgerald, "Technology and Agriculture in Twentieth-century America," in *A Companion to American Technology*, ed. Carroll Pursell (Malden, MA: Blackwell, 2005), 69–82.

17. See Daniel R. Headrick, *The Tentacles of Progress: Technology Transfer in the Age of Imperialism, 1850–1940* (New York: Oxford University Press, 1988), 209–258; and William K. Storey, "Plants, Power and Development: Founding the Imperial Department of Agriculture for the West Indies, 1880–1914," in *States of Knowledge: The Co-production of Science and Social Order*, ed. Sheila Jasanoff (London: Routledge, 2004), 109–130.

18. See Deborah Fitzgerald, "Beyond Tractors: The History of Technology in American Agriculture," *Technology and Culture* 32 (January 1991): 114–126; John H. Perkins, *Geopolitics and the Green Revolution: Wheat, Genes, and the Cold War* (New York: Oxford University Press, 1997); and Nicolas Rasmussen, "Plant Hormones in War and Peace: Science, Industry, and Government in the Development of Herbicides in 1940s America," *Isis* 92 (June 2001): 291–316.

19. Philip J. Pauly, *Fruits and Plains: The Horticultural Transformation of America* (Cambridge, MA: Harvard University Press, 2007), 3. See also Frieda Knobloch, *The Culture of Wilderness: Agriculture as Colonization of the American West* (Chapel Hill: University of North Carolina Press, 1996).

20. Wolf and Wolf, *Rubber*, 169; and David Fairchild, "Reminiscences of Early Plant Introduction Work in South Florida," *Proceedings of the Florida State Horticultural Society* 51 (1938): 11–33.

21. See Philip J. Pauly, *Biologists and the Promise of American Life: From Meriwether Lewis to Alfred Kinsey* (Princeton, NJ: Princeton University Press, 2000); and Jeffrey Jacob Jones, "The World Was Our Garden: U.S. Plant Introduction, Empire, and Industrial Agri(culture), 1898–1948" (PhD diss., Purdue University, 2004).

22. See Olga Elina, Susanne Heim, and Nils Roll-Hansen, "Plant Breeding on the Front: Imperialism, War, and Exploitation," in *Politics and Science in Wartime: Comparative International Studies on the Kaiser Wilhelm Institute*, ed. Carola Sachse and Mark Walker, *Osiris* 20 (2005): 161–178; Michael Flitner, *Sammler, Räuber, und Gelehrte: Die politischen Interessen an pflanzengenetischen Ressourcen, 1895–1995* (Frankfurt: Campus, 1995); and Susanne Heim, *Kalorien, Kautschuk, Karrieren: Pflanzenzüchtung und landwirtschaftliche Forschung in Kaiser-Wilhelm-Instituten 1933–1945* (Göttingen: Wallstein, 2003).

23. See Carroll W. Pursell Jr., "The Farm Chemurgic Council and the United States Department of Agriculture," *Isis* 60 (1969): 307–317; David E. Wright, "Waste Not, Want Not: The Michigan Roots of the Farm Chemurgic Movement," *Michigan History* 73 (1989): 32–38; William J. Hale, *Farmward March: Chemurgy Takes Command* (New York: Coward-McCann, 1939); Christy Borth, *Pioneers of Plenty: Modern Chemists and Their Work*, 2nd ed. (New York: New Home, 1942);

and Wheeler McMillen, *New Riches from the Soil* (New York: Van Nostrand, 1946).

24. See Daniel J. Kevles, "Patents, Protections, and Privileges: The Establishment of Intellectual Property in Plants and Animals," *Isis* 98 (June 2007): 323–331; and Glenn E. Bugos and Daniel J. Kevles, "Plants as Intellectual Property: American Practice, Law, and Policy in World Context," *Osiris* 7 (1992): 75–104.

25. See Jeffrey L. Meikle, *American Plastic: A Cultural History* (New Brunswick, NJ: Rutgers University Press, 1995); Williams Haynes, *The Chemical Front* (New York: Knopf, 1943); and James Phinney Baxter 3rd, *Scientists Against Time* (Boston: Little, Brown, 1946).

26. Clark A. Miller, "'An Effective Instrument of Peace': Scientific Cooperation as an Instrument of U.S. Foreign Policy, 1938–1950," in *Global Power Knowledge: Science and Technology in International Affairs*, ed. John Krige and Kai-Henrik Barth, *Osiris* 21 (2006): 133–160.

27. Gary R. Hess, *The United States' Emergence as a Southeast Asian Power, 1940–1950* (New York: Columbia University Press, 1987), 208–211, 270–272; and Andrew J. Rotter, *The Path to Vietnam: Origins of the American Commitment to Southeast Asia* (Ithaca, NY: Cornell University Press, 1987), 49–69, 151–154.

28. Clay, *World Agriculture and the Environment*, 333–345.

29. A few other potential rubber plants will be introduced in subsequent pages. These include the Colorado rubber plant, or pinguay (*Hymenoxys floribunda utilis* or *Hymenoxys richardsonii var. floribunda*); a number of milkweed species from the genus *Asclepias*; a few trees of the genus *Ficus* (some commonly known as rubber trees); the rubber rabbitbrush (*Chrysothamnus nauseosus*; also sometimes known as rabbit bush or rabbit brush), a wild weed native to the western United States; ocotillo (*Fourquieria splendens*), an unusual desert plant native to the American Southwest; a number of *Euphorbia* species, particularly the *Euphorbia instisy* of Madagascar; and a few others. As will be seen in chapter 6, the realization that rubber could be produced from alcohol as a feedstock, and thus derived from common agricultural crops, meant that wheat, corn, soybeans, sugarcane, and other common plants also could become sustainable and renewable domestic sources of rubber.

30. For the botany of guayule, see Francis E. Lloyd, *Guayule: A Rubber Plant of Chihuahuan Desert* (Washington, DC: Carnegie Institution, 1911); and National Academy of Sciences, *Guayule: An Alternative Source of Natural Rubber* (Washington, DC: National Academy of Sciences, 1977).

31. Charles S. Dolley, "Cryptostegia Grandiflora: A Possible Source of Commercial Rubber," *IRW* 71 (March 1925): 339–340; Charles S. Dolley, "Growing Cryptostegia in the United States as a Commercial Source of Rubber," *IRW* 76 (April 1927): 3–4; Dale W. Jenkins, "Cryptostegia as an Emergency Source of Rubber" (Board of Economic Warfare, technical bulletin no. 3) (n.p., 1943); Charles Morrow Wilson, *Trees and Test Tubes: The Story of Rubber* (New York: Holt, 1943), 15–20; Yvonne Baskin, *A Plague of Rats and Rubber Vines: The Growing Threat of Species Invasions* (Washington, DC: Island, 2002), 3, 86, 123; and D. J. Kriticos et al., "Climate Change and Biotic Invasions: A Case History of a Tropical Woody Vine," *Biological Invasions* 5 (September 2003): 147–165.

32. Walter M. Buswell, *Goldenrods of South Florida* (Coral Gables, FL: University of Miami, 1942); and Division of Rubber Plant Investigations, "Goldenrod—A Possible New Source of Rubber" (1943), box 163, EFWE, Fort Myers, Florida.

33. This species is often spelled kok-saghyz or koksaghyz, particularly in the botanic literature. I will use kok-sagyz, the more common colloquial spelling of the World War II era.

34. W. Gordon Whaley and John S. Bowen, compilers, *Russian Dandelion (Kok-Saghyz): An Emergency Source of Natural Rubber* (USDA, miscellaneous publication no. 618) (Washington, DC: Government Printing Office, 1947) (hereafter cited as *Russian Dandelion*).

Chapter 1 · **The American Dependence on Imported Rubber: The Lessons of Revolution and War, 1911–1922**

1. The Intercontinental Rubber Company was a Wall Street holding company that, at various times between 1906 and 1953, controlled the stock of several subsidiary companies: Cia Ganadora Textil do Cedros, which owned and operated the vast Cedros Ranch in northern Mexico; the Continental-Mexican Rubber Company, which at times owned and operated factories in Cedros, Torreón, Catorce, and Cedral; the Agricultural Products Company, a real estate operation that owned up to ten thousand acres, mainly in Arizona; the American Rubber Producers Inc., an entity responsible for guayule field and factory operations in California; the Continental Rubber Company, a New York–based sales organization; the Continental Plantation Company, which owned rubber plantation land in Sumatra; among others. For the sake of simplicity, however, I will use the terms "Intercontinental" and "IRC" to describe the activities of the Intercontinental Rubber Company as well as its subsidiaries. See "A Fifty-year Stretch: A History of the Intercontinental Rubber Company, 1903–1953," unpublished manuscript, June 1954, Texas Instruments Historical Collection, DeGolyer Library, Southern Methodist University, Dallas, Texas.

2. Memorial, docket 2483, *United States of America on Behalf of the Intercontinental Rubber Company vs. the United Mexican States*, file 5280, boxes 961–962; and Memorial, docket 2483, *United States of America on Behalf of the Continental-Mexican Rubber Company vs. the United Mexican States*, file 1733, box 318, E125a General Claims Commission, Case Files, RG76, Records of Boundary and Claims Commissions and Arbitrations, NA, College Park, Maryland (hereafter cited as E125a, RG76, NA).

3. W. S. Conduit to W. E. Giescke, 28 August 1912, in "A Fifty-Year Stretch," n.p. Conduit added, "Of course I don't want it to be generally known that we are offering this reward, but I would take great pleasure in paying it did I know for sure that it was Severo who was killed."

4. On the disruptions in the Torreón region, see, William K. Meyers, *Forge of Progress, Crucible of Revolt: Origins of the Mexican Revolution in La Comarca Lagunera, 1880–1911* (Albuquerque: University of New Mexico Press, 1994); and William K. Meyers, "Seasons of Rebellion: Nature, Organisation of Cotton Production, and the Dynamics of Revolution in La Laguna, Mexico, 1910–1916," *Journal of Latin American Studies* 30 (February 1998): 63–94.

5. The story of McCallum's tobacco tin may be apocryphal, but it is found in various sources, including: Peter Neushul, "Science, Technology, and the Arsenal of 'Democracy': Production, Research, and Development during World War II" (PhD diss., University of California, Santa Barbara, 1992), 136–137; Leslie E. Baird, "Guayule—The Viable Second Source of Natural Rubber," *Rubber World* 174 (March 1975): 44–50; and Burton Anderson, *America's Salad Bowl: An Agricultural History of the Salinas Valley* (Salinas, CA: Monterey County Historical Society, 2000), 117–120.

6. On some logistical and resource issues of World War I, see A. Joshua West, "Forests and National Security: British and American Forestry Policy in the Wake of World War I," *Environmental History* 8 (April 2003): 270–293; J. R. McNeill,

"Wood and Warfare in World History," *Environmental History* 9 (July 2004): 388–410; Peter Neushul, "Seaweed for War: California's World War I Kelp Industry," *Technology and Culture* 30 (July 1989): 561–583; and Peter A. Shulman, "'Science Can Never Demobilize': The United States Navy and Petroleum Geology, 1898–1924," *History and Technology* 19, no. 4 (2003): 365–385.

7. "New Mexican Rubber," *IRW* 24 (1 June 1901): 264; Rudolf Endlich, "The 'Guayule' Rubber Plant—I," *IRW* 32 (1 July 1905): 335–336; Rudolf Endlich, "The 'Guayule' Rubber Plant—II," *IRW* 32 (1 August 1905): 367–369; Henry C. Pearson, "The Production of Guayule Rubber," *IRW* 59 (March 1919): 289–291; Wolf and Wolf, *Rubber*, 271–273; and Meyers, *Forge of Progress*, 74–79.

8. "William Appleton Lawrence," *IRW* 49 (December 1913): 135–136; Meyers, *Forge of Progress*, 145–148; and Howard Page to Paul Morton, 30 June 1908, box 10, Records of the Intercontinental Rubber Company, DeGolyer Library, Southern Methodist University, Dallas, Texas (hereafter cited as IRC Records).

9. BMB, *Baruch: My Own Story* (New York: Holt, Reinhart, and Winston, 1957), 207–214; Margaret L. Coit, *Mr. Baruch* (Boston: Houghton Mifflin, 1957), 120–122; and John Mason Hart, *Empire and Revolution: The Americans in Mexico Since the Civil War* (Berkeley and Los Angeles: University of California Press, 2002), 184.

10. George D. Rogers to Thomas F. Ryan and Nelson W. Aldrich, 23 June 1904, Letterbooks, vol. 89, p. 113, cited in e-mail correspondence from Ken Rose of the Rockefeller Archive Center, 15 March 2002. Rogers's letter accompanied a check for twenty-five thousand dollars, "in payment of Mr. John D. Rockefeller Jr.'s subscription to the Continental Rubber Company." Although an otherwise valuable source, William Meyer's *Forge of Progress* is mistaken to repeatedly describe the IRC as "Rockefeller dominated."

11. Thomas F. Ryan et al. to Gentlemen, 24 January 1907, *The Papers of Nelson Aldrich* (Washington, DC: Library of Congress Photoduplication Service, 1972), reel 25.

12. *NYT*, 15 January 1906, 1; and Hart, *Empire and Revolution*, 184.

13. "The 'Guayule' Rubber Plant—III," *IRW* 33 (1 October 1905): 3–4; Intercontinental Rubber Company, *Annual Report to Stockholders for the Year Ending 31 July 1910*, Historic Corporate Report Collection, Baker Library, Harvard Business School (hereafter cited as IRC *Annual Report* and year); [Robert G. Cleland], "Memorandum on the Guayule Industry," box 4, Robert G. Cleland Papers, Huntington Library, San Marino, California; and Meyers, *Forge of Progress*, 139, 145–148.

14. Minutes of Advisory Committee of the Continental Rubber Company of America, 31 March 1906, *Papers of Nelson Aldrich*, reel 68.

15. Deed of agreement between J. G. Griner, John J. Foster, C. F. Young, and Charles D. Woodruff, 14 April 1906, *Papers of Nelson Aldrich*, reel 68.

16. J. Wesley White, "History of Guayule, with Reference to Texas," in box 28, E20 Records of the Emergency Rubber Project, RG95 United States Forestry Service, NA (hereafter cited as E20, RG95, NA); and "Guayule Interests," *IRW* 36 (1 August 1907): 332. See also W. H. Stayton to Nelson W. Aldrich, 26 February 1907; and Howard Page and A. Chester Beatty to Continental Rubber Co., 14 November 1907, *Papers of Nelson Aldrich*, reel 25. See also Stayton to Aldrich, 14 December 1909, *Papers of Nelson Aldrich*, reel 38.

17. "A Fifty-year Stretch," 21. See also L. C. Andrews to Howard Page, 4 February 1907, *Papers of Nelson Aldrich*, reel 25.

18. Memorandum attached to Daniel Shea to Nelson W. Aldrich, 28 March 1908, *Papers of Nelson Aldrich*, reel 26.

19. Baruch Brothers to E. A. Aldrich, 14 December 1909, box 10, IRC Records; and Baruch, *My Own Story*, 211.

20. Jerome L. Sternstein, "King Leopold, Senator Nelson W. Aldrich, and the Strange Beginnings of American Economic Penetration of the Congo," *African Historical Studies* 2, no. 2 (1969): 189–204. See also Hochschild, *King Leopold's Ghost*, 243–249.

21. William G. McGinnies, *Discovering the Desert: Legacy of the Carnegie Desert Botanical Laboratory* (Tucson: University of Arizona Press, 1981), 1–9.

22. "Francis E. Lloyd," *IRW* 58 (April 1918): 420; and "Francis Earnest Lloyd," in *Makers of North American Botany*, ed. Harry Baker Humphrey (New York: Ronald, 1961), 153–155.

23. The Colorado rubber plant, or pinguay, supplied a small, short-lived rubber mill in southern Colorado before World War I. See E. L. Dunbar to William H. Meadowcroft, 19 February 1928, box 4, Papers of the EBRC (hereafter cited as EBRC), ENHS; Wilmatte Porter Cockerell, "Note on a Rubber-producing Plant," *Science* 19 (19 February 1904): 314–315; "One Pound of Colorado Rubber," *IRW* 32 (1 July 1905): 334; and Charles P. Fox, "More Regarding American-grown Rubber," *IRW* 59 (1 October 1918): 32.

24. HMH to Francesco Franceschi, 2 January 1908, box 10, Francesco Franceschi Papers, Bancroft Library, University of California, Berkeley.

25. Lloyd, *Guayule*. Three decades later, Lloyd's book was considered required reading for the scientists who worked on the Emergency Rubber Project of World War II.

26. W. B. McCallum to Evan W. Kelley, 15 January 1943, in box 28, E20, RG95, NA. Additional biographical information from "Another Successful Graduate: Commercial Rubber Production in Mexico," *The OAC Review* 43 (April 1931): 428; and *London (Ontario) Free Press*, 21 November 1942.

27. Meyers, *Forge of Progress*, 76, 145–149; Meyers, "Seasons of Rebellion," 71; John Mason Hart, *Revolutionary Mexico: The Coming and Process of the Mexican Revolution* (Berkeley and Los Angeles: University of California Press, 1987), 96–100; and Edward Beatty, *Institutions and Investment: The Political Basis of Industrialization in Mexico Before 1911* (Stanford, CA: Stanford University Press, 2001), 168.

28. Meyers, *Forge of Progress*, 145–149; L. C. Andrews to Howard Page, 4 February 1907, *Papers of Nelson Aldrich*, reel 25; Ramón Eduardo Ruíz, *The Great Rebellion: Mexico, 1905–1924* (New York: Norton, 1980), 17; and Henry C. Pearson, "The Production of Guayule Rubber, II," *IRW* 60 (April 1919): 347–348.

29. Meyers, *Forge of Progress*, 86, 125–126; *NYT*, 9 October 1909; "Situation at Cedros," 31 December 1907, *Papers of Nelson Aldrich*, reel 25; and Guadalupe Villa Guerrero, "Riqueza en suelo eriazo: La industria guayulera y los conflictos interregionales de la elite norteña en México," *Secuencia* 46 (2000): 93–120.

30. [Henry C. Pearson], "A Journey Through Guayule Land—I," *IRW* 36 (1 March 1907): 173–177; and [Henry C. Pearson], "A Journey Through Guayule Land—II," *IRW* 36 (1 April 1907): 205–210.

31. Madero quoted in Meyers, *Forge of Progress*, 147. See also Friedrich Katz, "Rural Rebellions After 1810," in *Riot, Rebellion, and Revolution: Rural Social Conflict in Mexico*, ed. Friedrich Katz (Princeton, NJ: Princeton University Press, 1988), 546; and Hart, *Revolutionary Mexico*, 98–100.

32. Memorial, docket 2483, *United States of America on Behalf of the Intercontinental Rubber Company vs. the United Mexican States*, file 5280, boxes 961–962; and Memorial, docket 2483, *United States of America on Behalf of the Continental-*

Mexican Rubber Company vs. the United Mexican States, file 1733, box 318, E125a, RG76, NA.

33. G. C. Carothers "Events Transpiring Since the 23rd of April," to the honorable secretary of state, [May 1911?], decimal file 812.00/1968, Records Relating to the Internal Affairs of Mexico, RG59, U.S. Department of State, NA.

34. Cleland, "Memorandum on the Guayule Industry."

35. IRC, *Annual Report*, 1913 and 1914. See also Alan Knight, *The Mexican Revolution*, vol. 2, *Counter-revolution and Reconstruction* (New York: Cambridge University Press, 1986), 48, 132.

36. Untitled deposition, file 5820, box 961; and translation of clipping, *El Radical*, 23 September 1916, box 19, IRC Records; "Report Concerning Situation at the Cedros Estate" 22 February 1917; and untitled report, signed by GHC, [1919?], box 318, file 1733, E125a, RG76, NA.

37. *NYT*, 10 September 1912; Jehuda Reinharz, *Chaim Weizmann: The Making of a Statesman* (New York: Oxford University Press, 1993), 40–68; Edwin E. Slosson, *Creative Chemistry: Descriptive of Recent Achievements in the Chemical Industries* (Garden City, NY: Garden City Publishing, 1919), 151–153; S. P. Schotz, *Synthetic Rubber* (London: Benn, 1926), 124–126; and Jack Morrell, "W. H. Perkin Jr. at Manchester and Oxford: From Irwell to Isis," in *Research Schools: Historical Reappraisals*, ed. Gerald L. Geison and Frederic L. Holmes, *Osiris* 8 (1993): 104–126.

38. *NYT*, 8 September 1913.

39. Frank Robert Chalk, "The United States and the International Struggle for Rubber, 1914–1941" (PhD diss., University of Wisconsin, 1970), 6–9; and Riley Froh, *Edgar B. Davis and Sequences in Business Capitalism: From Shoes to Rubber to Oil* (New York: Garland, 1993).

40. Burton I. Kaufman, "The United States and Latin America: The Wilson Years," *Journal of American History* 58 (June 1991): 342–363. See also Elisabeth Glaser, "Better Late Than Never: The American Economic War Effort, 1917–1918," in *Great War, Total War: Combat and Mobilization on the Western Front, 1914–1918*, ed. Roger Chickering and Stig Förster (Cambridge: Cambridge University Press, 2000), 389–407.

41. David Jerome Rhees, "The Chemists' Crusade: The Rise of an Industrial Science in Modern America, 1907–1922" (PhD diss., University of Pennsylvania, 1987).

42. E. W. Townsend to TAE, 28 August 1914, Chemicals folder, Edison General File, 1914, ENHS.

43. Edison quoted in Neil Baldwin, *Edison: Inventing the Century* (New York: Hyperion, 1995), 344.

44. TAE to George W. Cope, marginal notes on Cope to TAE, 5 September 1914, Chemicals folder, Edison General File, 1914, ENHS. See also Byron M. Vanderbilt, *Thomas Edison, Chemist* (Washington, DC: American Chemical Society, 1971), 246–250; and Paul Israel, *Edison: A Life of Invention* (New York: Wiley, 1998), 452–454.

45. I am grateful to Paul Israel and Brian Shipley of the Edison Papers for their assistance with these issues, particularly Shipley's explanations of the chemical processes involved. Brian Shipley, personal communication to author, 26 May 2004 and 3 June 2004. See also Vanderbilt, *Thomas Edison, Chemist*, 253; David K. Van Keuren, "Science, Progressivism, and Military Preparedness: The Case of the Naval Research Laboratory, 1915–1923," *Technology and Culture* 33 (October 1992): 710–736; and Miller Reese Hutchinson to Josephus Daniels, 1 July 1916, box 1; and Miller Reese Hutchinson to Josephus Daniels and 25 June 1916,

box 2, Josephus Daniels Papers, Wilson Library, University of North Carolina, Chapel Hill.

46. Israel, *Edison*, 446–447; Robert D. Cuff, *The War Industries Board: Business-government Relations During World War I* (Baltimore: Johns Hopkins University Press, 1973), 15–16; and Paul A. C. Koistinen, *Mobilizing for Modern War: The Political Economy of American Warfare, 1865–1919* (Lawrence: University Press of Kansas, 1997).

47. Quoted in Grosvenor B. Clarkson, *Industrial America in the World War: The Strategy Behind the Lines, 1917–1918* (Boston: Houghton Mifflin, 1923), 13–14.

48. Cuff, *War Industries Board*, 20. See also Burton I. Kaufman, *Efficiency and Expansion: Foreign Trade Organization in the Wilson Administration, 1913–1921* (Westport, CT: Greenwood, 1974), 43–47; and Guy Alchon, *The Invisible Hand of Planning: Capitalism, Social Science, and the State in the 1920s* (Princeton, NJ: Princeton University Press, 1985), 23–25.

49. Marston T. Bogert, undated report [summer 1916?], reel 4; and [Hale?] to Charles H. Herty, [n.d.] May 1916, George Ellery Hale Papers, reel 18, Mount Wilson and Palomar Observatories Library, Pasadena, California. For more on Herty and the chemical industry's role in the preparedness effort, see Germaine M. Reed, *Crusading for Chemistry: The Professional Career of Charles Holmes Herty* (Athens: University of Georgia Press, 1995), 89–99; and Rhees, "The Chemists' Crusade." In fall 1916, the NRC was formed as the central location for scientific research on preparedness issues; its Botanical Raw Products Committee included rubber plants among its research subjects.

50. C. Paul Vincent, *The Politics of Hunger: The Allied Blockade of Germany, 1915–1919* (Athens: Ohio University Press, 1975), 27–59; Belinda J. Davis, *Home Fires Burning: Food Politics, and Everyday Life in World War I Berlin* (Chapel Hill: University of North Carolina Press, 2000), esp. 180–187; and Roger Chickering, *Imperial Germany and the Great War, 1914–1918* (Cambridge: Cambridge University Press, 1998), 36–46, 140–146.

51. For instance, an editorialist for the German rubber trade journal *Gummi-Zeitung* stated plainly that the industry's misery was "directly due to its total dependency upon foreign countries for its supplies in raw materials." Quoted in "The German Rubber Situation as Germans See It," *IRW* 51 (March 1915): 349.

52. *Times (London)*, 10 April 1915, 16 April 1915, and 5 November 1915; "The German Rubber Situation"; "Explaining High Prices to German Consumers," *IRW* 52 (May 1915): 349–350, 464; W. G. Max Müller, "The Economic Situation in Germany During July 1916, Being the Twenty-fourth Month of the War"; and W. G. Max Müller, "The Economic Situation in Germany During August 1916, Being the Twenty-fifth Month of the War," in *British Documents on Foreign Affairs: Reports and Papers from the Foreign Office Confidential Print, Series H, The First World War, 1914–1918* ([Frederick, MD]: University Press of America, 1989), 10:208–209, 251–252.

53. W. G. Max Müller, "The Economic Situation in Germany During April 1918, Being the Forty-fifth Month of the War"; and W. G. Max Müller, "The Economic Situation in Germany During September 1918, Being the Fiftieth Month of the War," *British Documents on Foreign Affairs: Reports and Papers from the Foreign Office Confidential Print, Series H, The First World War, 1914–1918*, 12:177, 370.

54. *NYT*, 10 September 1912, 14 December 1915, and 12 March 1916. See also Vernon Herbert and Attilio Bisio, *Synthetic Rubber: A Project That Had to Succeed* (Westport, CT: Greenwood, 1985), 27.

55. Georg Kassner, *Ist in Deutschland eine Production von Kautschuk möglich, gestützt auf den Anbau einheimischer Culturpflanzen?* (Breslau: Kern, 1885); "Eine einheimische

Kautschukpflanze, eine der kautschukreischten überhaupt," *Gummi-Zeitung* 30 (1916): 500–501; "The Rubber Trade in Germany," *IRW* 54 (May 1916): 440; "Rubber from German Weeds," *IRW* 54 (July 1916): 571; and "Rubber-producing Weeds in Germany," *IRW* 59 (January 1919): 201–202.

56. Slosson, *Creative Chemistry*, 157–158.

57. "S. A. Man Knows All About Guayule," *San Antonio Light*, 24 May 1942, clipping, box 28, E20, RG95, NA.

58. Alfred Lief, *The Firestone Story: A History of the Firestone Tire and Rubber Company* (New York: Whittlesey House, 1951), 90–91; and Chalk, "The United States and the International Struggle for Rubber," 5–23.

59. Chalk, "The United States and the International Struggle for Rubber," 12–14.

60. Charles McKinley Saltzman, "My Experience in the World War and Activities in the Organization of the OASW," in *Army Industrial College Lectures*, vol. 14, Special Collections, Archives and History Department, National Defense University, Washington DC. See also David M. Kennedy, *Over Here: The First World War and American Society* (Oxford: Oxford University Press, 1980), 113–116.

61. *NYT*, 7 February 1917.

62. Quoted in Clarkson, *Industrial America*, 30.

63. *NYT*, 14 April 1917.

64. "Final Report on Condition of the Rubber Industry, 1913–1918, and upon the Transactions of the Rubber Goods Section, War Industries Board, Relating to the Same and the War Program," series 2, vol. 13, part 2, Baruch Papers.

65. Emphasis in original. H. Stuart Hotchkiss to E. E. Pratt, 25 April 1917; and H. Stuart Hotchkiss to E. E. Pratt, 14 June 1917, box 552, E79, Bureau of Imports, Office of the Director, General Correspondence, 1917–1919, RG182, War Trade Board, NA (hereafter cited as E79, RG182, NA).

66. "Considerations, Presented by War Service Committee of the Rubber Industry for Elaboration and Discussion with Representatives of Government Departments in Connection with Plans for Regulating the Importation and Allocation of Rubber," 15 July 1918; "Memorandum to War Industries Board from the War Service Committee of Rubber Industry Regarding Brazil," 22 July 1918; and "Information and Suggestions Presented to the War Industries Board by the War Service Committee of the Rubber Industry, 11 September 1918," file 84, George N. Peek Papers, Western Historical Manuscript Collection, Columbia, Missouri (hereafter cited as Peek Papers).

67. HSF to Harry T. Dunn, 12 August 1918, folder: Fisk Rubber Company, 1914–1919; and HSF to Amos Miller, 21 June 1918, folder: Amos C. Miller, Firestone Tire and Rubber Company Records, University of Akron Archival Services, Akron, Ohio (hereafter cited as Firestone Records). Firestone himself used the word "czar" in the latter document.

68. B. G. Work to All Rubber Manufacturers, 2 August 1918, folder: War Service Committee (July–September 1918), Firestone Records.

69. J. M. Barnet to HSF, 2 August 1918, folder: War Service Committee (July–September 1918), Firestone Records; and George N. Peek to Bertram G. Work, 26 July 1918, box 553, E79, RG182, NA.

70. "War Industries Board Priorities Division Circular No. 24"; "Regulations Governing the Production of Rubber Products, Supplement to War Industries Board Circular No. 24"; and "Regulations Governing the Production of Rubber Products, Supplement to War Industries Board Circular No. 24, and Superceding W.S.C. Letter of September 23, 1918," file 84, Peek Papers.

71. Carnahan biography from *Who's Who in America*, vol. 13, 1924–1925 (Chicago: Marquis, 1924), 652; and "Protagonist of Guayule," *IRW* 79 (March 1929): 91. Obituaries in *NYT*, 21 March 1941; *RA* 49 (April 1941): 46; and *IRW* 104 (April 1941): 67.
72. GHC to Executive Committee of the IRC, 23 January 1917, box 10, IRC Records
73. See numerous diary entries, and esp. "Memorandum on the Mexican Situation," 24 May 1917, in "Diary of Chandler P. Anderson," Papers of Chandler P. Anderson, microfilm edition, series 1, vol. 1, Library of Congress. See also IRC, *Annual Report*, 1916 and 1917.
74. W. B. McCallum to Evan W. Kelley, 15 January 1943, box 28, E20, RG95, NA; and Blythe, "Taming the Wild Guayule."
75. G.E.P. Smith, *History of Pima County's Short-lived Rubber Industry* (Tucson: n.p., 1965), 1–8; GHC to R. B. Von Klein Smid, 6 July 1916; G.E.P. Smith to Smid, 18 July 1916; and GHC to Smith, 3 August 1916, box 11, George E. P. Smith Papers, University Archives, University of Arizona, Tucson.
76. GHC to William C. Potter, 30 June 1916, box 10, IRC Records.
77. Smith, *History of Pima County's*, 1–8; *The Tucson Citizen*, 18 July 1916; *The Tucson Citizen*, 30 September 1916; and *The Tucson Citizen*, 30 September 1918, clippings, box 31, E20, RG95, NA.
78. [Henry C. Pearson], "Guayule Cultivation in the United States: A Rubber Preparedness Suggestion," *IRW* 55 (December 1916): 133–135. For evidence that the IRC tried to shield its activities from the press, see GHC to J. M. Williams Jr., 25 July 1916; W. B. McCallum to Williams, 5 May 1917; and Williams to GHC, 11 July 1918, box 31, E20, RG95, NA.
79. "Machine-made Rubber," *Literary Digest* (14 December 1918), clipping, box 31, E20, RG95, NA.
80. W. B. McCallum to GHC, 19 November 1917 and 5 February 1918, E91, Records of the Intercontinental Rubber Company, Division of Rubber Plant Investigations, RG54, Bureau of Plant Industry, Soils, and Agricultural Engineering, NA (hereafter cited as E91, RG54, NA).
81. Cleland, "Memorandum on the Guayule Industry"; and GHC to TAE, 7 February 1919, folder: Rubber, 1919, Edison General Files, ENHS. According to Cleland, company officials had to remove human excrement from workers' doorways on a daily basis.
82. *Congressional Record*, 64th Congress, 1st sess., 22. See also memoranda dated 8 January and 31 January 1916, box 328, House General Correspondence of the Secretary, RG16, USDA, NA.
83. "The Ocotillo Rubber of Arizona," *IRW* 55 (November 1916): 75–76; and "Ocotillo Again to the Front," *IRW* 63 (1 October 1920): 9.
84. E. M. East to G. E. Hale, 19 July 1917; and "Report of the Chemistry Committee to the National Research Council," 10 July 1917, George Ellery Hale Papers, reel 50.
85. *Yearbook No. 17, 1918* (Washington, DC: Carnegie Institution, 1919).
86. Attachment to letter from University of California to Governor William D. Stephens, 27 March 1917, carton 1, State Council for Defense of California Papers, Bancroft Library, University of California, Berkeley.
87. For more on Hall, see Joel B. Hagen, "Clementsian Ecologists: The Internal Dynamics of a Research School," in *Research Schools: Historical Reappraisals*, ed. Gerald L. Geison and Frederic L. Holmes, *Osiris* 8 (1993): 178–195; and Robert E. Kohler, *Landscapes and Labscapes: Exploring the Lab-field Border in Biology* (Chicago: University of Chicago Press, 2002), 163–169.

88. *Caoutchouc* is the French word for rubber. Hall urged his wife to avoid using the word "rubber," particularly on postcards, for fear that enemies of the United States might learn of his work. See HMH to CCH, 3 November 1917; and HMH to Folks at Home, 4 November 1917, box 1, HMH Papers, Bancroft Library, University of California, Berkeley (hereafter cited as Hall Papers).

89. HMH to CCH, 5 March 1918, box 1, Hall Papers .

90. Hall's observations on variations among individual specimens of *Chrysothamnus* sparked a significant and decade-long project in "experimental taxonomy." See Kohler, *Landscapes and Labscapes*, 163–169; and HMH to Frederic E. Clements, 22 November 1918, Edith S. and Frederic E. Clements Papers, American Heritage Center, Laramie, Wyoming (hereafter cited as Clements Papers).

91. Based on several letters from HMH to CCH, 18 June 1917, 4 November 1917, 7 November 1917, folder: 1915–1917; and 15 February 1918 and 21 February 1918, box 1, Hall Papers. See also HMH, "Report of the Committee on Botany to the Pacific Coast Research Conference," 24 November 1917, carton 1, State Council for Defense of California Papers.

92. HMH to My Dear Peeved One [CCH?], 26 January 1918; and HMH to Folks at Home, 17 February 1918, box 1, Hall Papers.

93. Edith Clements, "Diary, 1918," box 52, Clements Papers.

94. For instance, he estimated that 832,000 acres of the shrub grew in the San Luis Valley of Colorado alone. HMH to CCH, [n.d.] August 1918, Hall Papers; and HMH, "Report on Rubber Investigations," 12 October 1918, carton 1, State Council for Defense of California Papers.

95. E. M. East to Frederic E. Clements, 7 October 1918; and East to Clements, 30 October 1918, box 1, Clements Papers.

96. HMH to gentlemen, 18 December 1918, carton 1, State Council for Defense of California Papers. On demobilization, see Koistinen, *Mobilizing for Modern War*, 281–285.

97. HMH to gentlemen, 18 December 1918, carton 1, State Council for Defense of California Papers.

98. Final report of the Committee on Botanical Investigations, 26 January 1919, carton 1, State Council for Defense of California Papers.

99. HMH and Thomas Harper Goodspeed, *A Rubber Plant Survey of Western North America* (Berkeley and Los Angeles: University of California Press, 1942), 251. This book, published amid the World War II rubber crisis, contains reprints of essays originally published in 1919 as a result of the World War I research expeditions.

100. Thomas Harper Goodspeed, "The Organization and Activities of the Committee on Scientific Research of the State Council of Defense of California," *Bulletin of the National Research Council* 5 (April 1923): 9–10.

101. Clarkson, *Industrial America*, 437; Chalk, "The United States and the International Struggle for Rubber," 23; and H. T. Dunn, "Final Report of the Chief of the Rubber Section," box 216, E152, Consolidated Files, RG61, War Industries Board, NA.

102. Clarkson, *Industrial America*, 436.

103. [Charles H. Herty], "National Self-Containedness," *Journal of Industrial and Engineering Chemistry* 10 (December 1918): 966. On total war, see Stig Förster, "Introduction," in Chickering and Förster, *Great War, Total War*, 1–15.

104. Rhees, "The Chemists' Crusade"; and Hugh R. Slotten, "Human Chemistry or Scientific Barbarism? American Responses to World War I Poison Gas, 1915–1930," *Journal of American History* 77 (September 1990): 476–498.

Chapter 2　*Domestic Rubber Crops in an Era*
of Nationalism and Internationalism

1. Emphasis in original. E. F. Koenig to "All Officers," 9 November 1922, file 336, box 34, E191, Assistant Secretary of War, War Planning Branch, RG107, Department of War, NA (hereafter cited as E191, RG107, NA).

2. [GHC] to directors of the IRC, 31 March 1920, box 10, IRC Records.

3. Carl P. Parrini, *Heir to Empire, United States Economic Diplomacy, 1916–1923* (Pittsburgh: University of Pittsburgh Press, 1969), 81–91; and Jordan A. Schwarz, *The Speculator: Bernard M. Baruch in Washington, 1917–1965* (Chapel Hill: University of North Carolina Press, 1981), 219–225.

4. "Final Report on Condition of the Rubber Industry, 1913–1918, and upon the Transactions of the Rubber Goods Section, War Industries Board, Relating to the Same and the War Program," series 2, vol. 13, part 2, Baruch Papers.

5. See P. Harvey Middleton, *Industrial Mexico: 1919 Facts and Figures* (New York: Dodd, Mead, 1919), esp. v–xiii, 95–97.

6. Joseph Brandes, *Herbert Hoover and Economic Diplomacy: Department of Commerce Policy, 1921–1928* (Pittsburgh: University of Pittsburgh Press, 1962). See also Alchon, *The Invisible Hand of Planning*, 3–4, 33–35; Edwin T. Layton Jr., *The Revolt of the Engineers: Social Responsibility and the Engineering Profession* (Baltimore: Johns Hopkins University Press, 1971).

7. Emphasis in original. Brooks Emeny, *The Strategy of Raw Materials: A Study of America in War and Peace* (New York: Macmillan, 1934), 1–11, quotation on p. 1.

8. Peter Mansfield Abramo, "The Military and Economic Potential of the United States: Industrial Mobilization Planning, 1919–1945" (PhD diss., Temple University, 1995), 1–55, quotation on p. 41. See also Paul A. C. Koistinen, *Planning War, Pursuing Peace: The Political Economy of American Warfare, 1920–1939* (Lawrence: University Press of Kansas, 1998).

9. Saltzman, "My Experience in the World War and Activities in the Organization of the OASW."

10. Charles McKinley Saltzman, "Commodity Production Study-Rubber," 27 September 1921, file 1082, box 100; and F. C. Brown to Office of the Assistant Secretary of War, 14 October 1921, file 336, box 34, E191, RG107, NA.

11. H. L. Rogers to the assistant secretary of war, 8 February 1922; and secretary of war to secretary of agriculture, 21 February 1922, file 336, box 34, E191, RG107, NA.

12. William A. Taylor to Dr. Ball, 3 March 1922, box 941, E17, General Correspondence of the Secretary, RG16, USDA, NA (hereafter cited as E17, RG16, NA); and Henry C. Wallace to secretary of war, 14 March 1922, file 336, box 34, E191, RG107, NA.

13. [Charles Holmes Herty], "Another Idol Shattered," *Journal of Industrial and Engineering Chemistry* 10 (November 1918): 880.

14. [Charles Holmes Herty], "National Self-Containedness," *Journal of Industrial and Engineering Chemistry* 10 (December 1918): 966; and Bernhard C. Hesse, "Our Preparations for After the War," *Journal of Industrial and Engineering Chemistry* 10 (November 1918): 881–887.

15. Julius Stieglitz, "Introduction," in Slosson, *Creative Chemistry*, n.p. See also Rhees, "The Chemists' Crusade"; Reed, *Crusading for Chemistry*, 156–218; and Kathryn Steen, "Confiscated Commerce: American Importers of German Synthetic Organic Chemicals, 1914–1929," *History and Technology* 12, no. 3 (1995): 261–284.

16. Stieglitz, "Introduction."

17. Slosson, *Creative Chemistry*, 145–163.

18. Edwin E. Slosson to William E. Ritter, 23 August 1921, and attached prospectus, n.d., box 19, William Emerson Ritter Papers, Bancroft Library, University of California, Berkeley.

19. Thomas H. Buckley, *The United States and the Washington Conference, 1921–1922* (Knoxville: University of Tennessee Press, 1970), 75–89; John Leyland, "The Washington Conference: Japan and the Powers," *The Nineteenth Century and After* 90 (September 1921): 393–402; and Valentine Chirol, "The Washington Conference and the Pacific Problem," *Contemporary Review* 121 (February 1922): 147–156.

20. Franklin D. Roosevelt, "Shall We Trust Japan?" *Asia* 23 (July 1923): 475–478, 526–558; and Christopher M. Bell, "Thinking the Unthinkable: British and American Naval Strategies for an Anglo-American War, 1918–1931," *International History Review* 91, no. 4 (1997): 789–808.

21. *Yearbook No. 19, 1920* (Washington, DC: Carnegie Institution, 1921), 365–366; and *Yearbook No. 20, 1921* (Washington, DC: Carnegie Institution, 1922), 398–399.

22. HMH to HSF, 20 April 1920, folder: Crude Rubber 1920, Firestone Records.

23. HMH to CCH 26 August 1919, box 1, Hall Papers. See also HHM to CCH, 20 July 1919; and HHM to CCH, 24 August 1919, Hall Papers.

24. HMW and Frances L. Long, *Rubber-content of North American Plants* (Washington, DC: Carnegie Institution, 1921). See also Harold A. Senn, "Early Studies of Milkweed Utilization in Canada," *The Canadian Field Naturalist* 58 (September–October, 1944): 177–180; and Charles P. Fox, "Ohio Grown Rubber, Crop of 1910," *Ohio Naturalist* 11 (January 1910): 271–272.

25. HMH to CCH, 4 Jan 1921; HMH to CCH, 6 Feb 1921; HMH to CCH, 24 Feb 1921, Hall Papers. See also *NYT*, 9 January 1921 and 29 May 1921.

26. For botanical information see, McGinnies, *Discovering the Desert*, 68, 198–203. For early investigations, see *Yearbook No. 10, 1910* (Washington, DC: Carnegie Institution, 1911), 65; and "The Ocotillo Rubber of Arizona," 75–76.

27. "Ocotillo Again in the Front," *IRW* 63 (October 1920): 9–10; Emmet S. Long, "Experiments with a New Cactus Rubber," *IRW* 62 (August 1920): 709–710; "Ocotillo as Rubber Substitute," *Pan American Magazine* 32 (March 1921): 219–200; and HMH to CCH, 4 January 1921, Hall Papers. On the delayed opening of the Salome factory, see Charles F. Day to Henry Ford, 3 July 1923, box 36, accession 94, Engineering-Dearborn Laboratories, Benson Ford Research Center, The Henry Ford, Dearborn, Michigan (hereafter cited as The Henry Ford.)

28. Henry Nehrling, *The Plant World in Florida*, collected and edited by Alfred and Elizabeth Kay (New York: Macmillan, 1933), 166, 174.

29. John Gifford, *The Everglades and Other Essays Relating to Southern Florida*, 2nd ed. (Miami: Everglade Land Sales, 1912), 117–122. The essay originally appeared under the title "Rubber in South Florida," in *Everglade Magazine*, February 1911. See also "200 Products Growing in Demonstration Farm," *Everglade Magazine* 4 (May 1913): 4; and Charles S. Dolley, "Rubber Planting [sic] Experimenting in Florida," *Florida Grower* 31 (17 January 1925): 19. For an introduction to efforts to develop the Everglades for agricultural purposes, see Christopher F. Meindl, "Past Perceptions of the Great American Wetland: Florida's Everglades During the Early Twentieth Century," *Environmental History* 5 (July 2000): 378–393. For more on cryptostegia, see Dolley, "Cryptostegia Grandiflora"; and Dolley, "Growing Cryptostegia in the United States."

30. David Fairchild, "Report of Southern Trip in 1913," 35–36; David Fairchild, "Report of Southern Trip, Spring of 1914," 44; David Fairchild, "Report of Southern

Trip in 1915," 32–32a; and David Fairchild, "Southern Trip, 1916," 9–11, Fairchild Papers.

31. J. E. Ingraham to F. J. Pepper, 14 August 1918, box 28, Model Land Company Collection, Archives and Special Collections, University of Miami, Coral Gables, Florida (hereafter cited as Model Land Company Collection).

32. J. E. Ingraham to Pepper and Potter, 23 November 1923; and Pepper and Potter to Ingraham, 31 May 1924, Model Land Company Collection.

33. FMNP, 14 December 1920.

34. Henry C. Wallace to Senator Duncan Fletcher, 28 April 1923; Pepper and Potter to J. E. Ingraham, 26 September 1923; Pepper and Potter to David Fairchild, 23 January 1924; and Pepper and Potter to Ingraham, 17 March 1924, box 28, folder 681, Model Land Company Collection. At least five articles promoting southern Florida rubber appeared in the Miami Herald in 1923 and 1924; quotation is from the Miami Herald, 18 January 1924. See also "Can Florida Become a Rubber Growing State?" Manufacturers Record 65 (8 November 1923): 87.

35. "Is Rubber Growing a Commercial Possibility in Florida?" Manufacturers Record 83 (15 March 1923): 59; Shelton S. Matlack, "Rubber Production a Florida Probability of the Near Future with Tremendous Possibilities," Florida Grower 30 (15 November 1924): 10; Mary Hayes Davis, "Rubber Plantation Started," Florida Grower 30 (12 July 1924): 3; and "Everglades Land Bought by Syndicate," Florida Grower 32 (3 October 1925): 21. See also various clippings and letters in box 1023, E17, RG16, NA.

36. An illustration of the volatility of rubber prices in the early twentieth century: in seventeen of the twenty-five years between 1905 and 1930, rubber prices either fell to less than half, rose to more than double their median price, or both. See James Cooper Lawrence, The World's Struggle with Rubber: 1905–1931 (New York: Harper, 1931), 1. For British fears, see "The Rubber Trade in Great Britain," IRW 65 (October 1921): 55; and "The Rubber Trade in Great Britain," IRW 65 (December 1921): 221. The United States and Britain consumed about the same amount of rubber in 1908. Thereafter, American demand grew at a much faster pace than did British demand. The U.S. share of world consumption reached a peak of about 75 percent in 1924. See William Woodruff, "Growth of the Rubber Industry of Great Britain and the United States," Journal of Economic History 15 (December 1955): 376–391.

37. Tensions grew again in July 1922, when evidence surfaced that American government loans issued to the Netherlands had been used to prop up accounts of Java rubber planters, which enabled the growers to withhold product from the market and drive up prices for American consumers. This apparent misuse of American capital sparked debates in Washington DC on how to retaliate. See Parrini, Heir to Empire, 201, 204.

38. Austin Coates, The Commerce in Rubber: The First 250 Years (New York: Oxford University Press, 1987), 207–216; Brandes, Herbert Hoover and Economic Diplomacy, 84–85; and "British Government Considers Plans" IRW 67 (October 1922): 5; NYT, 20 April 1924

39. See "Can Rubber Be Controlled—II," IRW 67 (November 1922): 69–71; and "Preliminaries of the International Plantation Company," IRW 67 (November 1922): 71–72.

40. Coates, The Commerce in Rubber, 216–219; and Miami Herald, 29 March 1925.

41. Chalk, "The United States and the International Struggle for Rubber," 27–34; Lawrence, The World's Struggle with Rubber, 27–36, 47; and "Stevenson Plan for Rubber Control Adopted," IRW 67 (November 1922): 122.

42. Ellis W. Hawley, "Herbert Hoover, the Commerce Secretariat, and the Vision of the 'Associative State,'" Journal of American History 61 (June 1974): 116–140;

Brandes, *Herbert Hoover and Economic Diplomacy*; and Kendrick A. Clements, *Hoover, Conservation, and Consumerism: Engineering the Good Life* (Lawrence: University of Kansas Press, 2000). For parallels between Baruch and Hoover, see Schwarz, *The Speculator*, 219–225.

43. A. L. Viles to HCH, 18 April 1921, box 533, Commerce Papers, Herbert Hoover Presidential Library, West Branch, Iowa (hereafter cited as HHPL).

44. Chalk, "The United States and the International Struggle for Rubber," 44.

45. HSF to C. H. Huston, 6 January and 9 January 1923; and HSF to Congressman Simeon Fess, 10 January 1923, Firestone Records.

46. Medill McCormick to HCH, 30 January 1923; HCH to McCormick, 31 January 1923 (draft letter, unsent); HCH to McCormick, 2 February 1923; and untitled press release, 3 February 1923, folder: Rubber 1921–1924, box 531, Commerce Papers, HHPL. Hoover's response went through two drafts that differed in tone and content. In the second version, for instance, Hoover deleted his prediction that high rubber prices would lead to the "ultimate demoralization . . . and ruin" of the rubber industry.

47. Hoover quoted in Claudius H. Huston, "America's Dependence on Britain for Rubber," *Saturday Evening Post* (26 May 1923): 22–23, 170–174.

48. *America Should Grow Its Own Rubber*, report of the Conference of Rubber, Automotive, and Accessory Manufacturers (Washington, DC, 27 February 1923), pamphlet, box 1023, E17, RG16, NA; McCormick quoted in Lief, *The Firestone Story*, 149; and HSF to W. S. Gilbreth, 19 February 1923, box 19: Rubber Growing, Firestone Records.

49. *America Should Grow Its Own Rubber*; and *NYT*, 28 February 1923.

50. *NYT*, 13 March 1923; "The Aftermath of Rubber Restriction," *IRW* 68 (April 1923): 421; and "Firestone Company Resigns from Rubber Association," *IRW* 68 (June 1923): 572.

51. HCH to William C. Redfield, 15 March 1923, box 520, E2, RG40, NA.

52. HCH to Henry C. Wallace, 19 March 1923, box 520, E2, RG40, NA. For more on Hoover's involvement in USDA policy, which led to complaints that he attempted to control two cabinet departments, see C. Fred Williams, "William M. Jardine and the Foundations for Republican Farm Policy, 1925–1929," *Agricultural History* 70 (Spring 1996): 216–232.

53. Chalk, "The United States and the International Struggle for Rubber," 71–74.

54. Memorandum on the Conference at Lotus Club, New York City, 20 May 1924, box 520, E2, RG40, NA; and Chalk, "The United States and the International Struggle for Rubber," 71–74.

55. G. M. Duncan to HSF, 10 November, 14 November, and 30 November 1923; W. D. Hines to HSF, 3 December 1923; and [G. Bernel?] to S. G. Carkhuff, 29 April 1925, folder: Florida Rubber Growing, Firestone Records; and F. E. Castleberry to HSF, 30 July 1925, folder: United States Rubber Growing, folder: Florida Rubber Growing, Firestone Records. It also is pertinent that Firestone reacted angrily to British press reports that Florida rubber experiments were a "joke" supported with "fake" photographs. See A. T. Hancock to Carkhuff, 22 May 1925; and HSF to TAE, 13 June 1925, box 7, EBRC, ENHS.

56. HSF to TAE, 2 April 1926, folder: Thomas A. Edison, 1926; and director, Botanic Gardens to managing director, Firestone Tire and Rubber Co. Ltd., 27 September 1927, folder: Edison 1927, Firestone Records.

57. Quoted in *Miami Herald*, 29 March 1925, clipping, folder: Florida Rubber Growing, Firestone Records.

58. William H. Smith to B. L. Graves, 20 April 1923; Graves to Smith, 23 April 1923; J. B. Groves to Smith, 11 May 1923; Groves to Smith, 6 September 1923;

and Ford Motor Company to George Carlson, 17 May 1923, box 101, accession 94, The Henry Ford.

59. For a sense of the growing power of American-based international business enterprises, see Emily S. Rosenberg, *Spreading the American Dream: American Economic and Cultural Expansion, 1890–1945* (New York: Hill and Wang, 1982), 7–13, 132–137.

60. Edison's copy of Hall and Long, *Rubber-content of North American Plants*, is in folder: Rubber, 1921, Edison General Files, ENHS.

61. HSF to TAE, 9 January 1923; and TAE to HSF, 16 January 1923, folder: Rubber, 1923, Edison General Files, ENHS.

62. TAE to E. G. Liebold, 19 July 1923, folder: Rubber, 1923, Edison General Files, ENHS. See also [Ford R. Bryan], "Edison Botanic Research Corporation," n.d., unpublished manuscript from The Henry Ford. Copy held at EFWE.

63. E. G. Liebold to TAE, 3 August 1923, folder: Rubber, 1923, Edison General Files, ENHS; and Reynold M. Wik, *Henry Ford and Grass-roots America* (Ann Arbor: University of Michigan Press, 1972), 146.

64. *FMNP*, 4 September 1923.

65. TAE to Henry Ford, 14 September 1923; TAE to Ford, 20 September 1923; and HSF to William H. Meadowcroft, 6 October 1923, folder: Rubber, 1923, Edison General Files, ENHS; and TAE to HSF, [September 1923?], folder: Thomas A. Edison, 1923, Firestone Records.

66. *(Miami) Metropolis*, 21 January 1924, clipping, box 28, Model Land Company Collection.

67. *FMNP*, 14 March 1924; James R. Weir to TAE, 13 March 1924; Richard H. Tingley to TAE, 19 March 1924; and Mark L. Felber to M. A. Cheek, 24 March 1924, folder: Rubber, 1923 [sic], Edison General Files, ENHS.

68. J. F. Menge to E. G. Liebold, 17 April 1924; TAE to Liebold, 22 April 1924; and David Fairchild to William H. Smith, 7 June 1924, box 101, accession 94, The Henry Ford.

69. David Fairchild to William H. Smith, 14 June 1924, box 101, accession 94; and Ford R. Bryan, "Henry's So-called Rubber Plantation in Florida," unpublished manuscript from, The Henry Ford, copy held at EFWE. For an introduction to Ford's interest in industrial by-products, see Tom McCarthy, "Henry Ford, Industrial Ecologist or Industrial Conservationist? Waste Reduction and Recycling at the Rouge," *Michigan Historical Review* 27 (Fall 2001): 53–88.

70. *Miami Herald*, 31 May 1924; Mary Hayes Davis, "Rubber Plantation Started," *Florida Grower* 30 (12 July 1924): 3; and *Detroit News*, 27 May 1927.

71. Pepper and Potter to J. E. Ingraham, 31 May 1924; and Ingraham to Pepper and Potter, 16 June 1924, Model Land Company Collection.

72. Sketch of Plots at Goodno, apparently attached to E. E. Goodno to Ernest G. Liebold, 16 July 1924, box 389, accession 285, Benson Ford Research Center, The Henry Ford. See also *FMNP*, 28 May 1924; and Bryan, "Henry's So-called Rubber Plantation."

73. See William Meadowcroft to Carters Tested Seeds Inc. et al., 3 October 1924, Edison General Files, folder: Rubber-1924, ENHS; and Bryan, "Henry's So-called Rubber Plantation in Florida."

74. A. C. Steger to William H. Smith, 30 August 1924; and Alfred Keys to Steger, 21 November 1924, box 101, accession 94; and Smith to Frank Campsall, 6 March 1925, box 31, accession 292, The Henry Ford.

75. *FMNP*, 19 March 1925; "Ford, Firestone, and Edison Unite to Break Rubber Monopoly Now Held by Great Britain," no citation available [probably *FMNP*, approx. April 1925]; *FMNP*, 1 May 1925; *FMNP*, 3 December 1925; E. E. Goodno

to Ernest G. Liebold, 30 October 1925, box 389, accession 285, The Henry Ford; and Tom Smoot, *The Edisons of Fort Myers: Discoveries of the Heart* (Sarasota, FL: Pineapple, 2004), 170.

76. Bryan, "Henry's So-called Rubber Plantation in Florida"; and *FMNP*, 7 November 1925.

77. D. N. Borodin to HSF, 7 October 1925, folder: Florida Rubber Growing, Firestone Records; and Ernest G. Liebold to A. C. Steger, 9 October 1925, box 427, accession 285, The Henry Ford.

78. E. H. Sykes to TAE, 17 September 1926, box 553, accession 285; and E. E. Goodno to Ernest G. Liebold, 20 January 1927, box 6, accession 572, The Henry Ford.

79. Harvey S. Firestone, *Men and Rubber: The Story of Business* (Garden City, NY: Doubleday, Page, 1926), 254. The Philippine government, which had some autonomy despite its colonial status, insisted that no company could control more than 2,500 acres in its territory. Rubber-industry leaders had been lobbying for changes in this Filipino law for some time. Some seized this issue in 1926 and called for the United States to remove the southern (and predominantly Muslim) islands from the rest of the Philippines so that they could be ruled directly under the American flag and developed for American economic interests. Neither proposal became law. See also B.J.C. McKercher, *The Second Baldwin Government and the United States, 1924–1929: Attitude and Diplomacy* (Cambridge: Cambridge University Press, 1984), 44–46; Howard T. Fry, "The Bacon Bill of 1926: New Light on an Exercise in Divide-and-Rule," *Philippine Studies* 26 (1978): 257–273; "A Setback to Filipino Progress" *IRW* 62 (June 1920): 555; "For American Owned Rubber," *IRW* 67 (March 1923): 343; and "Why Not Grow Our Own Rubber?" *Scientific American* 133 (December 1925): 404–406.

80. Hoover testimony discussed in *NYT*, 7 January 1926; and Brandes, *Herbert Hoover and Economic Diplomacy*, 76–80.

81. Chalk, "The United States and the International Struggle for Rubber," 96–98; and Brandes, *Herbert Hoover and Economic Diplomacy*, 122.

82. Chalk, "The United States and the International Struggle for Rubber," 108–110.

83. Undated clipping, box 532, Commerce Papers, HHPL.

84. Chalk, "The United States and the International Struggle for Rubber," 96–97, 106–111; and Brandes, *Herbert Hoover and Economic Diplomacy*, 95–105, 127–128.

85. For more on the work of first generation of USDA plant explorers, see David Fairchild, *The World Was My Garden: Travels of a Plant Explorer* (New York: Scribner's, 1938); Philip J. Pauly, "The Beauty and Menace of the Japanese Cherry Trees: Conflicting Visions of American Ecological Independence," *Isis* 87 (March 1996): 51–73; and Jones, "The World Was Our Garden," 11–35.

86. "Orator Fuller Cook," *National Cyclopedia of American Biography* (New York: White, 1953): 38:369–370. Before Cook began his thirty-nine-year career with the USDA, he spent several years in Liberia working to "repatriate the American Negro." Cook also gained attention in early in the twentieth century for his theory of "kinetic evolution," a provocative alternative to the Darwinian theory.

87. O. F. Cook to William A. Taylor, 10 August 1922, box 290, E2, RG54, NA; Taylor to secretary, 3 August 1923, box 941, E17, RG16, NA; and memorandum of war-planning diary, 2 June 1922, file 336, box 34, E191, RG107, NA.

88. O. F. Cook to Karl F. Kellerman, 13 March 1923, box 290, E2, RG54, NA. Cook had little regard, however, for Haiti's people. He also doubted that America could utilize the labor of "low-class Hindoos" employed in Trinidad and elsewhere in

the Caribbean because they too were "greatly inferior, physically, morally, and mentally."

89. F. C. Baker to O. F. Cook, 22 March 1924, box 290, E2, RG54, NA.

90. O. F. Cook to William A. Taylor, 4 June 1925; Cook to Taylor, 16 July 1925; and Cook to Taylor, 22 September 1925, box 290, E2, RG54, NA.

91. "Raising Rubber in California," *IRW* 71 (November 1924): 71–72; and R. E. Beckett, R. S. Stitt, and E. N. Duncan, "Growth and Rubber Content of Cryptostegia, Guayule, and Miscellaneous Plants Investigated at the United States Acclimatization Garden, Bard, California, 1923–1934," box 11, E90, Division of Rubber Plant Investigations, Subject Files, RG54, Bureau of Plant Industry, NA (hereafter cited as E90, RG54, NA).

92. Henry C. Wallace to Senator Lee S. Overman, 20 May 1924, quoted in *Congressional Record*, Senate, 68th Congress, 1st sess., 9282. See also Henry C. Wallace, "The Year in Agriculture," in USDA, *Yearbook of Agriculture, 1923* (Washington, DC: Government Printing Office, 1924), 49; Henry C. Wallace, "The Year in Agriculture," in USDA, *Yearbook of Agriculture, 1924* (Washington, DC: Government Printing Office, 1925), 74; and O. F. Cook, "Rubber Possibilities of Many Kinds Exist in the United States," in USDA, *Yearbook of Agriculture, 1927* (Washington, DC: Government Printing Office, 1928), 562–565.

93. GHC to William A. Taylor, 23 May 1923, box 16, IRC Records; and Taylor, confidential memorandum for the secretary, 29 May 1923, box 1023, E17, RG16, NA. At the May 1923 meeting, Carnahan insisted that his company's higher priority was to protect the research discoveries that they considered their "exclusive property."

94. William B. McCallum to GHC, 15 November 1923; and O. F. Cook, "Memorandum of Conference on Guayule," 16 November 1923, box 10, E91, RG54, NA.

95. Karl F. Kellerman to GHC, 20 May 1925, box 5, E91, RG54, NA.

96. Ibid.

97. HMH to W. M. Gilbert, 30 April 1923, file: Plant Biology—HMH, Archives of the Carnegie Institution of Washington, Washington, DC.

98. Ibid.

99. F. W. Bolzendahl to Governor J. G. Scrugham, 6 August 1925; Scrugham to Chrysil Rubber Association, 11 August 1925; and Scrugham to Bolzendahl, 15 September 1925, Governor James G. Scrugham Papers, Nevada State Library and Archives, Carson City.

100. HMH to Frederic E. Clements, 4 September 1925, box 5, Clements Papers; and HMH report attached to Samuel B. Doten to Governor James G. Scrugham, 11 September 1925, Governor James G. Scrugham Papers, Nevada State Library and Archives, Carson City.

101. Samuel B. Doten to HMH, 18 September 1925, file NUC 01/17, Agricultural Experiment Station Records, Special Collections, University of Nevada, Reno.

102. Quotations from [Samuel B. Doten], "Conclusions Reached after Season's Work on Chrysothamnus," 8 February 1926, file NUC 01/17, Agricultural Experiment Station Records, Special Collections, University of Nevada, Reno; and Samuel B. Doten to HMH, 10 November, box 45, College of Agriculture Records, 1889–1965, Special Collections, University of Nevada, Reno. See also J. M. Ryan to Samuel B. Doten, 27 September 1926, file NUC 01/17, Agricultural Experiment Station Records, Special Collections, University of Nevada, Reno; HMH to Samuel B. Doten, 7 November 1925; and O. F. Cook to HMH, 9 November 1925, box 45, College of Agriculture Records, 1889–1965, Special Collections, University of Nevada, Reno.

103. A summary of the 1926 report is described in S. B. Doten, *Rubber from Rabbit Brush (Chrysothamnus nauseosus)*, University of Nevada Agricultural Experiment Station Bulletin No. 157 (March 1942). At about the same time, the Carnegie Institution scaled back its rubber plant research at the Fallon, Nevada, facility because results had not been promising. See Carnegie Institution of Washington, *Yearbook No. 24, 1924–1925* (Washington, DC: Carnegie Institution, 1925), 342–343. Similarly, by 1925 and 1926, Hall had shifted his own research agenda to his interest in experimental taxonomy and botanic evolution.

104. "Rubber Grown Commercially in the United States," *IRW* 73 (October 1925): 7; John C. Treadwell, "Guayule Rubber from Texas," *RA* 20 (November 1926): 138–140; and J. Wesley White, "History of Guayule, with Reference to Texas," box 28, E20, RG95, NA.

105. *Bulletin of the National Association for the Protection of American Rights in Mexico*, no. 1 (1 July 1919): 1–2.

106. "Intercontinental Closes Torreon Factory," *IRW* 60 (July 1919): 596; William H. Yeandle Jr. to GHC, 22 April 1921, box 10, IRC Records; IRC *Annual Report*, 1920–1924; Shonnard and Company, "Intercontinental Rubber Company" (January 1922); and G.M.P. Murphy and Co., "Intercontinental Rubber Company" (January 1926), Harvard Corporate Report Collection. For annual guayule production data between 1905 and 1925, see "Cultivation to Enlarge Guayule Rubber Supply," *RA* 20 (November 1926): 127. See also the slightly different figures, 1911 to 1925, in "Guayule—A High Grade Rubber," *IRW* 72 (August 1925): 652–653.

107. William B. McCallum to GHC, 31 July 1922, box 1; and GHC to J. M. Williams, 14 January 1924, box 5, E91, RG54, NA. See also *Tucson Citizen*, 23 December 1924, clipping, box 31, E20, RG95, NA; and IRC, *Annual Report*, 1924.

108. "Guayule—A High Grade Rubber," 652–653; and GHC, "Method and Apparatus of Treating Gumlike Substances," U.S. Patent 1,671,570, issued 29 May 1928.

109. GHC to William B. McCallum, 17 March 1925, box 5, E91, RG54, NA.

110. Harry M. Blair to W. D. Hines, 20 January 1926, box 19: Rubber Growing, Firestone Records.

110. GHC to HCH, 26 January 1926; and GHC to HCH, 27 January 1926, folder: Guayule Rubber, box 532, Commerce Papers, HHPL. For nearly two decades, the National Committee on Plant Patents and other groups pushed for protective legislation before achieving success in 1930. See Bugos and Kevles, "Plants as Intellectual Property"; and Kevles, "Patents, Protections, and Privileges."

112. *NYT*, 13 February 1926. See also GHC, "Production of Guayule Rubber," *RA* 20 (November 1926): 136–137.

113. Murphy and Co., "Intercontinental Rubber Company"; and Investors' Prospectus, "Intercontinental Rubber Company" [March 1926], Harvard Corporate Report Collection.

114. See Chalk, "The United States and the International Struggle for Rubber," 126–127; Eugene Staley, *Raw Materials in Peace and War* (New York: Council on Foreign Relations, 1937), 131–135; and HCH to John J. Raskob, 19 June 1926; and Raskob to HCH, 23 June 1926, folder: Guayule Rubber, box 532, HHPL. Raskob was chair of the finance committee of General Motors, president of the National Automobile Chamber of Commerce, treasurer at E. I. du Pont de Nemours and Company, and a prominent power broker in national politics. In 1928, Raskob served as campaign manager for the Democratic presidential nominee, Alfred E. Smith, and thus directly opposed Hoover's presidential run.

115. William B. McCallum, "Botany and Cultural Problems of Guayule," *RA* 20 (November 1926): 129–132; David Spence, "The Chemistry of Guayule," *RA* 20 (November 1926): 133–135; and GHC, "Production of Guayule Rubber," 136–137.
116. GHC to H. A. Bighman, 15 September 1926, box 10, IRC Records.
117. Vincent Sauchelli, "Machine Grown Rubber in the United States," *IRW* 75 (November 1926): 67; and other articles in "Guayule Supplement," *RA* 20 (November 1926). Although no evidence has been found, it certainly seems feasible the IRC offered these journals some financial considerations in exchange for these remarkably favorable stories. Notably, *RA*'s twenty-page supplement featured many illustrations identical to those found in the company's prospectus for investors.
118. D. T. MacDougal to William McCallum, 18 October 1926, box 18, AZ 356, Records of the Desert Botanical Laboratory, University Archives, University of Arizona, Tucson.
119. HCH to secretary of war [Dwight F. Davis], 28 January 1926; and Hanford MacNider to HCH, 27 February 1926, folder: Guayule Rubber, box 532, Commerce Papers, HHPL.
120. Brandes, *Herbert Hoover and Economic Diplomacy*, 78–80, 96–100, 123–126. Between 1923 and 1927, the Firestone Tire and Rubber Company cleared about forty-nine million dollars, just behind similar profits for the Goodyear Tire and Rubber Company. See *Congressional Record*, House, 69th Congress, 1st sess., 5993.
121. F. R. Henderson to Col. Ferguson, 8 December 1926, file 336, box 34, E191, RG107, NA. The reports mentioned are Lt. Ben R. Morton, "Resource Study-Rubber," February 1924; and Lt. Col. James Regan, "Resource Study-Rubber," 16 April 1926, file 1082, box 100, E191, RG107, NA.
122. William Jardine to secretary of commerce [HCH], [n.d.] October 1926, box 1225, E17, RG16, NA. This report also suggests an approach of greater cooperation with Latin American governments in Costa Rica, Ecuador, and elsewhere rather than aggressive isolationism or economic nationalism.
123. Karl F. Kellerman to F. M. Russell, 9 July 1926; and William A. Taylor to Secretary Jardine, 18 October 1926, box 1225; and A. F. Woods to O. F. Cook, 8 January 1927, box 1301, E17, RG16, NA.
124. NYT, 23 December 1925.
125. William C. Redfield, *Dependent America* (Boston: Houghton Mifflin, 1926), quotations on pp. 9, 10, 13, 35.
126. Ibid., 252.
127. E. G. Holt, "Five Years of Restriction," *IRW* 77 (February 1928): 55–57; Chalk, "The United States and the International Struggle for Rubber," 151–156; and Staley, *Raw Materials in Peace and War*, 290–293.

Chapter 3   Thomas Edison and the Challenges of the New Rubber Crops

1. NYT, 27 February 1927, 1 March 1927, 20 March 1927; *FMNP*, 27 February 1927, 1 March 1927, 18 March 1927; and *Washington Herald*, 16 March 1927 and 18 March 1927. For accounts of Edison's premature publicity on his electric lighting project, see Israel, *Edison*, 168; and Thomas P. Hughes, *American Genesis: A Century of Invention and Technological Innovation* (New York: Penguin, 1989), 88–89. On the use of "Menlo Park archetypes" in media portrayal of Edison's rubber research, see Wyn Wachorst, *Thomas Alva Edison: An American Myth* (Cambridge, MA: MIT Press, 1981), 163–164.

2. Popular press reports touting Edison's rubber project include "Edison Absorbed in Question of Rubber," *RA* 21 (July 1927): 382; "Edison and Florida Rubber," *IRW* 76 (August 1927): 252; "Edison Hunting for Rubber in Weeds," *Literary Digest* 95 (26 November 1927): 9–10; and Frank Parker Stockbridge, "Rubber From Weeds—My New Goal—Edison," *Popular Science Monthly* 111 (December 1927): 9–11.

3. Matthew Josephson, *Edison: A Biography* (1959; repr., New York: Wiley, 1992), 470; and Baldwin, *Edison*, 380–381.

4. Correspondence with Rebecca Baker, archivist, Luther Burbank Homes and Gardens, 20 January 2001. Ms. Baker reports that "we have often heard this story from former visitors to Edison's winter home," but that she has "no information on this subject at all." Her archives contain no supporting evidence, and she reports that the Burbank Papers in the Library of Congress include no correspondence between Edison and Burbank. Preliminary searches at the ENHS also yielded no significant Edison-Burbank correspondence. HSF's account of the 1915 meeting with Burbank does not mention rubber. Firestone reports that he and Edison did discuss rubber on their 1919 camping trip, with the implication that this was the first time that he learned of Edison's interest in the subject. Firestone, *Men and Rubber*, 190–191, 226–227.

5. Israel, *Edison*, 446–447; and Van Keuren, "Science, Progressivism, and Military Preparedness."

6. Interview, General File, 1917, ENHS.

7. GHC to TAE, 4 February 1919; GHC to TAE, 7 February 1919; and TAE to William H. Meadowcroft, 13 October 1919, folder: Rubber, 1919, Edison General Files, ENHS. The first of these letters indicates that Edison first sought guayule samples in November 1918. Firestone's comment in Samuel Crowther and HSF, "What Vacations Have Taught Me About Business," *System* 50 (July 1926): 106. See also Vanderbilt, *Thomas Edison, Chemist*, 81–83; and Andre Millard, *Edison and the Business of Innovation* (Baltimore: Johns Hopkins University Press, 1990), 320.

8. Schimerka Laboratory Notes, November–December 1922, folder: Rubber, 1922; and Schimerka to TAE, 11 April 1923, folder: Storage Battery, 1923, Edison General Files, ENHS. For an introduction to the storage-battery issues, see Vanderbilt, *Thomas Edison, Chemist*, 203–233; and Israel, *Edison*, 410–421. Pontianak, or pressed Jelutong, is a rather inelastic rubber commonly obtained from the *Dyera costulata* plant of Borneo. See Harry L. Fisher, *Chemistry of Natural and Synthetic Rubbers* (New York: Reinhold: 1957), 59.

9. C. L. Fulbert to TAE, 12 December 1922; and TAE to Jos. Stokes Rubber Co., 14 December 1922, folder: Rubber, 1922, Edison General Files, ENHS; H. C. Egerton to TAE, 23 November 1923; Egerton to TAE, 13 December 1923; and J. P. Burke to TAE and Charles Edison, 17 April 1925, folder: Rubber, 1923 [sic, folder also contains later material], Edison General Files, ENHS; and J. P. Burke to Stringfellow, 4 December 1924, folder: Rubber 1924, Edison General Files, ENHS. Several engineers and experts studied the issue, an effort that culminated in a November 1923 report that concluded that the Edison Company could save $60,000 to $120,000 per year if it produced its own hard rubber. An explosion and other difficulties diminished the effectiveness of this plant, so it is not clear how much rubber was produced.

10. Terms from a confidential memorandum, P. L. Palmerton to Mr. Stokes, 23 October 1925, box 531, Commerce Papers, HHPL. For more on the Liberian episode, see Chalk, "The United States and the International Struggle for Rubber," 77–

95, 130–150, 176–195; *FMNP*, 6 March 1926; Lief, *The Firestone Story*, 147–170; and Tucker, *Insatiable Appetite*, 249–257.

11. Joseph A. Russell, "Alternative Sources of Rubber," *Economic Geography* 17 (October 1941): 391–408; Davis, *One River*, 338–340; and Ford R. Bryan, *Beyond the Model T: The Other Ventures of Henry Ford* (Detroit: Wayne State University Press, 1990), 151–162.

12. Smoot, *The Edisons of Fort Myers*; and Olav Thulesius, *Edison in Florida: The Green Laboratory* (Gainesville: University Press of Florida, 1997). A good Edisonian account is the transcript of an interview that describes his early impressions of Fort Myers, folder: Fort Myers, 1917, Edison General Files, ENHS.

13. Henry H. Wesener to HSF, 13 April 1926 and 29 April 1926, folder: Florida Rubber Growing, 1926, Firestone Records; H. E. Ellis to TAE, 2 July 1926; and Ellis to TAE, 14 August 1926, box 7, EBRC, ENHS; *FMNP*, 17 September 1926 and 27 February 1927; and *NYT*, 5 March 1927.

14. For example, the cipher "JULUJLEARD" was translated as "first of the month." P. W. Grandjeau to Ernest G. Liebold, 14 July 1926; William H. Meadowcroft to Liebold, 30 November 1926; Edwin Mayer and Co., copy of telegram dispatched, 31 December 1926; and Mayer to Henry Ford, 5 February 1927, file: Experimenter, box 6, accession 572, The Henry Ford; and "Reminiscences of E. G. Liebold," Oral History Section, accession 65, The Henry Ford. Typescript copy held at EFWE, p. 636.

15. E. G. Holt to Johannesburg Office, 27 July 1927, box 1165, E1, RG151, Bureau of Foreign and Domestic Commerce, NA; and Notebook (29–02–01), ENHS.

16. Untitled memorandum, 10 August 1929, in folder: Rubber Notes and Glossary; and "Rubber Experiment: Factory and Field Costs," 9 October 1929, box 25, EBRC; Notebook (28–11–05); Notebook (29–01–03); and Notebook (28–10–19), ENHS.

17. *NYT*, 18 June 1929, 3 January 1927, 5 March 1927; and *Washington Herald*, 16 March 1927.

18. "Edison and Florida Rubber," *IRW* 76 (August 1927): 252.

19. *NYT*, 27 February 1927 and 1 March 1927.

20. *FMNP*, 27 February 1927, 1 March 1927, and 18 March 1927. See also Stockbridge, "Rubber from Weeds," 9–11.

21. *NYT*, 20 March 1927; and *Washington Herald*, 16 March and 18 March 1927.

22. Charles Edison to TAE, 27 June 1927, folder: T. A. Edison, Work Outline, General Files, ENHS.

23. Notebook (27–08–15), ENHS; and William H. Meadowcroft to Frank M. Stout, 18 June 1927, box 17, EBRC, ENHS.

24. TAE to HSF, 9 May 1927, and accompanying notes, box 7, EBRC, ENHS.

25. Letters related to the procedure of collecting and preparing specimens include William H. Meadowcroft to John K. Small, 23 June 1927; Small to Meadowcroft, 24 June 1927; and Small to Meadowcroft, 25 July 1927, box 13, EBRC, ENHS.

26. John K. Small to TAE, 17 June 1927, box 13, EBRC, ENHS. See also Laura Zelasnic, "Thomas Edison, Botanist," *The Herbarium Sheet: Newsletter of the New York Botanical Garden*, no. 276 (29 February 2000).

27. Quotation from Notebook (27–08–00.1), ENHS. See also Notebook (27–08–00), ENHS.

28. "Edison Absorbed in Question of Rubber," 382.

29. TAE, [July 1927?], draft letter to various railway companies, box 18, EBRC, ENHS.

30. I estimate that Edison received approximately seven hundred unsolicited letters from private citizens with suggestions about rubber plants and the rubber industry.

31. A. W. Morrill to TAE, 8 April 1927, box 12, EBRC, ENHS.
32. E. L. Dunbar to William H. Meadowcroft, 19 February 1928, box 4, EBRC, ENHS.
33. W. Sam Clark to TAE laboratories, 19 July 1927; TAE to Clark, 29 July 1927; JVM to Clarkdota Fig Orchard, 26 September 1927; and Clark to EBRC, 4 October 1927, box 4, EBRC, ENHS.
34. Edison contacted several of the nation's leading professors of botany (including Arno Nehrling of Cornell University, son of the Florida botanist) for the names of those who might help in the search for rubber-bearing plants. These men had varying backgrounds and experience: Arthur P. Kelley was assistant professor of botany at Rutgers University; Robert Stratton was an assistant professor of botany and plant pathology at Oklahoma Agricultural and Mechanical College; G. Myron Shear was a fresh graduate of the University of Maryland but son of a well-respected USDA botanist; and Howard A. Barton and Seth Judson Ewer were still undergraduate students, at the New Mexico College of Agriculture and the Massachusetts Agricultural College, respectively.
35. TAE to Marshall A. Howe, 12 May 1927, box 13, EBRC, ENHS. Purchasing power determined at http://www.eh.net/ehresources/howmuch/dollarq.php.
36. Edison marginal notes, on J. Connie Walton and Edward E. Lanser to TAE, 27 July 1927, box 11, EBRC, ENHS. Tensions between the United States and Mexico reached a new peak in 1927, and war became a real possibility. In several cases, Mexican revolutionaries harassed, kidnapped, murdered, or otherwise terrorized the representatives of American economic interests. Tensions seemed to ease by September 1927, but botanist J. N. Rose ran into difficulties nevertheless.
37. On collection methods, see Edison quoted in JVM to John T. Reid, 12 July 1928, box 14, EBRC, ENHS. On traveling too quickly, see JVM to S. T. Moore, 27 November 1929, box 12, EBRC, ENHS. On hotel costs, see JVM to W. H. Hand, 21 November 1929, box 9, EBRC, ENHS. On railroad accommodations, JVM to R. A. Smith, 22 August 1928, box 20, EBRC, ENHS.
38. TAE to HSF, 6 September 1927, box 7, EBRC, ENHS. See also Karl F. Kellerman to TAE, 17 August 1927, box 18, ERBC, ENHS. See also TAE to Everett G. Holt, 27 June 1927; E. G. Holt to William H. Meadowcroft, 11 May 1927; and E. G. Holt to William H. Meadowcroft, 25 June 1927, box 18, EBRC, ENHS.
39. TAE to G. Myron Shear, 1 July 1927 and 8 July 1927, G. M. Shear Collection, Special Collections Department, Virginia Polytechnic and State University, Blacksburg (hereafter cited as Shear Collection).
40. Martin Jay Rosenblum, R. A., and Associates, *Historic Structures Report* (EFWE, 2002), I-86; and "Instructions for Collecting and Shipping Latex-bearing Plants for TAE, Orange, New Jersey," 1 August 1927, box 36, EBRC, ENHS.
41. Edison asked Rose to conduct rubber tests in the field: upon adding ten drops of denatured alcohol to each drop of plant latex, the elastic properties would become evident. These tests revealed Edison's growing awareness that rubber quantity and quality could not be determined visually. TAE to J. N. Rose, 29 September 1927, box 9, record series 221, Division of Plants, 1886–1928, Smithsonian Institution Archives, Washington DC. See also Notebook (27–09–01.1), ENHS.
42. For the South African difficulties, see Everett G. Holt to JVM, 6 April 1929, box 18, EBRC, ENHS. Edison forwarded three hundred dollars to his agents in South Africa to fund their search; two years later, they returned all but about four dollars. See Holt to Johannesburg Office, 21 January 1929; and Johannesburg Office to Rubber Division, 6 March 1929, box 1165, E1, RG151, NA.
43. In all, the thirteen plant collecting teams returned 423 specimens, or an average of 33 each. Precisely calculating the efficiency of this effort, Edison's staff

determined that the ratio of specimens collected to days employed was a disappointing 53.7 percent. In "Statistics Concerning Field Men," n.d., box 7, EBRC, ENHS. For difficulties, see Howard Barton to TAE, 29 July 1927, 10 October 1927, 23 October 1927, box 2; G. Myron Shear to TAE, 8 July 1927, box 16; and J. N. Rose to TAE, 26 November 1927, box 14, EBRC, ENHS.

44. Dr. Marlatt to William A. Taylor, 25 July 1927, folder: USDA-BPI, box 18, ERBC, ENHS. For more on Marlatt's politics and personal rivalries, see Pauly, "The Beauty and Menace of the Japanese Cherry Trees."

45. E. G. Holt to William H. Meadowcroft, 29 July 1927, box 18, EBRC, ENHS. Holt probably would not have pleased with the "Fighting British Control" headline that appeared two days later above an article that described Edison's research project. [New York?] *Daily News*, 1 August 1927, clipping, box 7, EBRC, ENHS.

46. *Washington Star*, 27 July 1927, clipping, Shear Collection.

47. E. G. Holt to Johannesburg Office, 27 July 1927, box 1165, E1, RG151, Bureau of Foreign and Domestic Commerce, NA.

48. William H. Meadowcroft to E. G. Holt, 18 July 1927, folder: United States Department of Commerce, box 18, EBRC, ENHS.

49. Numbers of species tested from Loren G. Polhamus, *Plants Collected and Tested by Thomas A. Edison as Possible Sources of Domestic Rubber* (Agricultural Research Service Publication 34–74) (Washington, DC: USDA, 1967); "NO RUBBER" stamp, box 14, EBRC, ENHS.

50. TAE to HSF, 6 September 1927; and A. T. Hancock to Firestone Tire and Rubber Company, 4 October 1927, folder: Firestone no. 1, box 7, EBRC, ENHS. The Fort Myers plots included *Landolphia* vines from Liberia, *Ceara* trees from Brazil, and *Parameria* vines and *Mascarenhasia* trees vines from the Dutch East Indies, among scores of other varieties.

51. List of plants from TAE notations on BJ memorandum, 27 May 1927, box 163, EFWE; "Report on Rubber Plants from March 10th 1927 to June 8th 1927"; and "Report on Rubber Plants from June 18th 1927 to July 20th 1927," box 17, EBRC, ENHS.

52. Notebook (27–12–15), ENHS. Although this notebook dates from December 1927, when Edison was in New Jersey, its contents suggest that they are summaries of work conducted earlier in the year in Fort Myers.

53. Notably, Edison had begun to test scores of plants in the thistle family, (*Compositeae*), which includes his eventual favorite, the goldenrod genus (*Solidago*). Initial results, however, were not very promising. "Rubber Experiments—List of Plants Planted at Glenmont, 1927," box 24, EBRC, ENHS; and "List of Best Plants—Nov. 28, '27," box 36, EBRC, ENHS.

54. "Reminiscences of E. G. Liebold," 637, Oral History Section, accession 65, The Henry Ford.

55. Examples include TAE to Stout, 5 July, 8 July, 9 July, 11 July 1927, box 17; TAE quoted in JVM to E. G. Holt, 17 September 1927, box 18; and TAE to John W. Martin, 25 August 1927, box 11, EBRC, ENHS.

56. HSF to TAE, 11 July 1927; and TAE to HSF, 16 July 1927, box 6, accession 572, The Henry Ford.

57. Ernest G. Liebold to William H. Meadowcroft, 22 July 1927; and Meadowcroft to Liebold, 25 July 1927, box 6, accession 572, The Henry Ford. See also [Ford R. Bryan], "Edison Botanic Research Corporation"; and Rosenblum, *Historic Structures Report*, I-86, I-87. Firestone's quotation is from HSF to Frank Campsall, 28 March 1928, box 785, accession 285, The Henry Ford. Edison and Firestone had proposed an initial capitalization of $40,000, but Ford countered that $75,000 was more appropriate. Additional financial reports also indicate that Edison, his

heirs, and TAE Inc. continued to match the contributions of Ford and Firestone through at least June 1934. In addition to their initial capitalization, each contributed roughly $90,000 to the budget of $269,115 that supported EBRC operations from 19 May 1928 to 30 June 1934. See financial documents in box 163, EFWE.

58. TAE to GHC, 9 August 1927; GHC to TAE, 29 September 1927 [two letters with this date]; GHC to TAE, 5 October 1927; and GHC to TAE, 11 November 1927, box 4, EBRC, ENHS.

59. TAE, sketch of letter, 19 March 1927, box 7. In a letter to IRC colleague Jim Williams, Carnahan indicated that the company hoped to "some day find occasion to utilize" Edison's note. See GHC to J. M. Williams, 29 March 1927, box 10, E91, RG54, NA.

60. Edison quoted in JVM to John T. Reid, 2 August 1927, box 14; and TAE to R. E. Beckett, 22 August 1927, box 2, EBRC, ENHS.

61. N.L. Britton to TAE, 11 November 1927, box 13, EBRC, ENHS. This is a response to Edison's count, reported in a letter of 8 November 1927; and TAE to J. N. Rose, 14 November 1927, box 14, EBRC, ENHS.

62. TAE to Mark McKernan, 27 July 1927, box 20, EBRC, ENHS; and GHC to members of the board of the IRC, 27 September 1927, series 6, vol. 18, Baruch Papers.

63. GHC to members of the board of the IRC, 27 September 1927.

64. Notebook (29–02–01), ENHS; and TAE to GHC, 13 January 1930, series 6, vol. 24, Baruch Papers.

65. GHC to TAE, 24 January 1930, series 6, vol. 24, Baruch Papers; and TAE to GHC, 10 February 1930, box 4, EBRC, ENHS.

66. TAE to HSF, 24 October 1927, box 7; and TAE to David Spence, 4 November 1927; and Spence to TAE, 30 November 1927, box 4, EBRC, ENHS.

67. HSF to TAE, 3 January 1928, Firestone Records; W. A. Benney to JVM, 2 May 1928, box 2; and Benney to TAE, 30 October 1928; and TAE to GHC, 9 November 1928, box 4, EBRC, ENHS.

68. GHC to TAE, 10 December 1928; GHC to TAE, 17 December 1928; and TAE marginalia response on the 17 December letter, box 4, EBRC, ENHS. Firestone shipped additional sets of guayule tires and tubes to Fort Myers in 1929 and again in 1930, and Edison asked his employees to take great care with the rare specimens. At least one of the guayule tires used on the Ford Model A truck employed in southern Florida plant collecting was still in satisfactory condition after some fourteen thousand miles. Trials continued until at least 1933, when the tires were placed in the Fort Myers vault. In the aftermath, a Firestone scientist concluded that although overall results were "very poor," guayule tires nevertheless could match, and probably surpass, the quality of fabric tires used during the World War I tire emergency. JVM to Walter N. Archer, 4 November 1930, box 1, EBRC, ENHS; Archer to JVM, 9 February 1933, box 2, EBRC, ENHS; TAE to GHC, 10 February 1930, series 6, vol. 24, Baruch Papers; and J. H. Doering, "Guayule in Tires and Tubes: Service Tests in Which the Rubber Was Exclusively Guayule," *Journal of Industrial and Engineering Chemistry* 26 (May 1934): 541–543.

69. W. A. Benney to TAE, 4 December 1927; and Benney to JVM, 22 January 1928, box 2, EBRC, ENHS; and *FMNP*, 13 January, 14 January, 16 January 1928.

70. Baldwin, *Edison*, 385.

71. *Fort Myers Tropical News*, 7 January 1928; W. A. Benney to JVM, 22 January 1928, box 2, EBRC, ENHS; Notebook (27–12–28), ENHS; and Mina Edison to HSF, 25 March 1928, folder: TAE, 1928, Firestone Records.

72. *Collier City (Florida) News*, 1 February 1928; and *NYT*, 13 February 1928 and 19 February 1928. Information about snake protection found in W. A. Benney to JVM, 12 December 1927; and Benney to JVM, 18 January 1928, box 2, EBRC, ENHS. Machetes mentioned in Walter N. Archer to JVM, 22 January 1930, box 1, EBRC, ENHS. Other clues about these trips derived from a summary of the notebooks copied onto microfilm, and "Inventory of Material Supplies and Equipment, Edison Research Laboratory, Fort Myers, Florida, June 1935," box 178, EFWE.

73. Notebook (28–03–28.1), ENHS; and W. A. Benney to JVM, 2 May 1928, box 2, EBRC, ENHS.

74. TAE to unidentified recipient, 12 April 1928, box 163, EFWE.

75. TAE to "Gentlemen. Chamber of Commerce," 13 July 1928, box 4, EBRC, ENHS.

76. TAE to W. A. Benney, 9 August 1928; Benney to TAE, 12 August 1928; JVM to Benney, 22 August 1928; and TAE to Benney, 4 September 1928, box 2, EBRC, ENHS.

77. "Hullmann Ga. 252" indicated that it was the 252nd plant that collector Walter Hullmann had sent to the EBRC during his meanderings through Georgia.

78. W. A. Benney to TAE, 11 September 1928; Benney to TAE, 28 October 1928; W. A. Benney to TAE, 11 November 1928; and Benney to JVM, 1 December 1928, box 3, EBRC, ENHS; and W. H. Hand to TAE, 12 November 1928 and 30 November 1928, box 9, EBRC, ENHS.

79. TAE to W. A. Benney, 17 November 1928, box 3, EBRC, ENHS. Similarly, Edison's letter to Jonas explained that he planned to test about fifty specimens of the same species for rubber content and use the seeds from the best yielder as the basis for the next year's planting. See marginal note on BJ to TAE, 28 October 1928, box 10, EBRC, ENHS.

80. Despite a few articles in the popular press that used the word "crossbreeding," there is no evidence that Edison made any effort to integrate the genetic material of different plant lines. See also Walter N. Archer to JVM, 22 June 1929; and Archer to TAE, 16 August 1929, box 1, EBRC, ENHS; and *Fort Myers Press*, 11 April 1930.

81. Emphasis in original. TAE to William Benney, [June 1928?], box 3, accession 1630, The Henry Ford.

82. Edison Notebook no. 1, 1 April 1928 to 26 December 1929, box 186, EFWE.

83. Ibid.; and *NYT*, 14 February 1928, 11.

84. Notebook (27–12–15), ENHS.

85. Notebook (29–02–01); and Notebook (29–02–04.1), ENHS.

86. *The Intellect* (1927), 7–8, box 163, EFWE. Jonas's satire also claimed that Edison wanted more rubber because, although automobiles killed two thousand pedestrians per year, that was not enough to keep pace with the rising population.

87. BJ to Henry Ford, with *The Intellect* attached, 24 February 1928, box 827, accession 285, The Henry Ford.

88. The Justice Department's Bureau of Investigation had investigated Jonas and his radical publications during the Red Scare that followed World War I. The agent found Jonas living near Toledo, Ohio, in a dirty shanty with no bed. Although his anarchist philosophies seemed extreme, the agent concluded that Jonas was "absolutely harmless" and likely to soon be in a "lunatic asylum." See case no. 376601, Report of 22 April 1920, Investigative Case Files of the Bureau of Investigation, series: Old German Files, roll 826, NA Microfilm Publication M1085. After he left Edison's project, Jonas remained in the Fort Myers area, where he created a community named "Fort Intellect" that claimed to study intellectual

issues. Jonas published his rants and philosophies in newspaper columns and in the form self-published, handwritten, mimeographed books entitled *Feast of the Cannibals* and *The Roof of the Palace*, among others. See also BJ, "We Become Realtors"; and BJ to TAE, 5 June 1928, box 10, EBRC, ENHS; and BJ, "Wanted: A New Science," *FMNP*, 19 October 1929.

89. BJ to JVM, 13 April 1928, box 10, EBRC, ENHS.
90. BJ to JVM, 13 April 1928; and BJ to TAE, 5 June 1928, 10 August 1928, 28 November 1928, box 10, EBRC, ENHS.
91. BJ to TAE, 21 June 1928, box 10, EBRC, ENHS.
92. BJ to TAE, 22 August 1928, 29 August 1928, 25 September 1928, 14 October 1928, 7 November 1928, 16 November 1928, 23 November 1928, 30 October 1928, and TAE marginalia on the same letter, box 10, EBRC, ENHS. See also untitled memo, [17 November 1928?], box 25, EBRC, ENHS.
93. W. A. Benney to TAE, 30 October 1928; Benney to JVM, 1 December 1928; and Benney to TAE, 28 December 1928, box 3, EBRC, ENHS.
94. Edison Notebook no. 1, 1 April 1928 to 26 December 1929, box 186, EFWE.
95. TAE to W. A. Benney or BJ, 7 September 1928; and Benney to TAE, 7 October 1928, box 2, EBRC, ENHS; and TAE to Benney, 17 November 1928, box 3, EBRC, ENHS.
96. Notebook (28–11–05), ENHS.
97. Emphasis in original. Notebook (28–11–05), ENHS. This page is dated "Nov 7/28," but 7 December 1928 seems much more likely. Pages dated November 5 through November 8 are located between the pages dated December 3 and December 13 and appear in the notebook well after other pages that begin with a November 5 date.
98. Notebook (29–06–02), ENHS.
99. Notebook (29–01–03), ENHS. The entry also states, "So far it appears that *Pterocaulin* [sic] . . . and *Artemisia lindheimeriana* are the best plants for us. They grow in all states, both a perennial, have long roots."
100. Unsigned memorandum, "Method of Collecting, Testing, and Recording Plants" 30 January 1929, box 36, EBRC, ENHS.
101. Notebook (29–02–01), ENHS.
102. *FMNP*, 12 February 1929; and *NYT*, 12 February 1929.
103. Samuel Crowther, "T.A.E.—A Great National Asset, An Interview with Thomas A. Edison," *Saturday Evening Post* 201 (5 January 1929): 128.
104. Notebook (29–04–19), ENHS; and J. K. Small to TAE, 4 June 1929, box 13, EBRC, ENHS.
105. TAE quoted in JVM to Herbert Stirling Wilson, 21 June 1929, box 19, EBRC, ENHS.
106. William G. Sutlive, to TAE, 17 July 1929, box 16, EBRC, ENHS. The Georgia rumors also reached the *NYT* (see 4 August 1929). By the spring of 1930, several news reports suggested that Henry Ford was ready to begin goldenrod production on a large scale on his twenty-five thousand-acre property in Bryan County, Georgia. Not coincidentally, Edison also hinted in the spring of 1930 that he thought that goldenrod would soon be ready for a twenty-five-thousand-acre test project. For a rather clear indication that the planting had not taken place, see *NYT*, 12 April 1930. For more on paving roads with rubber, see David O. Whitten, "Rollin' on Rubber: 120 Years of Rubber Road Surfacing," *Essays in Economic and Business History* 12 (1994): 414–427.
107. Notebook (28–07–06); and Notebook (29–06–24), ENHS.
108. Notebook (29–06–24), ENHS; and JVM to Walter Hullmann, 15 September 1929, box 25, EBRC, ENHS.

109. Walter Hullmann to JVM, 21 September 1929, box 1, EBRC, ENHS.
110. JVM to Edwin E. Slosson, 3 October 1929, box 16, EBRC. ENHS. The letter includes a draft of Slosson's article "A Conversation with Edison on Home-grown Rubber." Slosson died on 15 October 1929, and it appears that this article was never published. Meanwhile, a hurricane hit the Fort Myers facility in October 1929 and caused considerable damage. Significantly, however, Edison's Florida employees understood that only the *Solidago leavenworthii* plants were a high priority. They spent several days trying to save them and built special cradles to help realign the blown-over plants in an erect position. By the end of the year, the situation had returned to normal and nearly one acre of *leavenworthii* had been planted. See Walter N. Archer to JVM, 5 October 1929; Archer to JVM, 8 October 1929; and Archer to JVM, 28 December 1929, box 1, EBRC, ENHS.
111. Patent available at http://edison.rutgers.edu/patents/01740079.pdf.
112. U.S. Patent 1,740,079; "Edison's Rubber Process," *IRW* 81 (February 1930): 55; *NYT*, 13 February 1930 and 12 April 1930; and *FMNP*, 11 April 1930. See also preliminary notes in Notebook (28–12–28), ENHS.
113. Notebook (29–01–03), ENHS.
114. W. A. Benney to TAE, 4 December 1927; and Benney to TAE, 31 July 1928, box 2; and Walter N. Archer to JVM, 11 September 1931, box 1, EBRC, ENHS.
115. N. A. Shepard to HSF, 9 November 1928, folder: Edison, 1938, Firestone Records.
116. Schimerka explained to Edison the cost advantages of using benzol rather than carbon tetrachloride as the principle solvent. He further showed the precise calculations necessary to ensure a continuous operation and maximum benefit from the benzol. In brief, Schimerka's strategy was to develop a leaching series of four tanks. Each batch of powdered leaf material would be treated in four successive tanks, with three of the tanks using recycled rather than fresh benzol. At the end of the process, chemists would recover as much benzol as possible, leaving the exhausted, rubber-less powder behind. Schimerka to TAE, 22 November 1929, box 163, EFWE; Schimerka to TAE, 16 January 1930, 21 January 1930, 6 February 1930, box 16; and Schimerka to TAE, box 37, EBRC, ENHS.
117. Notes from interview of TAE, 28 March 1930, document available at http://www.eBay.com, closing date 26 November 2006, accessed 20 November 2006.
118. Wheeler McMillen, "Edison Hunts a *Rubber Crop* for Your Farm," *Farm and Fireside* 52 (June 1928): 10–11, 47–48; and *NYT*, 18 June 1929. See also Wheeler McMillen to TAE, 8 February 1928, box 18, EBRC, ENHS; and *FMNP*, 5 June 1928. McMillen met with Henry Ford that day as well, which may have played a role in Ford's subsequent decision to become a significant promoter of chemurgy on the national level.
119. Fred Scheffler to William H. Meadowcroft, 25 March 1930 and 14 April 1930, box 16, EBRC, ENHS.
120. Bureau of Standards, "Report on Paper-making Quality of Goldenrod Stalks"; and S. W. Stratton to Charles Edison, 20 January 1930, box 163, EFWE. Sample of the goldenrod paper is found in the same location.
121. *NYT*, 3 December 1929 and 5 December 1929; and *Fort Myers Press*, 2 November 1929 and 20 November 1929.
122. *Fort Myers Press*, 16 December 1929 and 28 December 1929.
123. S. T. Moore to TAE, 15 December 1929, box 12, EBRC, ENHS.
124. S. T. Moore to JVM, 13 January 1930, box 12, EBRC, ENHS. Edison responded with an offer to reward the plant hunters with a twenty-five-dollar bonus for any new crop yielding over 5 percent rubber. Note by TAE, box 37, ENHS; and Walter N. Archer to JVM, 18 March 1930, box 1, EBRC, ENHS.

125. TAE to Walter N. Archer, 23 September 1930; TAE to Archer, 30 September 1930 (all three letters with this date); and TAE to Archer, 1 October 1930, box 1, EBRC, ENHS; TAE to Miss Caroline Dorman, 2 February 1931, box 4, EBRC, ENHS; and *FMNP*, 24 September 1930.
126. Mina Edison to Charles Edison, 28 April 1931, box 1, Charles Edison Fund Collection, ENHS.
127. Layoffs and salary reductions discussed in JVM to Walter N. Archer, 13 July 1931, box 2, EBRC, ENHS.
128. Earl B. Babcock to JVM, 28 October 1931, box 7, EBRC, ENHS.
129. *NYT*, 16 June 1931, 11 July 1931, 22 October 1931; and "Edison's Last Experiment," *IRW* 85 (November 1931): 52
130. *Plant Patents*, report to accompany S 4015, 71st Congress, 2nd sess., 2 April 1930.
131. John G. Townsend Jr. to TAE, telegram 25 February 1930; and John G. Townsend Jr., "The Importance of Plant Patents to Agriculture," undated press release, [1930?], box 17, EBRC, ENHS; and Edison telegram of 26 February 1930 quoted in *Congressional Record*, Senate, 71st Congress, 2nd sess., 6764. See also TAE to W. Sam Clark, 29 July 1927, box 4, EBRC, ENHS.
132. *NYT*, 23 February 1930 and 20 May 1930; "Patents on Plants," *Science* 71, supplement 14 (25 April 1930): xiv; "Millions for Plant Inventors," *Popular Mechanics* 54 (November 1930): 763–765; and "Patenting of Plants Promises Big Profits—and Big Problems," *Business Week* (16 August 1931): 26–29. For background analysis, see Cary Fowler, "The Plant Patent Act of 1930: A Sociological History of Its Creation," *Journal of the Patent and Trademark Office Society* 82 (2000): 621–644; Bugos and Kevles, "Plants as Intellectual Property"; and Kevles, "Patents, Protections, and Privileges."
133. Ukkelberg, quoted in Vanderbilt, *Thomas Edison, Chemist*, 316.
134. The Department of Labor did not calculate unemployment figures on a monthly basis in 1931. Estimate based on correspondence with labor historian Judson MacLaury, 12 September 2005.
135. Draft of Annual Report for Year Ending 31 December 1931, box 6, EBRC, ENHS. On early 1930s progress in synthetic-rubber research, see Peter J. T. Morris, *The American Synthetic Rubber Research Program* (Philadelphia: University of Pennsylvania Press, 1989), 7–16; and John K. Smith, "The Ten-year Invention: Neoprene and Du Pont Research, 1930–1939," *Technology and Culture* 26 (January 1985): 34–47.

Chapter 4   The Nadir of Rubber Crop Research, 1928–1941

1. Dwight Eisenhower, "Guayule Diary," in *Eisenhower: The Prewar Diaries and Selected Papers, 1905–1941*, ed. Daniel D. Holt (Baltimore: Johns Hopkins University Press, 1998), 114–125. See especially entries of 21 and 27 April 1930.
2. Emphasis in original. Major Gilbert Van B. Wilkes and Major Dwight Eisenhower, "Report of Inspection of Guayule Rubber Industry," in Holt, *Eisenhower*, 126–138. Also in Gilbert Van B. Wilkes and Dwight Eisenhower, "Report of Inspection of Guayule Rubber Industry," 6 June 1930, box 81, E135f, Division of Plant Exploration and Introduction, General Correspondence, 1900–1940, RG54, Bureau of Plant Industry, NA (hereafter cited as E153f, RG54, NA).
3. GHC to Patrick J. Hurley, 26 December 1930, box 34, E191, Records of the Assistant Secretary of War, War Planning Branch, RG107, Department of War, NA (hereafter cited as E191, RG107, NA).
4. Lawrence, *The World's Struggle with Rubber*, 67.

5. O. F. Cook to Karl. F. Kellerman, telegram, 29 October 1931; and Kellerman to Cook, 30 October 1931, box 291, E2, RG54, NA.

6. Memorandum, 30 April 1932, 838.6176/30, decimal file, 1930–1939, RG59, U.S. Department of State, NA.

7. Walter D. Marcuse, *Through Western Madagascar in Quest of the Golden Bean* (London: Hurst and Blackett, 1914).

8. Charles F. Swingle, "Across Madagascar by Boat, Auto, Railroad, and Filanzana," *National Geographic* 56 (August 1929): 178–211.

9. Walter T. Swingle to Charles F. Swingle, 20 April 1928, box 1, Charles Fletcher Swingle Papers, Hunt Institute for Botanical Documentation, Carnegie Mellon University, Pittsburgh, Pennsylvania (hereafter cited as C. F. Swingle Papers).

10. Knowles A. Ryerson to Charles F. Swingle, 11 May 1928 and 22 May 1928, box 1, C. F. Swingle Papers.

11. Charles F. Swingle, "Plant Exploration Trip to Madagascar, 1928," box 1, C. F. Swingle Papers. Swingle's published report on his trip, "Across Madagascar by Boat, Auto, Railroad, and Filanzana," covers much the same material, although in somewhat less frank and less revealing language. For more on natives' resistance to foreign penetration, see Jeffrey C. Kauffman, "*La Question des Raketa*: Colonial Struggles with Prickly Pear Cactus in Southern Madagascar, 1900–1923," *Ethnohistory* 48 (Winter–Spring 2001): 87–121.

12. Charles F. Swingle, "The Anatomy of *Euphorbia Intisy*," *Journal for Agricultural Research* 40 (April 1930): 615–625; and "Planting the Seeds for a Possible Revolution in the American Rubber Industry," *St. Louis Post-Dispatch*, 18 February 1934, clipping, box 18, EBRC, ENHS.

13. Charles F. Swingle and Eugene May Jr., "Experiments in Propagating *Euphorbia Intisy*," draft manuscript; and Charles F. Swingle to Henri Humbert, 12 June 1930, box 1, C. F. Swingle Papers.

14. Charles F. Swingle, "The Present Status of Euphorbia Intisy," 16 June 1930; Charles F. Swingle, "Memorandum for Dr. Magness Regarding Euphorbia Intisy Propagation," 16 June 1930; and O. F. Cook to [?], 24 June 1930, box 1, C. F. Swingle Papers; and R. E. Beckett, "Lost Plant Material, Bard, California, March 1934," box 39, E90, RG54, NA.

15. GHC to BMB, 23 April 1927; and GHC to C. Bakker, 18 November 1927, series 6, vol. 18, Baruch Papers.

16. GHC to BMB, 29 November 1927, attached to GHC to BMB, 20 December 1927, series 6, vol. 18, Baruch Papers.

17. GHC to Claudius H. Huston, 8 January 1927, box 16, IRC Records. Huston was a close ally of President Hoover. He served as assistant secretary of commerce from 1921 to 1923, went to Singapore in 1922 in an effort to forestall Britain's plans to raise rubber prices, and wrote an article that embraced Hoover's challenge to the Stevenson Plan. See C. H. Huston, "America's Dependence on Britain for Rubber," *Saturday Evening Post* 191 (26 May 1923): 22–23. Huston later played a leading role in Hoover's presidential campaign of 1928, and he was Hoover's choice to take over the Republican Party in September 1929. Connected with several financial improprieties, he resigned in embarrassment in 1930. Carnahan had also pushed the guayule agenda with John Raskob, national chairman of the Democratic Party.

18. At about the same time, Congress rejected HCH's efforts to overturn the Webb-Pomerene Act, a move that prevented formation of a conglomeration of rubber buyers dominated by American industrialists.

19. Ampar was short for "American parthenium," a moniker derived from the botanical name of the guayule plant, *Parthenium argentatum*.

20. GHC to E. G. Liebold, 12 January 1928, box 757, accession 285, Henry Ford Office Correspondence, The Henry Ford; and GHC to Claudius H. Huston, 7 December 1927, box 16, IRC Records.
21. "Ampar Balloon Tires," *IRW* 77 (January 1928): 65–66.
22. Cartoon from *Oakland Tribune*, 13 January 1928, clipping, box 10; and GHC to David Spence, telegram, 23 January 1928, box 1, E91, RG54, NA.
23. GHC to S. H. Starr, 17 February 1928; and GHC to Otis Woodward, 25 June 1928, box 7; GHC to George E. Murrell, 28 December 1926; GHC to Murrell, 7 February 1927; and GHC to J. M. Williams, 29 March 1927, box 10, E91, RG54, NA.
24. R. B. Smith to J. M. Williams, 18 August 1928; Williams to GHC, 26 May 1931; GHC to chief, Bureau of Chemistry, 25 May 1934; and GHC to Williams, 15 November 1934, box 11, E91, RG54, NA.
25. GHC to P. P. Thomas, 21 August 1930, box 10, E91, RG54, NA.
26. GHC to David Spence, 16 April 1928; GHC to B. L. Atwater, 17 January 1929; and A. B. Perlin to GHC, 7 February 1929, box 5; and GHC to Millard L. Caldwell, 23 October 1931; Caldwell to GHC, 29 October 1931; Caldwell to GHC, 12 December 1931; E. R. Pickert to IRC, 30 December 1931; Pickert to IRC, 6 January 1932; and GHC to Caldwell, 7 January 1932, box 10, E1, RG54, NA.
27. *NYT*, 20 March 1929 and 31 March 1929. Kidnappers demanded thirty thousand pesos ransom for T. L. Carnahan. Several days later, after federal troops paid ten thousand pesos, they stormed the house where Carnahan had been held and secured his release.
28. GHC to Claudius H. Huston, 5 November 1929; Huston to GHC, 15 November 1929; and GHC to Charles Hayden, 12 December 1929, box 16, IRC Records.
29. GHC to David Spence, 27 March 1930, box 16, IRC Records.
30. Dwight Eisenhower to General Mosely, 22 March 1930, file 336-Commodity Committees, Rubber, box 34, E191, RG107, NA. Recall that the Plant Patent Bill, discussed in chapter 3, did not become law until May 1930.
31. Paul A. C. Koistinen, *Planning War, Pursuing Peace: The Political Economy of American Warfare, 1920–1939* (Lawrence: University Press of Kansas, 1998), 123. I have found no archival evidence that elucidates the War Department's reaction to Eisenhower and Wilkes's glowing report. It is evident, however, that most officials ignored Carnahan's pleas for action. See GHC to Patrick J. Hurley, 26 December 1930, box 34, E191, RG107, NA; and GHC to Dwight Eisenhower, 26 December 1930, box 16, IRC Records.
32. "Guayule Extraction Mill," *IRW* 83 (1 March 1931): 53–55; Carnahan and Manning, "American-grown Rubber Produced from Guayule," *Chemical and Metallurgical Engineering* 38 (March 1931): 128; and Blythe, "Taming the Wild Guayule," 28, 30, 106, 109–110.
33. GHC to William B. McCallum, 31 December 1931, box 6, E91, RG54, NA.
34. IRC, *Annual Report*, 1930–1932. See also *Salinas Index-Journal*, 5 April 1941, Guayule File, Monterrey County Historical Society, Salinas, California.
35. GHC to BMB, 6 December 1931, series 6, vol. 26, Baruch Papers; and GHC to Patrick J. Hurley, 4 May 1932, box 34, E191, RG107, NA.
36. GHC to HCH, 1 May 1931; GHC, "Memorandum for the President Re: American Crude Rubber," [1931?]; GHC to Theodore Joslin, 30 April 1932; and GHC to Theodore Joslin, 7 May 1932, Presidential Papers, box 276, HHPL. See also GHC to Grayson Murphy, 15 October 1931, box 16, IRC Records.
37. Nikolai Ivanovich Vavilov, *Five Continents*, trans. Doris Löve (Rome: International Plant Genetic Resources Institute, 1997), 147. According to Löve, Vavilov

intended to devote sections of this work to his study of guayule in Mexico and California, and to his meetings with Edison, Fairchild, and other American rubber plant experts, but it appears that major portions of the manuscript have been lost.

38. *Congressional Record*, Senate, 29 May 1933, 73rd Congress, 1st sess., 2963, 4454. The two resolutions were referred to Senate committees, where they stalled.

39. William B. McCallum to GHC, 18 October 1933; GHC to J. M. Williams, 21 October 1933; GHC to Williams, 9 December 1933; GHC to Rexford G. Tugwell, 6 June 1934; and GHC to William D. Lupe, 26 February 1935, box 16, IRC Records.

40. *Salinas Index-Journal*, 15 September 1934, Guayule File, Monterrey County Historical Society, Salinas, California.

41. GHC to HAW, 18 September 1934; GHC to BMB, 3 October 1934; BMB to GHC, 4 October 1934; and GHC to BMB, 9 October 1934, series 6, vol. 33, Baruch Papers.

42. For an excellent analysis of the International Rubber Regulation Committee, see Gibson Bell Smith, "Rubber for Americans: The Search for an Adequate Supply of Rubber and the Politics of Strategic Materials" (PhD diss., Bryn Mawr College, 1972).

43. HAW, *New Frontiers* (New York: Reynal and Hitchcock, 1934), 72.

44. GHC to Frederick Osborn, 18 October 1934; GHC to Frederick P. Garvan, 18 October 1934; and GHC to William B. McCallum, 16 January 1936, box 16, IRC Records.

45. J. H. Linxweiler, "Riensch and Held," 1 August 1947, box 17, IRC Records.

46. Earl B. Babcock to Harvey S. Firestone Jr., 16 September 1938, folder: Pirelli-Societa-Italiana, Firestone Records.

47. Mario Parducci, "Il Problema della Gomma Elastica le Piante Secondarie," *L'Ingegnere* 16 (January 1938): 2–7.

48. *Salinas Index-Journal*, 4 February 1941; and IRC, *Annual Report* 1935 and 1937. Figures for 1941 from HGA to BMB, 18 July 1941, series 12, box 193, General Correspondence, Baruch Papers.

49. Based on analysis of the 1931 crop results maps, flat file 9, EFWE; and Walter N. Archer to JVM, 26 October 1931, Incoming Rubber Reports, 18 July 1931 to 31 December 31, box 37, EBRC, ENHS.

50. [JVM], draft of report to stockholders, 31 December 1931, box 6, EBRC, ENHS.

51. JVM to Walter N. Archer, 14 January 1932; JVM to Archer, 19 April 1932, JVM to Archer, 13 June 1932; JVM to Archer, 15 June 1932; and Archer to JVM, 28 June 1932, binder: July 23 1931 to August 29 1932, box 168, EFWE; and JVM to Walter M. Buswell, 18 May 1932; and Buswell to JVM, 18 July 1932, box 3, EBRC, ENHS.

52. Earl B. Babcock to Harvey S. Firestone Jr., 26 July 1932, folder: Edison Botanical Research Corporation, 1931–1934, Firestone Records.

53. JVM to E[lvin] C[harles] Stakman, 15 July 1932, 1932 binder, item 181, EFWE; and "Reminiscences of H. G. Ukkelberg," accession 65, Oral Histories, The Henry Ford. Stakman had already trained several of the men who directed Firestone's rubber operations in Liberia.

54. JVM to HGU, 19 August 1932, 1932 binder, item 181, EFWE.

55. O. F. Cook, "Saint Luther: A Burbank Cult, with an Account of his Wonderworking Methods of Plant Breeding," *Journal of Heredity* 20 (1929): 309–318. Cook had briefly been one of Burbank's colleagues in 1905. See Peter Dreyer, *A Gardner Touched with Genius: The Life of Luther Burbank*, rev. ed. (Berkeley and Los Angeles: University of California Press, 1985), 132–133.

56. Paolo Palladino, "Wizards and Devotees: On the Mendelian Theory of Inheritance and the Professionalization of Agricultural Science in Great Britain and the United States, 1880–1930," *History of Science* 32 (1994): 409–444. On Burbank's continued resonance among the American public, see Katherine Pandora, "Knowledge Held in Common: Tales of Luther Burbank and Science in the American Vernacular," *Isis* 92 (September 2001): 484–516.

57. HGU to JVM, 9 November 1932, binder 1932, item 181, EFWE; HGU to JVM, 11 November 1933, binder 1933, item 182, EFWE; and HGU, "The Propagation and Breeding of Goldenrod," [April 1936?], box 173, EFWE.

58. To combat the threat, the young scientist initiated a number of precautions, such as treating all pups and cuttings in "Semesan," a common mercurial fungicide.

59. HGU to JVM, 16 December 1932, 1932 binder, item 181, EFWE; JVM to HGU, 16 January 1933; and HGU to JVM, 21 January 1933, 1933 binder, item 182, EFWE; and HGU to JVM, 2 December 1933, binder 5, box 170, EFWE.

60. HGU to JVM, 12 November 1932, 14 November 1932, 22 December 1932; and Walter N. Archer to JVM, 22 December 1932, binder: July 23, 1931 to August 29, 1932, box 168, EFWE. The natural habitat of *Solidago leavenworthii* was swampy lowlands, and it did not thrive as well in the sandy, dry soils of the Fort Myers facility. Walter N. Archer to JVM, 6 April 1933, binder 4, box 170, EFWE; and [JVM?] to Henry Ford, 18 November 1932, box 8, ENHS. For more information on the water issues, see Rosenblum, *Historic Structures Report*, I-88–I-90.

61. William A. Taylor to JVM, 23 October 1932; JVM to HGU, 23 October 1932; JVM to HGU, 3 November 1932; HGU to JVM, 9 November 1932; C. B. Doyle to HGU, 23 November 1932; and Doyle to HGU, 3 December 1932, 1932 binder, item 181, EFWE. Taylor and Miller first discussed cooperating on goldenrod research in May 1932. See also William A. Taylor to JVM, 21 May 1932, box 18, EBRC, ENHS.

62. E. G. Liebold to JVM, 30 November 1932, 1932 binder, item 181, EFWE.

63. HGU to JVM, 22 April 1933, box 170, EFWE; and HGU, "Report of Work at Edison Botanic Gardens, Ft. Myers, Fla.," [April 1933], binder 4, box 170, EFWE. One hundred twelve beds showed a higher average rubber yield than the year before; only twenty-six showed lower yields. E.P.C. 573-A yields rose especially rapidly—its highest rubber yield was 4.5 percent in 1931, but it was 9.69 percent in 1932.

64. Analysis of 1933 crop maps, flat file 9, EFWE. HGU, "Report of Work at Edison Botanic Gardens, Ft. Myers, Fla."; HGU to JVM 19 January 1933; and HGU to JVM, 12 April 1933, binder 4, box 170, EFWE.

65. Results for the related variety, E.P.C. 573, showed only a small decline—from 67 percent of the samples testing over 7 percent rubber in 1932 to 65 percent in 1933—but the lack of progress was nevertheless a disappointment. Results for the Moore, Florida, 502 variety followed a similar pattern—from 10 percent testing over 7 percent rubber in 1932 down to 4 percent in 1933. See "Tabulation of Number of Plants Tested by Years," box 176, EFWE.

66. HGU to JVM, 22 October 1932; and C. B. Doyle to HGU, 3 December 1932, 1932 binder, item 181, EFWE; HGU to JVM, 20 May 1933; and HGU to JVM, 2 December 1933, binder 1933, item 182, EFWE; and Mildred Pladeck Jr. to HGU, 15 August 1933, box 165, EFWE. See also Ernest Brown Babcock and Roy Elwood Clausen, *Genetics in Relation to Agriculture*, 2nd ed. (New York: McGraw-Hill, 1927), 384. Ukkelberg also urged Miller to consult with HSF's friend (and Ukkelberg's mentor), plant pathologist E. C. Stakman, for confirmation. See also HGU to JVM, 6 October 1933; HGU to JVM, 20 October 1933; HGU to JVM,

2 December 1933; and "General Conclusion Regarding E.P.C. 573A" [December 1933], binder 5, box 170, EFWE.

67. Harvey S. Firestone Jr. to JVM, 11 April 1933; and Earl B. Babcock to Harvey S. Firestone Jr., 8 January 1934 (two letters that date), folder: Edison Botanical Research Corporation, 1931–1934, Firestone Records.

68. EBRC, "General Report on Research on Season of 1933," box 179, EFWE.

69. EBRC, "Report for First Half of 1934"; and HSF to Harvey S. Firestone Jr., 5 April 1934, box 163, EFWE.

70. JVM to HGU, 16 April 1934, binder 1934, item 183, EFWE; and JVM to Henry Ford, 17 April 1934; and JVM to Henry Ford, 26 April 1934, binder 2, box 172, EFWE.

71. David Fairchild, *The Barbour Lathrop Plant Introduction Garden* (Washington, DC: n.p., 1928), 11–16; W. H. Hodge, H. F. Loomis, Lloyd E. Joley, and John L. Creech, "Federal Plant Introductions Gardens," *National Horticultural Magazine* 35 (April 1956): 86–92; David Fairchild, *The World Was My Garden: Travels of a Plant Explorer* (New York: Scribner's, 1938), 449; *SMN*, 30 November 1930; *Savannah Evening Press*, 26 May 1932; and *NYT*, 11 March 1930.

72. Mina Edison still had a sentimental attachment to her husband's final project and hoped to make the facility "a sort of permanent thing, an object of interest to the people in Fort Myers." In her own search for an alternative to transferring the experiments to Savannah, she asked if it might be possible to secure more government funding in the next budget year, so that the government would keep the experimental work in Fort Myers. See [JVM], "Notes from Talk with Mrs. Edison," 1 May 1934; and JVM to Charles Edison, 1 May 1934, binder 2, box 172, EFWE.

73. JVM to Karl G. Pearson, 13 July 1934; and HGU to JVM, 28 November 1934, 1934 binder, item 183, EFWE; and HGU to JVM, 6 April 1935, 1935 binder, item 184, EFWE.

74. "Rubber Growing in the United States," *RA* 36 (October 1934): 23–24.

75. JVM to Mrs. Thomas A. Edison, 25 July 1934, box 5, EBRC, ENHS; and Vanderbilt, *Thomas Edison, Chemist*, 302. The EBRC sent two of these pounds to the U.S. Bureau of Standards for testing, and it kept two pounds on hand as "exhibit specimens."

76. HAW to Franklin D. Roosevelt, 31 January 1935, box 2236, E17H, RG16, NA.

77. Franklin D. Roosevelt to HAW, 19 February 1935, quoted in Smith, "Rubber for Americans," 41.

78. Mina Edison to Charles Edison, [8 March 1935?], box 1, Charles Edison Fund Collection, ENHS.

79. HAW to president, [February 1936]; Paul H. Appleby to Malvina T. Schneider, 12 February 1936; HAW to [George H. Dern], 14 May 1936; and M. L. Wilson to [Cordell Hull], 12 August 1936, box 2418, E17H, General Correspondence of the Office of the Secretary, RG16, NA.

80. In several cases the plots yielded rubber at a rate of two hundred, and in some cases even three hundred, pounds per acre. HGU, "Quarterly Report of the Edison Botanic Garden, Ft. Myers, Fla. April 1, to July 1, 1934," box 5, EBRC, ENHS; and [HGU], "General Report of Experimental Work at the Edison Botanic Garden. Fort Myers Florida. July 1 to December 31 1934," 1935 binder, item 184, EFWE. See also Anne Jenkins and HGU, "Scab of Goldenrod by Elsinoe," *Journal of Agricultural Research* 51 (June 1935): 515–525.

81. HGU to JVM, 15 February 1935, 1935 binder, item 184, EFWE.

82. Handwritten notes by JVM regarding discussion with Charles Edison on 28 May 1936 and with Mrs. Mina Edison Hughes on 29 May 1936, box 5, EBRC, ENHS.

83. Minutes of meeting of the board of directors of EBRC, 1 July 1936, binder 1936, item 185, EFWE.

84. McMillen, *New Riches from the Soil*, 29.

85. *Proceedings of the Dearborn Conference of Agriculture, Industry, and Science* (New York: Chemical Foundation, 1935), 30–39; McMillen, *New Riches from the Soil*, 33; and *NYT*, 8 May 1935. See also Pursell, "The Farm Chemurgic Council"; Wright, "Waste Not, Want Not"; Hale, *Farmward March*; Borth, *Pioneers of Plenty*; and Mark R. Finlay, "Old Efforts at New Uses: A Brief History of Chemurgy and the American Search for Biobased Materials," *Journal of Industrial Ecology* 7, nos. 3–4 (2004): 33–46.

86. Ways Station is now the city of Richmond Hill. JVM to HGU, 16 April 1934; HGU to JVM, 28 November 1934; and JVM to HGU, 30 November 1934, 1934 binder, item 183, EFWE; and JVM to HGU, 6 June 1935, 1935 binder, item 184, EFWE.

87. HGU to JVM, 25 May 1936; and HGU to JVM, 13 June 1936, binder 1936, item 185, EFWE.

88. HGU to E. G. Liebold, 6 June 1936, box 1897, accession 285, The Henry Ford.

89. "Reminiscences of H. G. Ukkelberg."

90. "Goldenrod Research Reports, 1936–1942 and 1944," Frank S. McCall Papers, Richmond Hill Historical Society and Museum, Richmond Hill, Georgia; and HGU to E. G. Liebold, 4 January 1937, box 2015, accession 285, The Henry Ford. See also Franklin Leslie Long and Lucy Bunce Long, *The Henry Ford Era at Richmond Hill, Georgia* (n.p.: Long and Long, 1998); Buddy Sullivan, *From Beautiful Zion to Red Bird Creek: A History of Bryan County, Georgia* (n.p.: Bryan County Board of Commissioners, 2000), 303–309; and Bryan, *Beyond the Model T*, 186–198.

91. Earl B. Babcock to Harvey S. Firestone Jr., 28 February 1940, folder: Rubber Development Corporation, 1940–1944, Firestone Records; "Summary of Goldenrod Investigations at U. S. Plant Introduction Garden, Savannah, Georgia, 1934–1940"; and "Goldenrod as an Emergency Source of Natural Rubber," box 5, E90, RG54, NA; and *SMN*, 31 January 1940 and 27 March 1940.

92. R. E. Beckett, R. S. Stitt, and E. N. Duncan, "Rubber Content and Habits of a Second Desert Milkweed (*Asclepias erosa*) of Southern California and Arizona" (USDA, technical bulletin no. 604, 1938). See also *NYT*, 22 May 1932; and "Rubber Content of Desert Milkweed," *The Scientific Monthly* 46 (June 1938): 574–575.

93. David W. Figart to JVM, 19 March 1934, 3 April 1934, and 2 May 1934, box 7, EBRC, ENHS. The Luling Foundation had cooperated with the EBRC in 1929 by planting milkweed and goldenrod. See Jack Shelton to JVM, 13 May 1929, box 11, EBRC, ENHS. See also Zona Adams Withers, *A History of the Luling Foundation, 1927–1982* (n.p.: Luling Foundation, 1982).

94. Chester Kennison to Governor David Sholtz, 14 April 1933 and 28 November 1934; Wilmon Newall to J. P. Newell, 27 December 1934; and other documents in an extensive set of correspondence on the poinsettia issue on microfilm reel 17.3, Administrative Policy Records, 1905–1962, Office of the Director, Institute of Food and Agricultural Sciences, Subseries 90A, Public Records Collection, University of Florida Archives, Gainesville. See also "Claims for Rubber from Poinsettia Unwarranted," *IRW* 97 (December 1937): 63.

95. *Washington Star*, 6 November 1934 and 20 January 1935, clippings; and Loren G. Polhamus to W. L. Finger, 27 July 1940, box 80, E135f, RG54, NA. Pitman claimed poinsettia yielded 14 percent rubber, when the USDA's data in fact showed an average of 0.14 percent.

96. Press release from Science Service, 8 November 1934, box 80, E135f, RG54, NA.
97. Abramo, "The Military and Economic Potential of the United States," 148–150; and Koistinen, *Planning War, Pursuing Peace*, 253–304.
98. Henry Latrobe Roosevelt was a sixth cousin of the president and the fourth member of the Roosevelt family to serve as assistant secretary of the navy. During his tenure in this post, Henry Roosevelt frequently urged Americans to improve military and naval preparedness. See *NYT*, 16 March 1933, 25 September 1935, and 23 February 1936. For Edison's role in this post, see John D. Venable, *Out of the Shadow: The Story of Charles Edison* (East Orange, NJ: Charles Edison Fund, 1978), 120–163. Also pertinent, Charles Edison had been a leader in the American-Soviet Chamber of Commerce, an organization that embraced friendly trade relations with the Soviet Union in the 1930s and touted Soviet successes with kok-sagyz and other botanic rubber sources.
99. BMB to Charles Edison, 8 March 1937, cited in Smith, "Rubber for Americans," 153.
100. See Staley, *Raw Materials in Peace and War*, 84, 173–178. Yet even Brooks Emeny, author of an important text that warned of America's dependence on imported raw materials, suggested, "It is impossible to conceive of a complete severance of imports." See Emeny, *The Strategy of Raw Materials*, 137.
101. Alonzo Hamby, *For the Survival of Democracy: Franklin Roosevelt and the World Crisis of the 1930s* (New York: Free Press, 2004), 98–99. For DuPont's announcement and its aftermath, see *NYT*, 4 November 1931 and 15 November 1931.
102. *NYT*, 24 April 1940, 13 September 1940, and 13 December 1940.
103. Fleming MacLeish and Cushman Reynolds, *Strategy of the Americas* (New York: Duell, Sloan, and Pearce, 1941), 21.
104. See P. H. Groggins, "Report on 'The Rubber Situation in the United States, 1938,'" box 85, File 40281, E32, General Correspondence, 1935–1939, RG97, Bureau of Agricultural and Industrial Chemistry, NA.
105. Marshall, *To Have and Have Not*, 33–35; and Smith, "Rubber for Americans," 139–181, Fish quoted on p. 178.
106. Charles Edison and Louis Johnson to HAW, 25 August 1939; and HAW to Charles Edison and Louis Johnson, 25 August 1939, reel 21, HAW Papers at the University of Iowa, microfilm edition.
107. MacLeish and Reynolds, *Strategy of the Americas*, 22. These authors predicted, rather accurately, that the six-month stockpile could be stretched long enough to last until synthetic rubber came online once a war emergency actually arrived.
108. Smith, "Rubber for Americans," 112, 142, 149; "Rubber Growing in the United States," 23–24; "Rubber Plant Experiments by Federal Scientists," *Scientific American* 152 (July 1935): 79; and *NYT*, 2 April 1939. Chapman Field also had success with acclimatization and tapping experiments on *Castilloa* rubber, a species immune from the rubber blight. Some promoters saw this as a sign that southern Florida could become the center of an American *Castilloa* rubber industry. See HAW to Pat Cannon, 15 May 1939; and [Colonel Robert H. Montgomery] to Cannon, 22 May 1939, Bertram Zuckerman Garden Archives, Fairchild Tropical Garden Research Center, Miami, Florida.
109. Earl N. Bressman to HAW, 8 February 1938, box 2869, E17J, General Correspondence of the Office of the Secretary, 1938, RG16, NA.
110. HAW to BMB, 29 September 1939, series 6, vol. 46, Baruch Papers.
111. Smith, "Rubber for Americans," 178.
112. Hull quoted in Marshall, *To Have and Have Not*, 1.

113. A. L. Viles, "The Rubber Industry," lecture delivered at the Army Industrial College, 16 November 1939, http://www.ndu.edu/library/ic7/L40a-013.pdf, accessed 20 March 2005.
114. *NYT*, 24 April 1940.
115. Marshall, *To Have and Have Not*, 130.
116. Signed editorial by S. C. Stillwagon, "Rubber, A Strategic Material," *IRW* 102 (July 1940): 52.
117. Marshall, *To Have and Have Not*, 67–68.
118. W. L. Finger to Mr. Stettinius, 29 July 1940, box 1891, E1 Policy Documentation Files, RG179, WPB, NA (hereafter cited as E1, RG179, NA). See also Herbert and Bisio, *Synthetic Rubber*, 49–51; and Jesse Jones, *Fifty Billion Dollars: My Thirteen Years with the RFC (1932–1945)* (New York: Macmillan, 1951), 404–405.
119. See Chalk, "The United States and the International Struggle for Rubber," 222–273.
120. Smith, "Rubber for Americans," 182–242. For a defense of his stockpiling program, see Jones, *Fifty Billion Dollars*, 396–401. See also Neushul, "Science, Technology, and the Arsenal of Democracy," 69–73.
121. E. W. Brandes to Eubanks Carsner, 26 August 1940, box 38, E90, RG54, NA. See also M. A. McCall to P. S. Burgess, 30 July 1940, box 208, E135f, RG54, NA; and [Chester Davis] to W. Averill Harriman, 19 August 1940, box 1899, E1, RG179, NA.
122. Dean, *Brazil and the Struggle for Rubber*, 87–88; and Edward O. Guerrant, *Roosevelt's Good Neighbor Policy* (Albuquerque: University of New Mexico Press, 1950), 198–199. See also *NYT*, 22 February 1941. Following the Dutch surrender, the Germans officially occupied the Netherlands on 17 May 1940.
123. Several files in the papers of University of Michigan botanist H. H. Bartlett describe these efforts. See for instance, "Memorandum to Rubber Scientists," 16 September 1940, box 6, Harley Harris Bartlett Papers, Bentley Historical Library, University of Michigan, Ann Arbor. For an especially engaging account, see Davis, *One River*.
124. For reactions to efforts to cut the rubber crop budget, see Earl B. Babcock to Harvey S. Firestone Jr., 28 February 1940; and T. S. Markey to Firestone, 25 March 1940, folder: Rubber Development Corporation, 1940–1944, Firestone Records. See also "Reductions Made by House Bill Below the Budget Estimates for 1940," series 9, subseries B, Richard B. Russell Jr. Papers, Richard B. Russell Library, University of Georgia, Athens; and Harold Loomis to David Fairchild, 20 May 1940, Fairchild Papers.
125. "Rubber Possibilities in the United States" [April 1940?], box 104, E90, RG54, NA. The most likely author of this report was Loren Polhamus, the USDA employee who worked most extensively on rubber issues from the 1930s to the 1960s.
126. *Salinas Index-Journal*, 29–30 April 1940.
127. Anderson quoted in *Congressional Record*, House, 13 June 1940, 76th Congress, 3rd sess., 8522–8524. See similar arguments in *Congressional Record*, House, 22 June 1940, 76th Congress, 3rd sess., 9082. Anderson delivered one of these congressional speeches on guayule on the very day that France surrendered much of its territory to German control. It is pertinent to note that Anderson framed the issues in terms of the politics of New Deal farm relief, rather than the rhetoric of war preparedness. Guayule, he suggested, offered the opportunity to develop new crops for arid lands and "a real chance to assist the American farmer" as part of a solution to unemployment and commodity crop surpluses.

128. *Defense of the United States and Other Nations in the Western Hemisphere (Rubber), Hearings Before the Committee on Military Affairs, United States Senate, 76th Congress, 3rd Sess., on S 4082* (Washington, DC: Government Printing Office, 1940), 26–29.

129. BMB to HAW, 4 April 1940; HAW to BMB, 9 April 1940; and BMB to HAW, 15 April 1940, series 6, vol. 50, Baruch Papers.

130. *NYT*, 14 June 1940.

131. BMB to Sheridan Downey, 14 June 1940, series 6, vol. 47, Baruch Papers.

132. "Notes on Meeting with Intercontinental Rubber Company, 9 July 1940," box 1901, E1, RG179, NA; GHC to Clarence Francis, 13 July 1940, series 6, vol. 47, Baruch Papers; and GHC to William McCallum, 25 July 1940, box 16, IRC Records.

133. *Congressional Record*, House, 13 June 1940, 76th Congress, 3rd sess., 8522–8524.

134. HAW to John Z. Anderson, 10 July 1940, quoted in *Congressional Record*, House, 16 April 1941, 77th Congress, 1st sess., 3139. See also Anderson to HAW, 24 June 1940, *Congressional Record*, House, 77th Congress, 1st sess., 3139.

135. H. O. Chute to HAW, 26 June 1940, box 208, E2, RG54, NA.

136. BMB to E. R. Stettinius Jr., 1 August 1940, box 1901, E1, RG179; BMB to Stettinius, 16 August 1940, quoted in BMB, *Baruch: The Public Years* (New York: Holt, Reinhart, and Winston, 1960), 301; and GHC to BMB, 21 October 1940, box 16, IRC Records.

137. "Notes on Meeting with Intercontinental Rubber Company, 9 July 1940"; and W. L. Finger to D. P. Morgan, C. C. Monrad, and E. W. Reid, 21 August 1940, box 1901, E1, RG179, NA.

138. W. L. Finger to Mr. Stettinius, 29 July 1940, box 1890, E1, RG179.

139. Henry Granger Knight Diary, 14 December 1940, Special Collections, National Agricultural Library, Beltsville, Maryland (hereafter cited as Knight Diary).

140. *Salinas Index-Journal*, 15 April 1941.

141. *Congressional Record*, House, 16 April 1941, 77th Congress, 1st sess., 3137–3142.

142. HGA to William B. McCallum, 15 April 1941; HGA, "Suggested Basis for Determination of a Subsidy for Domestic Production of Guayule Rubber," 8 May 1941; HGA to C. L. Baker, 10 May 1941; HGA to Baker, 17 May 1941; and HGA to Baker, 11 June 1941, box 16, IRC Records.

143. Claude Wickard to president, 28 April 1941, box 81, E135f, RG54, NA.

144. "Guayule Legislation Introduced in House," *IRW* 104 (July 1941): 54.

145. E. W. Brandes "Cultivation of Improved, Domesticated Guayule as an Emergency Source of Rubber," 6 June 1941, box 16, IRC Records; and Brandes to E. C. Auchter, 13 June 1941, and attachments, box 49, E90, RG54, NA.

146. Loren G. Polhamus, "Guayule as an Emergency Source of Crude Rubber, 16 July 1941," box 16, IRC Records.

147. James S. Adams to Clifford Townsend, 20 June 1941, box 1901, E1, RG179, NA.

148. HGA to Howard J. Klossner, 18 June 1941, series 6, vol. 51, 1941, Baruch Papers; and HGA, memorandum for the file, 20 June 1941; HGA, memorandum for the file, 27 June 1941; HGA, memorandum for the file, 24 July 1941; and HGA, memorandum for the advisory committee, 28 July 1941, box 16, IRC Records.

149. See Carl Hayden to secretary of agriculture, 29 August 1941; and [Claude Wickard] to Hayden, 3 September 1941, file 11645, box 119, E2, RG54, NA.

150. Earl B. Babcock, "Guayule as an Emergency Source of Rubber," 16 June 1941; and Harvey S. Firestone Jr., press release, 23 June 1941, folder: Guayule, 1941–1948, Firestone Records. Interestingly, the Firestone Tire and Rubber Company chose

to not revive its slogan from the 1920s: "Americans Should Produce Their Own Rubber."

151. HGA to BMB, 30 September 1941, series 6, vol. 51, Baruch Papers.

152. The U.S. Tariff Commission reported that for a twenty-million-dollar investment, it would be possible to obtain one hundred thousand tons of additional rubber. It would require one hundred million dollars to obtain the same amount from synthetic rubber. See U.S. Tariff Commission, *Rubber: Possibilities of Producing Rubber in the United States and Rubber Conservation* (Washington, DC: n.p., 1941).

153. Hauser had been a leading colloid chemist in Germany in the 1920s, where he developed a revolutionary method to concentrate and store the latex in natural rubber. In 1933, he fled to Austria, and in 1935, while on a world tour of rubber-industry sites, MIT president Karl Compton offered him the chance to join that faculty. Convinced that Germany soon would conquer Austria and that "the Nazis certainly would have done everything to liquidate me," Hauser accepted. His family emigrated soon thereafter. See [Ernst A. Hauser], "Twenty Years in the United States," unidentified speech, January 1955, in Ernst A. Hauser biographical file, MIT General Collections, MIT Museum, Cambridge, Massachusetts; and Hauser to A. N. Spanel, 6 September 1942, box 107, AC 4, Office of the President, 1930–1959, Compton-Killian, Institute Archives and Special Collections, MIT Libraries, Cambridge, Massachusetts (hereafter cited as Compton-Killian Papers).

154. Ernst A. Hauser, "Home-grown and Home-made Rubber," *IRW* 104 (September 1941): 27–29; and "Hits Synthetic Rubber Claims," *RA* 49 (July 1941): 272.

155. *NYT*, 12 September 1940. Hauser and IRC officials guessed that General Tire's guayule promotion campaign was an effort to divert attention from the real crisis; after all, the company had not purchased any guayule since the early 1930s. See GHC to John R. Morron, 20 September 1940; and Ernst A. Hauser to HGA, 2 January [1942], box 16, IRC Records.

156. *Salinas Index-Journal*, 25 November 1941; *DMN*, 17 November 1941; and "Jap War Spotlights Guayule," *Tires* (December 1941): 36.

157. *DMN*, 13 January 1941, 24 February 1941, and 3 March 1941.

158. Waldemar Kaempffert, "The World Has Just Begun," *American Magazine* 129 (January 1940): 42, 129–131; and "Rubber 'Grown' in Factories," *Popular Mechanics* 74 (December 1940): 846–847, 134A–135A.

159. In fact, synthetic-rubber production in 1942 was merely 22,434 long tons, about 3 percent of national demand. Herbert and Bisio, *Synthetic Rubber*, 127.

160. *DMN*, 2 November 1941.

Chapter 5   **Crops in War: Rubber Plant Research on the Grand Scale**

1. *NYT*, 7 December 1941.

2. Examples include Blythe, "Taming the Wild Guayule"; "American Rubber at Last," *Literary Digest* 110 (19 September 1931); "Rubber Plant Experiments by Federal Scientists," 79; "Rubber on Bushes," *Business Week* 579 (5 October 1940): 44–45; "Rubber Shrub," *Newsweek* (10 March 1941): 48–49; S. R. Winters, "Guayule," *Buick Magazine* 7 (May 1941): 3, 14; and Frank J. Taylor, "Uncle Sam's Rubber Farmer," *Country Gentleman* 65 (June 1941): 57.

3. See Neushul, "Science, Technology, and the Arsenal of Democracy," 158.

4. John Collyer, "Olin Hall Dedication Talk" (3 October 1942), box 20, John Lyon Collyer Papers, Department of Rare and Manuscript Collections, Cornell University, Ithaca, New York. See also *New York Herald Tribune*, 4 October 1942.

5. Neushul, "Science, Technology, and the Arsenal of Democracy," 158–160.

6. HGA to Howard J. Klossner, 19 December 1941, box 16, IRC Records.

7. N. H. Hunt to Loren G. Polhamus, 16 December 1941, box 80, E135f, RG54, NA; and A. D. Jackson, "Progress Report—Texas Agricultural Experiment Station—A and M College of Texas," (5 March 1942), Economic Botany Collection, Harvard University Libraries, Cambridge, Massachusetts. The latter group also retraced the work of soil scientist Hugh Hammond Bennett, who in 1927 spent several weeks trekking through lonely mesas and prairie dog towns in western Texas in search of guayule's preferred soil habitat. See entries of 19 July and 8 August 1927, in Field Diaries, box 17, Hugh Hammond Bennett Collection, Special Collections, Iowa State University, Ames.

8. Bayard L. Hammond to B. Y. Morrison, 3 January 1942, box 53, E90, RG54, NA.

9. Paul H. Appleby to Hampton P. Fulmer, 16 December 1941, file: Guayule Rubber, 1941–1948, Firestone Records.

10. Anderson quoted in *Hearing Before the Committee on Agriculture, U.S. House of Representatives, 77th Congress, 2nd Sess., on HR 6299 (Guayule Rubber)* (Washington, DC: Government Printing Office, 1942), 48 (hereafter cited as *Hearings on HR 6299*).

11. *Final Report: Emergency Rubber Report: A Report on Our War-time Guayule Rubber Program* (Washington, DC: USDA, Forest Service, 1946), 17–28 (hereafter cited as *Final Report*); and *NYT*, 11, 18, and 23 December 1941.

12. *Strategic and Critical Materials (Guayule Rubber), Hearings Before the Committee on Military Affairs, U.S. Senate, 77th Congress, 1st Sess., on S 2152* (Washington, DC: Government Printing Office, 1941). Jones quoted on p. 30. See also HAW to Jesse Jones, 27 December 1941, quoted in Neushul, "Science, Technology, and the Arsenal of Democracy," 113.

13. *Hearings on HR 6299*; and *Congressional Record*, House, 5 February 1942, 77th Congress, 2nd sess., 1056.

14. *Hearings on HR 6299*, 26.

15. *Congressional Record*, House, 5 February 1942, 77th Congress, 2nd sess., 1055–1085, Fulmer quoted on p. 1063.

16. "Extension of Remarks of John Z. Anderson in the House of Representatives, 4 February 1942," box 1901, E1, RG179, NA.

17. Ernst A. Hauser to Hampton P. Fulmer, 23 January 1942, box 16, IRC Records.

18. *Congressional Record*, House, 5 February 1942, 77th Congress, 2nd sess., 1065.

19. HGA to C. L. Baker, 16 February 1942, IRC Records.

20. *SMP*, 16 January 1942; *NYT*, 6 February 1942; *Salinas Index-Journal*, 9 February 1942; and *SMP*, 10 February 1942.

21. *NYT*, 18 and 19 February 1942; and *SMP*, 18 February 1942.

22. Downey quoted in *SMP*, 18 February 1942.

23. Sheridan Downey to HAW, 1 March 1942, HAW Papers at the Franklin D. Roosevelt Presidential Library, microfilm edition, reel 13.

24. *SMP*, 1 March 1942.

25. *Salinas Index-Journal*, 5 March 1942; *Final Report*, 9–13; and Neushul, "Science, Technology, and the Arsenal of Democracy," 143.

26. *SMP*, 22 February 1942, 5 March 1942, 6 March 1942, and 7 March 1942.

27. *Final Report*, 87–91.

28. Hallie Leyda to Mr. Wickard, 14 January 1942, file 44814, box 207, E2, RG54, NA.

29. Elizabeth Hobson to Jesse Jones, 28 May 1942, box 4, E91, RG54, NA.

30. Marlene Leda to Mr. Roosevelt [June 1942], box 206, E2, RG54, NA.

31. *DMN*, 15 December 1941. Schoffelmeyer's first series of articles on guayule appeared in Dallas's *Semi-weekly Farm News* in June 1923. Clippings, box 36, accession 94, The Henry Ford.

32. *DMN*, 22 December 1941, 29 December 1941, and 5 January 1942.

33. This episode is also significant because it made Waco's congressman, William R. "Bob" Poage, aware of the guayule issue. As will be seen below, Poage soon became one of guayule's most sustained and dedicated supporters. See Dewitt T. Hicks to W. R. Poage, 30 December 1941; Victor Schoffelmeyer to Poage, 3 January 1942; Poage to Schoffelmeyer, 8 January 1942; and Poage to Hicks, 10 January 1942, box 9, William Robert Poage Papers, Collections of Political Materials, Baylor University, Waco, Texas (hereafter cited as Poage Papers).

34. "Rabbit Brush Funds Asked," *Reno Gazette*, 6 February 1942; "Rabbit Brush Rubber Test Is Scrugham Goal," *Review Miner*, 26 February 1942; and Mark M. Butler to George W. Malone, 29 January 1942, Governor Edward P. Carville Papers, Nevada State Library and Archives, Carson City.

35. "Rubber in the Desert," transcript of radio broadcast, KGO San Francisco, 7 February 1942, Governor Edward P. Carville Papers, Nevada State Library and Archives, Carson City.

36. "Rubber from Rabbit Brush," *Chemurgic Digest* 1 (31 January 1942): 14; based on "Rabbit-brush Rubber," *San Francisco Chronicle* (22 January 1942).

37. Doten, *Rubber from Rabbit Brush*, 6–10.

38. Carl Pfaender to Ross A. Gortner, 14 July 1940; and Gortner to W. C. Coffey, 21 January 1942, box 1, collection 583, Division of Agricultural Biochemistry, Elmer L. Anderson Library, University of Minnesota Archives, Minneapolis.

39. R. A. Gortner to W. C. Coffey, 21 January 1942; and Gortner to H. K. Wilson, 8 January 1942, box 1, collection 583, Division of Agricultural Biochemistry, Elmer L. Anderson Library, University of Minnesota Archives, Minneapolis.

40. *(Sioux Falls) Daily Argus*, 21 January 1942, clipping; *Minneapolis Morning Tribune*, 24 January 1942, clipping; R. A. Gortner to editor, *Minneapolis Star Journal*, 23 January 1942; E. W. Brandes to F. R. Immer, 5 March 1942; and undated report by W. W. Benton, box 1, collection 583, Division of Agricultural Biochemistry, Elmer L. Anderson Library, University of Minnesota Archives, Minneapolis.

41. "Conference on World Rubber Situation," 12 March 1942, Edmund E. Day Papers, box 65, Division of Rare and Manuscript Collections, Cornell University, Ithaca, New York; L. H. MacDaniels to E. C. Auchter, 6 April 1942, box 205, E2, RG54, NA; and "Research for Rubber," *Chemurgic Digest* 1 (30 May 1942): 79.

42. [Herbert Hice Whetzel], "Apios Tubersoa Moench., a Possible Source of Commercial Rubber," unpublished notes; E. A. B[ates] to Whetzel, 8 June 1942; and L. M. Black to Whetzel, 10 July 1942, box 15, Herbert Hice Whetzel Papers, Division of Rare and Manuscript Collections, Cornell University, Ithaca, New York (hereafter cited as Whetzel Papers).

43. Lewis Knudson, "Rubber Investigations at Cornell University" [January 1943?], box 24, Carl E. Ladd Papers, Division of Rare and Manuscript Collections, Cornell University, Ithaca, New York (hereafter cited as Ladd Papers).

44. On Mitchell, see "Rubber Investigations," *Chemurgic Digest* 1 (30 May 1942): 77. On Jones, see William R. Jones, "Economic Consideration of Goldenrod for Rubber Production" (BS thesis, Virginia Polytechnic University, 1943).

45. Alvin L. Moxon and E. I. Whitehead, "The Rubber Content of Some South Dakota Plants," *Proceedings of the South Dakota Academy of Science* 23 (1943): 53–55; and "Oregon Rubber Study," *Chemurgic Digest* 1 (30 June 1942): 89.

46. Franklin Stewart Harris to Alfred Atkinson, 24 April 1942, box 1, University of Arizona Records of the Office of the President, 1937–1947, University Archives, University of Arizona, Tucson.

47. For soybeans, see *NYT*, 30 September 1942 and 20 December 1942; for cotton, see *NYT*, 30 September 1942; for pine trees, see *DMN* 8 June 1942; and *Atlanta Constitution*, 26 December 1942.

48. J. H. Foss to F. F. Baldwin, 5 August 1942, box 1901, E1, RG179, NA; and Bruce K. Brown to L. D. Thompkins, 30 November 1942; and Robert S. Herbert to Thompkins, 9 December 1942, box 1899, E1, RG179, NA. Early efforts to establish a rubber industry in Hawaii are described in W. A. Anderson, "Plantation Rubber in Hawaii," *Hawaiian Agricultural Experiment Station, Press Bulletin*, no. 44 (1913).

49. *NYT*, 22 July 1942.

50. For example, see Isaac Jones to George Washington Carver, 31 July 1942; Mrs. B. A. King to Carver, 6 August 1942; Mrs. Kork Kelly to Carver, 24 September 1942; and Charles Greenleaf to Carver, 10 September 1942, reel 42, George Washington Carver Papers, Tuskegee Institute, Tuskegee, Alabama. On Carver's mythic status, see Barry Mackintosh, "George Washington Carver: The Making of a Myth," *Journal of Southern History* 42 (November 1976): 507–528.

51. For biographical information, see Christopher M. Granger, "Forest Management in the United States Forest Service, 1907–1952," interview by Amelia R. Fry (Berkeley, CA: Regional Oral History Office, Bancroft Library, 1974); Evan W. Kelley, "The Making of a Regional Forester," interview by Amelia R. Fry (Berkeley, CA: Regional Oral History Office, Bancroft Library, 1974); and "Major Evan W. Kelley," unidentified manuscript, box 6, Paul H. Roberts Papers, Nebraska State Historical Society, State Archives and Manuscript Division, Lincoln.

52. "Memorandum for Historical File," 22 February 1942, box 28, E20, RG95, NA. See also *SMP*, 15 February 1942; and *Final Report*, 19–35.

53. According to a February 1941 report, the IRC's warehouses included 105 drums of guayule seed destined for Italy, but shipment had been prevented by a British blockade. Because that blockade remained in effect a year later, it seems likely that those same seeds ended up in American fields after the ERP took over the IRC's resources in March 1942. See *Salinas Index-Journal*, 4 February 1941. This assumption, however, cannot be confirmed from 1942 reports and news articles. Such sources presented the IRC's work in a profoundly patriotic light, focusing on its interest in the resource needs of the United States rather than those of Italy.

54. *Final Report*, 19–21; *SMP*, 1 March 1942, 10 March 1942, 3 April 1942, and 1 May 1942; and Lee McCardell, "Tire of Future Growing in U. S. 'Salad Bowl,'" *(New York) Daily Worker*, 20 March 1942, clipping, box 28, E20, RG95, NA. Here is another indication of the importance of guayule in the national consciousness in the summer of 1942: the news that a driver in Salinas had veered off the road and destroyed 150 guayule plants appeared in the *NYT*, 19 July 1942.

55. Neushul, "Science, Technology, and the Arsenal of Democracy," 145; and *SMP*, 26 May 1942, 30 May 1942, and 5 June 1942. For the shift to chemical and incendiary weeding technologies, see *Final Report*, 125–126.

56. Standard histories of the bracero program highlight its connection with national defense initiatives. See Richard B. Craig, *The Bracero Program: Interest Groups and Foreign Policy* (Austin: University of Texas Press, 1971), 36–43. For hints that guayule rubber added to this pressure, see *Los Angeles Daily News*, 25 February 1943; and *Final Report*, 46–50. I am indebted to Margaret Crowl MacArthur for her reminiscences of the social upheavals associated with her father's work for the

ERP. E-mail communication with author, 31 May 2004. (ERP officials later relied increasingly on labor from German and Italian POWs.)

57. James Bonner, "Physiology and Chemistry of Guayule," in *An International Conference on the Utilization of Guayule, November 17–19, 1975*, ed. William G. McGinnies and Edward F. Haase (Tucson: University of Arizona, 1975), 78.

58. F. W. Went to E. C. Auchter, 9 July 1942, reel 27, microfilm edition of the Robert Andrews Millikan Papers at the California Institute of Technology, Center for the History of Physics, American Institute of Physics, New York, New York (hereafter cited as Millikan Papers); and "Interview with James F. Bonner," Caltech Archives Oral Histories Online, http://oralhistories.library.caltech.edu, accessed 15 March 2004.

59. F. W. Went to IRC, 6 August 1940, box 16, IRC Records.

60. Greg Robinson, *By Order of the President: FDR and the Internment of Japanese Americans* (Cambridge, MA: Harvard University Press, 2001), 90.

61. See Eugene Rabinowitch, "Robert Emerson, 1903–1959," *Plant Physiology* 34 (May 1959): 178–184. Millikan's views from Julie L. Benton, "Interview with Hugh Harris Anderson," Pasadena Oral History Project (Pasadena, CA: Pasadena Historical Society, 1982), 36; HHA, "Unpublished Autobiography," 162, 174; and HHA to friends, 25 January 1945, HHA Collection, Pasadena Historical Museum, Pasadena, California (hereafter cited as Anderson Collection).

62. See E. W. Kelley to Robert Emerson, 30 April 1942, file 56.101, box 322, E16 Subject-classified General Records, RG210, War Relocation Authority, NA (hereafter cited as E16, RG210, NA). See also Robert A. Millikan to P. V. Cardon, 20 April 1942; J. Bonner, "Report on Guayule Investigations to December 15, 1941"; Millikan to E. C. Auchter, 13 February 1942; and Cardon to Millikan, 8 April 1942, reel 27, Millikan Papers; and "Report of the Guayule Investigations Carried Out at the California Institute of Technology, October 1940–April 1942," box 45, E90, RG54, NA.

63. Both men had unusual access to the ERP. Emerson had contact with the Salinas group because his brother worked as a mycologist for the ERP. Because Anderson was stricken with polio in the spring of 1943, he was able to secure additional gas coupons, which permitted trips among Pasadena, Manzanar, Salinas, and his workplace at the internment camp in Poston, Arizona. See "Report on Guayule Project at the Manzanar Relocation Area, for Submission to the U.S. Forest Service and the Bureau of Plant Industry, February 1943," reel 378, American Evacuation and Resettlement Records, Bancroft Library, University of California, Berkeley (hereafter cited as Berkeley Japanese American Records); Neushul, "Science, Technology, and the Arsenal of Democracy," 151; Irvin Ashkenazy, "As the Guayule Ball Bounces," *Westways* 69 (September 1977): 56–60; interview with Hugh Harris Anderson, OH 1473, Center for Oral and Public History, California State University, Fullerton; and HHA, "Unpublished Autobiography," 156–211.

64. Robert Emerson, "Objectives of the Guayule Project at Manzanar," manuscript report, 28 April 1942; and Robert Emerson, "Objectives of the Guayule Project in the Japanese Relocation Areas, and Recommendations for the Furtherance of These Objectives" manuscript report, 27 May 1942, box 1, Grace Nichols Pearson Collection, Hoover Institution Archives, Stanford, California (hereafter cited as Nichols Papers).

65. *Manzanar Free Press*, 20 May 1942 and 8 October 1942; and "Monthly Report of Guayule Experimental Project, Manzanar California, June 1942," box 14, collection 122, Manzanar War Relocation Center Records, Department of Special

Collections, Young Research Library, University of California, Los Angeles (hereafter cited as UCLA Manzanar Records).

66. Nichols, "Guayule Research Project," manuscript (1 August 1942), reel 157, Berkeley Japanese American Records. There also were attempts to grow cryptostegia at the Poston internment camp in Arizona. See Frank M. Kramer to Donald R. Longman, 8 September 1942, box 1899, E1, RG179, NA.

67. I am grateful to Dr. Glenn H. Kageyama and his father, Frank Akira Kageyama, for their help with this section. E-mail correspondence to author, 6 May 2007; and interview with Frank Akira Kageyama, 1 October 2007, Pomona, California. See also "Monthly Report of the Guayule Experimental Project, Manzanar, California, June 1942"; and Robert Emerson to Grace Nichols, 16 June 1942, 5 July 1942, 22 July 1942, and 3 August 1942, box 1, Nichols Papers; "Personnel" [1943?], box 14, UCLA Manzanar Records; and *Manzanar Free Press*, 8 October 1942 and 10 November 1942.

68. Nichols, "Guayule Research Project"; and "Report on the Guayule Project at the Manzanar Location Area, for Submission to the U. S. Forest Service and Bureau of Plant Industry, February 1943," reel 157, Berkeley Japanese American Records; Paul H. Roberts, memorandum for the record, 20 June 1942; and Robert Emerson to Major Kelley, 2 July 1942, box 16, E20, RG95, NA.

69. "Japanese Gardeners Work at Propagation of Guayule," *Science News Letter* 41 (9 May 1942): 296.

70. Linda Gordon and Gary Y. Okihiro, eds., *Impounded: Dorothea Lange and the Censored Images of Japanese American Internment* (New York: Norton, 2006). See also Brain Lain, "Different Archives, Different Values: Photographic Realism and Panoramic Perspectives on America's Concentration Camps," http://www.comm.unt.edu/research/different_archives.htm, accessed 29 November 2006.

71. Nichols, "Guayule Research Project." The censorship is evident in John C. Baker to Grace Nichols, 13 October 1942, file 56.101, box 322, E16, RG210, NA. Like Anderson and Emerson, Nichols was a member of the Quaker pacifist organization, the American Friends Service Committee.

72. Transcription of *Washington Post* story, attached to Fred McCargar to J. Edgar Hoover, 29 September 1942, box 8, Sheridan Downey Papers, Bancroft Library, University of California, Berkeley (hereafter cited as Downey Papers).

73. Fred S. McCargar to J. Edgar Hoover, 29 September 1942, box 8, Downey Papers.

74. Glenn H. Kageyama to author; Robert Emerson to Grace Nichols, 1 October 1942; Emerson to Nichols, 5 November 1942; Emerson to Nichols, 10 November 1942; and John C. Baker to Nichols, 24 November 1942, box 1, Nichols Papers.

75. See Dillon P. Myer to Robert Emerson, 28 October 1942, E16, RG210, NA; A. C. Hildreth to Robert A. Millikan, 18 December 1942; F. W. Went to Millikan, 16 January 1943, and E. C. Auchter to Millikan, 28 January 1943, reel 27, Millikan Papers. See also Emerson to Hildreth, 20 November 1942, box 16; and Gordon Salmond to Evan Kelley 13 November 1942, box 21, E20, RG95, NA.

76. Williams Haynes and Ernst A. Hauser, *Rationed Rubber and What to Do About It* (New York: Knopf, 1942), v.

77. Ibid., v.

78. Karl T. Compton, personal notebook, "Rubber Survey, August 1942," box 188, Compton-Killian Papers.

79. See J. R. Killian Jr. to Ernst A. Hauser, 21 August 1942; and Hauser to Killian, 30 August 1942, box 107, Compton-Killian Papers. On fears of bottlenecks, see Karl T. Compton, "Rubber Survey, August 1942"; and David Spence to A. C. Brett 9 August 1942, box 1901, E1, RG179, NA. On Baruch's lobbying to remove

those bottlenecks, see BMB to Sheridan Downey, 26 August 1942, box 8, Downey Papers.

80. *Report of the Rubber Survey Committee, September 10, 1942* (Washington, DC: n.p., 1942); and Ernst A. Hauser to Malcolm Davis, 20 October 1942, box 67, Compton-Killian Papers.

81. *Report of the Rubber Survey Committee*, 54–55; *Final Report*, 230–31; and Sheridan Downey to BMB, 24 September 1942, series 6, vol. 55, Baruch Papers. Quotation from Congressman Roy Woodruff of Michigan. See *Congressional Record*, House, 8 October 1942, 77th Congress, 2nd sess., 7951–7953.

82. Based on *(Indio, California) Date Palm*, [8?] October 1942, clipping; and *Los Angeles Times*, 8 December 1942 and 16 January 1943, clipping, scrapbook of John M. Crowl, copies provided to me by his daughter, Ms. Margaret Crowl MacArthur (hereafter cited as Crowl Scrapbook).

83. *Final Report*, 68, 97.

84. "Soviet Prizes for Synthetic Rubber," *IRW* 75 (November 1926): 105; "Soviet Planting Fails," *IRW* 82 (July 1930): 88; and *NYT*, 2 October 1930, 26 January 1932, 24 October 1932, and 18 March 1934.

85. A. I. Kuptsov, *Kok-sagyz v zapadnoi Sibiri* [The Kok-saghyz in West Siberia] (Ogiz, Novosibirsk: Oblastnoe gosudarstvennoe izdatel'stvo, 1942), 3.

86. Paul Appleby to J. W. Pincus, 30 December 1941 (draft of letter sent 7 January 1942), box 16, E90, RG54, NA. See also Pincus to Milislav Demerec, 4 March 1936, box 18, Milislav Demerec Papers, American Philosophical Society, Philadelphia.

87. Kolachov came to the United States in dramatic fashion. After six months of imprisonment during the Russian Civil War, he escaped from the Red Army into Bulgaria, then France, and then Czechoslovakia, before he became a vocal defender of American economic independence. Because of these experiences with the Soviet Union, Kolachov recognized that the Soviets would not supply him with kok-sagyz seeds directly, so he appealed to the government officials for help. See B. Y. Morrison to files of the Division of Plant Exploration and Introduction, 10 February 1942, box 80, E135f, RG54, NA; and "Chemurgic Personalities: Paul J. Kolachov," *Chemurgic Digest* 1 (15 September 1942): 135.

88. *NYT*, 28 December 1941; *DMN*, 5 January 1942; and Paul Kolachov, "Kok-Sagyz, Family 'Compositae,' as Practical Source of Natural Rubber for the United States," speech presented at Symposium on the Florida Rubber Situation, 11 February 1942, box 1, Lewis Knudson Papers, Division of Rare and Manuscript Collections, Cornell University, Ithaca, New York (hereafter cited as Knudson Papers). See also Kolachov, "Rubber for Peace and War," speech presented at the Eighth Annual Chemurgic Conference, 26 March 1942, in box 11, E90, RG54, NA.

89. *NYT*, 27 March and 7 June 1942; and A. T. Steele, "Kok-sagyz Scores Again," *Chemurgic Digest* 1 (30 March 1942): 41.

90. C. E. F. Guterman to R. G. Wiggins, 2 April 1942; and Guterman to H. H. Love, 2 April 1942, box 3, Carl E. F. Guterman Papers, Division of Rare and Manuscript Collections, Cornell University, Ithaca, New York; and "Memo Regarding Species of Taraxacum (Dandelion) Used in Russia as a Source of Rubber," 19 March 1942, box 24, Ladd Papers.

91. Secretary of agriculture [Claude Wickard] to secretary of state [Cordell Hull], [March 1942?]; Maxim Litvinoff to Jesse Jones, [April 1942?]; and E. W. Brandes to R. M. Salter, 1 September 1943, box 16, E90, RG54, NA.

92. E. W. Brandes to files of Rubber Plant Investigations, 4 May 1942; Dreyfuss to secretary of state, 1 May 1942; and Mattison to secretary of state, 13 May 1942, box 38, E90, RG54.

93. J. W. Pincus, "The USSR Grows Its Own Rubber," *Soviet Russia Today* 10 (June 1941): 14–15, 34. In subsequent months, a traveling exhibit that connected Soviet rubber crops with food relief toured the New Jersey State Fair, the American Seed Trade Association convention in Chicago, and elsewhere.

94. E. W. Brandes to E. C. Auchter, 5 June 1942, box 16, E90, RG54, NA; and A. M. Troyer to Walter T. Swingle, 8 September 1934, box 27, Walter Tennyson Swingle Collection, University of Miami, Department of Archives and Special Collections, Coral Gables, Florida; *Russian Dandelion*, 9.

95. Vavilov met with intisy-rubber expert Charles Swingle and invited him to study the next year in the Soviet Union. See I. G. Loskutov, *Vavilov and His Institute: A History of the World Collection of Plant Genetic Resources in Russia* (Rome: International Plant Genetic Resources Institute, 1999), 47. Further, American plant scientists readily invited Soviet scientists to visit the Savannah plant introduction station during the peak of the goldenrod harvest. JVM to HGU, 4 November 1935, binder 1935, item 184, EFWE; and David A. Bisset to B. Y. Morrison, 19 October 1935, box 84, E135f, RG54, NA.

96. Daniel T. MacDougal to N. Maximov, 23 February 1929, box 18, Records of the Desert Botanical Laboratory of the Carnegie Institution, 1903–1985, University Archives, University of Arizona.

97. T. D. Lyssenko [sic], "More Data Concerning the Production of Good Stands of Kok-saghyz"; T. D. Lysenko, "Good Germination of Kok-saghyz as the Best Assurance of Good Crops"; H. Altukhov, "Our Rubber Is Not Inferior to the Imported Product"; and E. Milner, "Stakhanov's Agricultural Technique for Raising Kok-Saghyz," Administrative Policy Records, 1905–1962, Office of the Director, Institute of Food and Agricultural Sciences, subseries 90A, Public Records Collection, University of Florida Archives, Gainesville.

98. E. W. Brandes to cooperators making fall (1942) test plantings of the Russian dandelion, kok-sagyz, 15 August 1942, Administrative Policy Records, 1905–1962, Office of the Director, Institute of Food and Agricultural Sciences, subseries 90A, Public Records Collection, University of Florida Archives, Gainesville. Brandes also warned his colleagues: "Publicity on this crop should be avoided. . . . Correspondence resulting from premature press notices is usually voluminous and needless."

99. For a review of the Lysenko and Vavilov controversy, see David Javorsky, *The Lysenko Affair* (Cambridge, MA: Harvard University Press, 1970); Mark Popovsky, *The Vavilov Affair* (Hamden, CT: Archon, 1984); Nils Roll-Hansen, "A New Perspective on Lysenko?" *Annals of Science* 42 (1985): 261–278; and Loskutov, *Vavilov and his Institute*, 106.

100. LeRoy Barnett, "An American Need for a Russian Weed," *Michigan History Magazine* 89, no. 2 (2005): 29–33.

101. "Memo Regarding Species of Taraxacum."

102. Robert W. Henderson to C. E. Steinbauer, 6 December 1942, Robert W. Henderson Papers, Oregon State University Archives, Corvallis (hereafter cited as Henderson Papers).

103. J. W. Wilson to Sumner Welles, 13 March 1942; and Wilson to Donald M. Nelson, 17 April 1942, box 1899, E1, RG179, NA. See also Leslie N. Gooding to T. D. Mallory, 9 June 1942, box 45, E90, RG54, NA.

104. "Kok-sagyz Seed Arrives," *Chemurgic Digest* 1 (15 May 1942): 65; and *NYT*, 7 January 1943.

105. *Russian Dandelion*, 9; *Summary of Kok-saghyz Investigations Spring and Summer 1942, Rubber Plant Field Laboratory, University Farm, St. Paul, Minnesota*

(Beltsville, MD: Bureau of Plant Industry, 1943); and "Directory of Research Personnel and Plantings of Taraxacum Kok-Sagyz," box 16, Whetzel Papers.

106. "Kok-sagyz in Canada," *Chemurgic Digest* 2 (15 July 1943): 110–111. See also Raphael Zon to E. W. Kelley, 5 May 1942, box 1, collection 583, Division of Agricultural Biochemistry, University of Minnesota Archives, Elmer L. Anderson Library, Minneapolis.

107. Arnold White, "Progress Report on Kok-saghyz Trial Planting in Chile," 25 February 1943, box 6, Harley Harris Bartlett Papers, Bentley Historical Library, University of Michigan, Ann Arbor.

108. Heim, *Kalorien, Kautschuk, Karrieren*, 125–171; and Olga Elina, Susanne Heim, and Nils Roll-Hansen, "Plant Breeding on the Front: Imperialism, War, and Exploitation," 161–178.

109. Dolley, "Cryptostegia Grandiflora," 339–340; and Dolley, "Growing Cryptostegia in the United States," 3–4.

110. Charles S. Dolley to David Fairchild, 2 May 1934, Fairchild Papers. Dolley complained he received "never a word of acknowledgement from either" Edison or Ford.

111. H. L. Trumball, "Confidential: Rubber from Domestic Sources: Summary" (May 1942), box 11, E90, RG54, NA; and Trumball to Eugene C. Auchter, 1 June 1942, box 26, E1 General Correspondence, 1942–43, RG310, Agricultural Research Administration, NA.

112. *NYT*, 22 May 1942.

113. *DMN*, 25 August 1941.

114. Drew Pearson and Robert S. Allen, "Washington Merry-Go-Round," *Philadelphia Record*, [June 1942?], clipping, box 28, E20, RG95. See also *DMN*, 8 July 1942. Wallace's views from Earl N. Bressman, "Rubber Diary," 22 June 1942, file 2/5, Earl N. Bressman Papers, Special Collections, Iowa State University, Ames (hereafter cited as Bressman Rubber Diary); "Launch Planting of Cryptostegia," *RA* 52 (November 1942): 146. See also clipping from *Chicago Daily Tribune* (14 October 1942), box 46, Baruch Papers. For its part, the Chicago paper invested in a process that promised to obtain grain alcohol rubber through the sulfites disposed in the newsprint production process.

115. Wilson, *Trees and Test Tubes*, 15–20.

116. Ibid., 15–17.

117. *DMN*, 13 July 1942. See also "Domestic Rubber Plant Investigation Launched by United States Rubber Co.," *RA* 51 (June 1942): 213–214; and *Baltimore Sun*, 19 July 1942, clipping, box 26, E20, RG95, NA.

118. "Report of the Technical Advisory Committee," 10 August 1942, box 11, E90, RG54, NA.

119. *(Washington DC) Times–Herald*, 11 September 1942, clipping, box 11, E90, RG54, NA.

120. Charles Dolley to HRH the duke of Windsor, 19 August 1942, series 6, vol. 55, Baruch Papers; and Charles Dolley to HRH duke of Windsor, 26 November 1942, series 12, box 191, Baruch Papers.

121. *Report of the Rubber Survey Committee*, 55.

122. Bressman Rubber Diary, 11 September, 16 September, 17 September, and 23 September 1942; HAW to Franklin D. Roosevelt, 30 October 1942, series 9, box 43, Baruch Papers; and "Launch Planting of Cryptostegia," *RA* 52 (November 1942): 146. Some considered the sudden enthusiasm for cryptostegia as premature. See E. G. Holt to Jesse Jones, 28 September 1942, box 1895, E1, RG179, NA.

123. SHADA was an unusual organization. In spring 1941, USDA official TAF proposed the basic concept: Haiti needed an organization to help its economic

development, particularly through the promotion of crops that did not compete with U.S. agriculture. At first, *Hevea*, rather than cryptostegia, was a centerpiece of the project. See TAF to Earl N. Bressman, 17 March 1941, 838.51/Cooperative Program/1, decimal files, 1940–1944, RG59, U.S. Department of State, NA (hereafter cited as decimal files, RG59, NA); and Bressman Rubber Diary, 18 April and 23 September 1942. Although the Haitian government owned all of the shares in the corporation, its funds came from the U.S. Department of Agriculture, Department of State, and the Export-Import Bank of Washington. In November 1942, the Rubber Reserve Company, an entity of the U.S. government, agreed to pay all expenses for the cryptostegia project, plus a management fee for SHADA. See "Haitian Rubber Timeline," http://www.webster.edu/~corbetre/haiti/misctopic/leftover/rubber.htm, accessed 3 January 2007.

124. For biographical information and personal reminiscences, I am indebted to Thomas Dudley Fennell, particularly an interview of 7 February 2008.

125. Intelligence report from J. E. Kerley, 16 November 1942, 838.51/Cooperative Program/44, decimal files, RG59, NA.

126. M. D. Knapp, "Activities of the SHADA Cryptostegia Program," [1943?], 838.51/Cooperative Program/59, decimal files, RG59, NA. See also Morris S. Rosenthal to BMB, 21 September 1942; Rosenthal to BMB, [late September 1942?]; HAW to Jesse Jones, 12 October 1942; Milo Perkins to BMB, 15 October 1942; Perkins to William Jeffers, 15 October 1942; and C. Reed Hill to Robert Herbert, 23 October 1942, series 9, box 43, Baruch Papers. On an "urgent" request to the Boyce Thompson Institute, a leader in the study of plant hormones, to develop a chemical method for the rapid regeneration of root cuttings of cryptostegia for the planting millions of plants, see Earl N. Bressman to William Crocker, 30 September 1942, and Crocker to Bressman, 8 October 1942; for the decision to shift the project to Haiti, see Bressman to Nelson Rockefeller, 13 October 1942, reel 23, microfilm edition, HAW Papers at the University of Iowa Library.

127. Knapp, "Activities of the SHADA Cryptostegia Program." See also Jenkins, "Cryptostegia as an Emergency Source of Rubber"; and "Crypto Wins Again," *A propos de SHADA* 1, no. 2 (July 1943): 1–3.

128. J. McGavack and Sam R. Hoover, "A Brief Study of the Cryptostegia Problem in Haiti" (15 June 1943), box 11, E90, RG54, NA.

129. Charles S. Dolley to David Fairchild, 4 August 1943, Dolley file; and Dolley to Fairchild, 3 December 1943, Rubber Subject file, Fairchild Papers.

130. "Expansion of Guayule Program Forecast by Officials," *RA* 53 (September 1943): 537–538; "Synthetic Rubber Can't Win Alone," *A propos de SHADA* 1 (July 1943): 5; and "Lescot Tours SHADA," *A propos de SHADA* 1, no. 3 (August 1943): 1. See also Matthew Jordan Smith, "Shades of Red in a Black Republic: Radicalism, Black Consciousness, and Social Conflict in Postoccupation Haiti, 1934–1957" (PhD diss., University of Florida, 2002), 84.

131. "SHADA's Job Is Rubber," *A propos de SHADA* 1, no. 1 (June 1943): 3; TAF, "In the Family," *A propos de SHADA* 1, no. 2 (July 1943): 4; and TAF, "Cryptostegia Rubber—Its Rebirth and Development," *RA* 54 (January 1944): 329–332.

132. Ironically, the facility faced problems in retaining and recruiting farm laborers due to competition from a farm operation just a few miles away—the Henry Ford Farms. Given the chance to work for Henry Ford, many of the agricultural workers in this predominantly African American district chose not to work on the government's goldenrod research. David A. Bisset to J. L. Mahoney, 8 June 1936, box 84, E135f, RG54, NA.

133. NYT, 7 May 1939; SMN, 31 January and 27 March 1940; and Loren G. Polhamus to R. H. Gerke, 2 September 1937, box 81, E135f, RG54, NA.

134. David A. Bisset to B. Y. Morrison, 6 January 1942, box 48, E90, RG54, NA; Bisset to Morrison, 14 February 1942; and Bisset to J. L. Mahoney, 27 February 1942, box 81, E135f, RG54, NA; and "Reminiscences of H. G. Ukkelberg." Ukkelberg turned down an offer to direct the USDA's goldenrod project, alleging later that the government did not have enough experienced personnel to make the project a success.

135. Charles Melchior, "The Production of Goldenrod as a Source of Emergency Rubber," unpublished manuscript (May 1945), box 6, E90, RG54, NA; and David A. Bisset to J. L. Mahoney, 3 March 1942; and Bisset to E. W. Brandes, 6 March 1942, box 81, E135f, RG54, NA. See also *SMN*, 23 October 1942; and "Goldenrod—A Possible New Source of Rubber," manuscript report from Bureau of Plant Industry, Soils, and Agricultural Engineering, June 1943, in Rubber Laboratory Documents, box 163, EFWE.

136. "Extraction, Characterization, and Utilization of Goldenrod Rubber: Work at the Southern Regional Research Laboratory Under the Emergency Rubber Project, U.S. Forest Service, July 1942 to June 1944," box 5; and John McGavack to LGP, 23 July 1943, box 11, E90, RG54, NA.

137. "Goldenrod as an Emergency Source of Natural Rubber," unpublished manuscript (16 January 1943), box 5, E90, RG54, NA. Evidently, the unnamed authors included E. A. Gastrock and E. D. Gordon of the Bureau of Agricultural Chemistry and Engineering, B. L. Hammond, of the Bureau of Plant Industry, and F. A. Cossitt of the Forest Service. See Melchoir, "The Production of Goldenrod."

Chapter 6 *Sustainable Rubber from Grain: The Gillette Committee and the Battles over Synthetic Rubber*

1. Rubber consumption data from press release of the Rubber Manufacturers Association, 29 July 1945, box 10, Poage Papers; Davis, *One River*, 297–300; and http://www.taphilo.com/history/WWII/Production-Figures-WWII.shtml, accessed 26 March 2007.

2. Morris, *The American Synthetic Rubber Research Program*, 7–16; and Smith, "The Ten-year Invention," 34–47.

3. Tuttle, "The Birth of an Industry," focuses on internal squabbling among bureaucrats, while Bruce Catton, *The War Lords of Washington* (New York: Harcourt Brace, 1948), 151–167, focuses on the timidity of Washington DC politicians in the face of the difficult rubber situation. In addition, many of the biographies and autobiographies of the participants situate their subject into a debate over who was to blame for the rubber mess. Examples include John C. Culver and John Hyde, *American Dreamer: A Life of Henry A. Wallace* (New York: Norton, 2000); Donald M. Nelson, *Arsenal of Democracy: The Story of American War Production* (New York: Harcourt Brace, 1946); Jones, *Fifty Billion Dollars*; Bascom N. Timmons, *Jesse H. Jones: The Man and the Statesman* (New York: Henry Holt, 1956); BMB, *Baruch*; James B. Conant, *My Several Lives: Memoirs of a Social Inventor* (New York: Harper and Row, 1970); and T. H. Watkins, *Righteous Pilgrim: The Life and Times of Harold Ickes, 1874–1952* (New York: Holt, 1990). A prime example of the more triumphant narratives is Herbert and Bisio, *Synthetic Rubber.* An outstanding dissertation and a strong undergraduate paper also offer solid introductions to this topic. See Neushul, "Science, Technology, and the Arsenal of Democracy"; and Robert A. Seeman, "The Corn Rubber Controversy of Guy M. Gillette," unpublished paper, January 1967, box 4, Del Stelck Collection of the Papers of Guy M. Gillette, Special Collections Department, University of Iowa Libraries, Iowa City (hereafter cited as Gillette Papers).

4. *NYT*, 26 September 1938, 20 January 1939, 25 January 1939, 6 February 1939, and 10 February 1939.

5. William J. Hale, *Prosperity Beckons: Dawn of the Alcohol Era* (Boston: Stratford, 1936). In this 201-page book, rubber is mentioned only on p. 6.

6. J. F. Fiege, letter to the editor, *DMR*, 2 August 1942.

7. Max Weinberg to HAW, 7 July 1938; and W. W. Skinner to Paul Appleby, 13 July 1938, box 2869, E17J General Correspondence of the Office of the Secretary, 1938, RG16, NA.

8. Mark R. Finlay, "The Industrial Utilization of Farm Products and By-Products: The USDA Regional Research Laboratories," *Agricultural History* 64 (Spring 1990): 41–52.

9. It appears that Knight requested a report on the synthetic-rubber situation in 1938 but that there was no more mention of the issue until 1940. See diary entries of 24 and 28 September 1938, Knight Diary.

10. Knight Diary, 9 May, 25 May, and 31 May 1940.

11. [William J. Sparks], "Proposed Program for Rubber Research Work in the Northern Regional Laboratory," n.d., cover letter dated 19 July 1940, box 163, E17K, General Correspondence of the Office of the Secretary, 1940, RG16, NA.

12. R. E. Buchanan to HAW, 22 July 1942, HAW Papers at the Franklin D. Roosevelt Presidential Library, microfilm edition, reel 6. See also Buchanan testimony, U.S. Senate, *Utilization of Farm Crops: Industrial Alcohol and Synthetic Rubber, Hearings Before a Subcommittee of the Committee on Agriculture and Forestry*, 77th Congress, 2nd sess., 1942 (hereafter cited as Gillette Hearings), 79–87.

13. Herbert and Bisio, *Synthetic Rubber*, 55; and *NYT*, 1 February 1942.

14. For example, a twenty-paragraph article under the headline "Synthetic Rubber from Petroleum Seen as New Opportunity for Texas" devoted only one paragraph to the reality that significant production was still at least three years away. *DMN*, 13 January 1941.

15. *NYT*, 13 September 1940. See also W. L. Finger to E. R. Stettinius, 29 July 1940, box 1890, E1, RG179, NA, which endorsed an August 1940 article in *Fortune* that emphasized the unlikelihood of the nation needing a synthetic-rubber program.

16. *NYT*, 1 February 1941.

17. *NYT*, 22 March 1941.

18. *NYT*, 9 September 1941. Additional optimistic reports found in *DMN*, 15 September 1941 and 3 November 1941.

19. Goodyear advertisement in *DMN*, 2 November 1941.

20. Ernst A. Hauser, "Science and Democracy," commencement address at the University of Wyoming, 29 August 1940 (Laramie: n.p., 1940).

21. Ernst Hauser, "Fundamental Problems in the Chemistry, Economy, and Processing of Rubber-like Elastomers," abbreviated text of speech, 7 July 1941, series 12, box 193, Baruch Papers; Hauser, "Home-grown and Home-made Rubber"; and "Hits Synthetic Rubber Claims."

22. The *New York Times Index* lists over five hundred articles on rubber in the year 1942, twenty-three of them on p. 1. See also "Current News of the Month," *RA* 50 (January 1942): 286–287.

23. Herbert and Bisio, *Synthetic Rubber*, 15; and *NYT*, 8 December, 14 December, and 21 December 1941.

24. See Herbert and Bisio, *Synthetic Rubber*, 63–69. In late 1940, the Reconstruction Finance Corporation planned for 10,000 tons of synthetic rubber. By 1941, the synthetic-rubber program was slated to produce 40,000 tons. In December, war planners raised the target to 120,000 tons; by January 1942, they raised the goal

to 400,000 tons; by spring, there were plans to produce 800,000 tons of synthetic rubber per year.

25. HAW to Henry Granger Knight, 24 January 1942, HAW Papers at the Franklin D. Roosevelt Presidential Library, microfilm edition, reel 26. Seven days later, Wallace complained to WPB chair Donald Nelson that "somebody is holding out on us" on rubber possibilities.

26. Knight Diary, 7 February 1942. When he met Iowa State College's delegation, Knight persuaded them to visit the NRRL in Peoria in order to learn the complexities of the synthetic-rubber problem.

27. Knight Diary, 20 January 1942 and 22 January 1942. See also *NYT*, 10 January 1942.

28. U.S. Senate, *Investigation of the National Defense Program, Hearings Before a Special Committee Investigation of the National Defense Program,* 77th Congress, 1st sess., 1942. For the IG Farben and Standard Oil agreements, see Peter Hayes, *Industry and Ideology: IG Farben in the Nazi Era* (Cambridge: Cambridge University Press, 1987), 335–336; and Joseph Borkin, *The Crime and Punishment of I. G. Farben* (New York: Free Press, 1978), 76–94. Also pertinent, Borkin charges that as SONJ and IG Farben finalized their agreements in December 1939, they decided to backdate the document to 1 September 1939, thus before the war had officially begun. For background on the Truman committee, see Donald H. Riddle, *The Truman Committee: A Study in Congressional Responsibility* (New Brunswick, NJ: Rutgers University Press, 1964).

29. For consistency and concision, I will refer to the Subcommittee of the Committee on Agriculture and Forestry as the "Gillette committee."

30. For brief biographies of committee members, see John A. Garraty and Mark C. Carnes, general eds., *American National Biography* (New York: Oxford University Press): McNary, 15:167–169; Norris, 16:499–501; Thomas, 21:505–506; and Wheeler, 23:130–131; and n.a., *Dictionary of American Biography* (New York: Scribner's, 1928–1958), Gillette, Supplement 9, 310–312; McNary, Supplement 3, 496–497; Norris, Supplement 3, 557–561; Thomas, Supplement 7, 738–739; and Wheeler, Supplement 9, 859–860. See also Jerry Harrington, "Senator Guy Gillette Foils the Execution Committee," *Palimpsest* 62, no. 6 (1981): 170–180; and Elmer Thomas, *Forty Years a Legislator,* ed. Richard Lowitt and Carolyn G. Hanneman (Norman: University of Oklahoma Press, 2007).

31. Christensen testimony, Gillette Hearings, 21–38; *DMN*, 18 August 1941, II, 3; and Leo Christensen, "Chemurgy in the News," *The Nebraska Alumnus,* February 1943, 7–8, Christensen Papers, Special Collections, Iowa State University Library, Ames. See also Leo M. Christensen's "Chemurgy Project Progress Report, No. 2," 1 December 1941, file 10/5/5, Records of the Agricultural Experiment Station, University Archives and Special Collections, University Libraries, University of Nebraska-Lincoln. Although it is thin on the World War II years, further information on Christensen is available in David Ray Duke, "Leo Christensen: A Biography of the Founder of Ethyl Alcohol in the United States" (PhD diss., St. Louis University, 1992).

32. Johnson testimony, Gillette Hearings, 2–21; and Johnson to director of priorities, Office of Price Administration, 30 March 1942; and Johnson to Guy M. Gillette, 22 April 1942, box 39, Senate 78A-F1, Committee on Agriculture and Forestry, 78th Congress, RG46, Records of the U.S. Senate, NA (hereafter cited as Senate 78A-F1, RG46, NA). For more on Johnson, see George E. Johnson II, *The Nebraskan* (Winter Park, FL: Anna, 1981). For Christensen, see Leo M. Christensen, "Synthetic Rubber Can Eliminate Need for Rationing," [approx. June 1942], box 36, Senate 78A-F1, RG46, NA. On Moffett, see "Rubber from

the Farm," *The Nation*, 16 May 1942, clipping, box 40, Senate 78A-F1, RG46, NA.

33. Knight Diary, 28 August 1940.

34. Basic elements of the Szukiewicz drama are found in testimony by Rosten, Knight, May, Henderson, Marks, Szukiewicz, and Lacy, Gillette Hearings, 119–123, 123–128, 201–207, 247–259, 259–261, 261–263; and "An Account of the So Called Polish Formula for the Manufacture of Butadiene," Leon Henderson Papers, box 35, Rubber-Congressional Committee, Franklin D. Roosevelt Presidential Library, Hyde Park, New York. The implication that British spies played a role is found in Hal Bernton, William Kovarik, and Scott Sklar, *The Forbidden Fuel: Power Alcohol in the Twentieth Century* (New York: Boyd Griffin, 1992), 31.

35. Knight Diary, 14 November 1941.

36. See David King Rothstein, "Report on the Economics and Technology of Synthetic Rubber," September 1942, Policy Documentation Files, box 1893, E1, RG179, NA.

37. John M. Weiss to Ferdinand Eberstadt, 1 May 1942, box 1891, E1, RG179, NA. Gillette quoted in *NYT*, 20 May 1942. Gillette's views may have been influenced by a letter from George Johnson, who wrote that he suspected that Szukiewicz was being kept "behind closed doors [as] simply an attempt to prohibit the use of grain for rubber production." George Johnson to Guy M. Gillette, 22 April 1942, box 39, Senate 78A-F1, RG46, NA.

38. Chaim Weizmann, *Trial and Error: The Autobiography of Chaim Weizmann* (New York: Harper, 1949), 423–430. See also Selman A. Waksman, "Chaim Weizmann as Bacteriologist," in *Chaim Weizmann: A Biography by Several Hands*, ed. Meyer W. Weisgal and Joel Carmichael (New York: Atheneum, 1963), 107–113. According to Knight, the British government sent Weizmann to the United States that spring explicitly to "find out why America was not producing rubber." Knight Diary, 23 April 1942.

39. Knight Diary, 23 April, 24 April, 4 May, 5 May, 6 May, and 27 May 1942. Quotation from 6 May 1942. See also Chaim Weizmann to Ernst David Bergmann, 8 May 1942; Weizmann to Bergmann, 6 June 1942; and Weizmann to Viscount Halifax, 19 June 1942, *The Letters and Papers of Chaim Weizmann*, ed. Michael J. Cohen (Jerusalem: Israel Universities Press, 1979), series A, 20:289–291, 302–303, 309–310.

40. Bressman Rubber Diary, 24 April, 1 May, 5 May, and 6 May 1942.

41. Bressman Rubber Diary, 6 May 1942.

42. H. T. Herrick, "Memorandum of Conference Held on 12 May 1942," box 27, E1, General Correspondence, 1942–43, RG310, Agricultural Research Administration, NA.

43. Bressman Rubber Diary, 1 May and 8 May 1942; and Carl Hamilton, "The Reminiscences of Carl Hamilton," transcript of interviews for the Oral History Research Office of Columbia University, 1953, 509–515.

44. Gillette Hearings, 103–119. Quotations on pp. 105 and 118. See also *NYT*, 20 May 1942.

45. *NYT*, 26 June 1942. See also *NYT*, 5 July 1942.

46. Gillette Hearings, 535–540. Quotation on p. 537. See also *NYT*, 4 February 1942. After months of delay, government officials finally awarded USSC access to the necessary critical materials in March 1942 but through an odd contract that presented additional roadblocks.

47. Gillette Hearings, 533.

48. *NYT*, 25 June 1942.

49. Draft of "Statement of Thurman Arnold, Assistant Attorney General in Charge of Antitrust Division" before the Senate subcommittee, [May 1942?]; and Thurman Arnold to Guy M. Gillette, 1 June 1942, box 36; *Washington Post*, 8 May 1942, clipping, box 40; and C. J. Farr to Isador Lubin, 8 May 1942, box 39, Senate 78A-F1, RG46, NA. See also Gillette Hearings, 285–303.
50. I. F. Stone, *The War Years, 1939–1945* (Boston: Little, Brown, 1988), 116–118. Stone's column, "Handcuffing Thurman Arnold," originally appeared on 4 April 1942. Other examples of the criticism include clippings from *NYT*, 25 March 1942; *Kansas City Star*, 13 May 1942; and *The Nation*, 16 May 1942, boxes 38 and 40, Senate 78A-F1, RG46, NA.
51. *NYT*, 10 May 1942.
52. *NYT*, 27 May 1942.
53. Gillette Hearings, 221.
54. *NYT*, 10 April 1942. See also Tuttle, "The Birth of an Industry," 35. Roosevelt's press conference response quoted in Neushul, "Science, Technology, and the Arsenal of Democracy," 70.
55. Knight Diary, 9 April 1942.
56. Bressman Rubber Diary, 21 April and 7 May 1942.
57. Quoted in Tuttle, "The Birth of an Industry," 38; and in Neushul, "Science, Technology, and the Arsenal of Democracy," 72.
58. Unsigned report on alcohol process, sent under a cover letter from Isador Lubin to Donald Nelson, 25 May 1942, box 1891, E1, RG179, NA.
59. *NYT*, 27 May 1942.
60. Gillette Hearings, 533.
61. Ibid., 435–436. Elsewhere, Wheeler had attacked bureaucrats as "asleep at the switch" in their failure to invest in rubber-from-grain research. *NYT*, 9 May 1942.
62. Gillette Hearings. 531–535. All cited comments except Wheeler's made on 28 May 1942.
63. *NYT*, 25 June 1942.
64. Roosevelt quoted in Catton, *War Lords of Washington*, 162. See also Drew Pearson and Robert S. Allen's nationally syndicated column, "A Warning to F. R.," *DMR*, 11 June 1942. Roosevelt's radio message of 12 June did focus on rubber, especially the scrap-rubber drive.
65. *Penny-a-Pound: The President's Rubber Drive, June 15–July 10, 1942* (Washington, DC: The Petroleum Industry War Council, 1943). See also Smith, "Rubber for Americans," 256.
66. *NYT*, 6 June 1942.
67. *DMR*, 12 July 1942.
68. D. Howard Doane, "Agriculture and Rubber" [1942?], in D. Howard Doane Papers, collection 3121, Western Historical Manuscript Collection-Columbia, University of Missouri, Columbia. This speech is undated, but similar rhetoric can be found in other Doane speeches dated May, June, and November 1942.
69. "Open Letter to Henry A. Wallace from Edward Breen of Fort Dodge, Iowa," published in the *Fort Dodge (Iowa) Messenger*, 10 July 1942, box 5, Edward Breen Papers, Special Collections Department, University of Iowa Libraries, Iowa City (hereafter cited as Breen Papers).
70. Fulton Lewis Jr., transcripts of radio broadcasts on the Mutual Broadcasting Systems, and WHN, in New York City, 19, 22, 23, 24, 25, 26 June 1942, box 5, Breen Papers; and Bressman Rubber Diary, 18 June 1942. Portions of Lewis's broadcast also appeared in the *Congressional Record*, 77th Congress, 2nd sess., A2746 and

A2833. For a summary of Lewis's career, see *Current Biography Yearbook* (New York: Wilson, 1942): 505–511; and "The Winner," *Time*, 15 March 1943, 50, 52–53. For more on Lewis's politics, see George Seldes, *Facts and Fascism* (New York: In Fact, 1943), 184–197.

71. Jay N. Darling to Carl Miner, 30 June 1942, Jay N. Darling Papers, Special Collections Department, University of Iowa Libraries, Iowa City. Miner's letter to Darling, 28 June 1942, also hints at the possibility of a government cover-up of the "horrible mess."

72. George Johnson to Otto Knudsen, 2 July 1942, box 5, Breen Papers.

73. Catton, *War Lords of Washington*, 163.

74. George H. Halbert to Guy M. Gillette, 23 July 1942, reel 19, HAW Papers at the Franklin D. Roosevelt Presidential Library.

75. Rex Price to Secretary Wickard, 22 May 1942, box 26, E208, Office of Agricultural Defense Relations/Office of Agricultural War Relations, RG16, NA.

76. J. Jacobs to Guy M. Gillette, 25 May 1942, box 39, Senate 78A-F1, RG46, NA.

77. M. I. Browne to the Senate Agricultural Committee, 17 June 1942, box 39, Senate 78A-F1, RG46, NA.

78. J. C. Cranmer and T. W. Reddington to Senator Carter Glass, 5 July 1942, box 39, Senate 78A-F1, RG46, NA.

79. E.B.V. Aldrich to BMB, 13 August 1942, series 9, box 46, Baruch Papers. The Oregonians called for Gillette, Baruch, Nelson, and various USDA officials to attend their September meeting in Pendleton, although none were able to make the trip. See Carl Hamilton to E.B.V. Aldrich, 14 August 1942, box 26, E1, General Correspondence, 1942–43, RG310 Agricultural Research Administration, NA.

80. *DMR*, 30 June 1942, 7.

81. Mary Lundell to Claude Wickard, 8 July 1942, box 26, E1 General Correspondence, 1942–43, RG310, NA. Lundell was secretary of the Morrow County Pomona Grange in Willows, Oregon. See also B. F. Johnson and Mariam Bagley to Senator M. C. Wallgren, 21 October 1942, box 26, E208, RG16, NA; and Harriet Heyworth to BMB, 20 July 1942, series 6, box 46, Baruch Papers. Heyworth was secretary of the Garcia Grange, Manchester, California.

82. *Congressional Record*, 24 July 1942, 77th Congress, 2nd sess., A2947–2948.

83. "Report of the Nebraska Agricultural and Industrial Conference," 17 July 1942, box 2, Records of the Chancellor: C. S. Boucher, University Archives and Special Collections, University Libraries, University of Nebraska, Lincoln.

84. Roy Bender to Lyle H. Boren, 8 July 1942, box 11, Lyle Boren Papers, Carl Albert Center of Congressional Research and Studies, Norman, Oklahoma.

85. Chaim Weizmann to Franklin D. Roosevelt, 8 July 1942; and Weizmann to Berl Locker, 15 July 1942, *Letters and Papers of Chaim Weizmann*, series A, 20:326–332.

86. *DMR*, 13 July 1942.

87. "Developments in the Synthetic Rubber Situation," *RA* 51 (August 1942): 408–413; and *NYT*, 10 July 1942. Early samples of the corn and soybean oil rubber demonstrated a "stretch" of 200 percent and tensile strength of five hundred pounds per square inch, compared to 600 percent elasticity and tensile strength of three thousand pounds per square inch for natural rubber. By January 1943, USDA scientists in Peoria continued to make progress with this product, dubbed "Norepol," and two companies began to produce small amounts on a commercial scale. See *NYT*, 31 January 1943.

88. Full-page Houdry advertisements found in *Washington Evening Star*, 7 July 1942; and *NYT*, 8 July 1942.

89. John Morris Weiss to F. Eberstadt, 2 July 1942, box 13, Resource Series, Elmer Thomas Papers, Carl Albert Center of Congressional Research and Studies, Norman, Oklahoma; *Washington Times-Herald*, 7 July 1942; and *Philadelphia Inquirer*, 11 July 1942, clippings, box 40, Senate 78A-F1, RG46, NA. See also "Outline of Scope of Presentation by Mr. Nelson Before Senate Agricultural Subcommittee on Tuesday July 14, 1942," box 1894, E1, RG179, NA; and "Seek Legislation to Centralize Production of Synthetic Rubber," *RA* 51 (July 1942): 314–317.

90. HAW, "Wallace Warns Against a 'New Isolationism,'" *NYT*, 12 July 1942.

91. "Nelson Fights Use of Grain for Rubber," *Washington Times-Herald*, 7 July 1942, clipping, box 40, Senate 78A-F1, RG46, NA; Drew Pearson, "Closed Door Session on Rubber," *DMR*, 22 July 1942; and "Rubber—The Official Football of the Amateur League," *IRW* 106 (August 1942): 466. See also Neushul, "Science, Technology, and the Arsenal of Democracy," 78–79.

92. *DMR*, 7 July 1942. Some WPB officials remained opposed to the alcohol scheme; one official claimed he would not have approved the Nebraskans' plan even if they had accumulated 98 percent of the needed strategic materials. See Robertson to Harrop, 15 July 1942, quoted in Richard Lowitt, *George W. Norris: The Triumph of a Progressive, 1933–1944* (Urbana: University of Illinois Press, 1978), 375. See also Catton, *War Lords of Washington*, 172–173.

93. Guy M. Gillette, "Synthetic Rubber," Vital Speeches of the Day, undated clipping, box 1, Gillette Papers. The clipping is text of Gillette's speech on the Blue Network, 21 July 1942.

94. *DMR*, 23 July 1942; and U.S. Senate, *Expediting the Prosecution of the War by Making Provision for the Increased Supply of Rubber Manufactured from Alcohol Produced from Agricultural or Forest Products*, report delivered by Mr. Thomas, 77th Congress, 2nd sess., 1942, report no. 1516.

95. U.S. Senate, *Rubber Manufactured from Alcohol*, veto message from the president of the United States, 77th Congress, 2nd sess., 1942, document 243. See also Seeman, "The Corn Rubber Controversy of Guy M. Gillette."

96. "Editorials," *RA* 51 (August 1942): 405; "President Vetoes the Farm Rubber Bill; Appoints New Fact-finding Committee," *RA* 51 (August 1942): 406–407; and *NYT*, 23 July and 28 July 1942.

97. Weizmann was among those called, and he again linked the Zionist cause with his expertise in the realm of rubber chemistry. To a colleague, Weizmann wrote, "Whatever else happens, the scientific work has opened many avenues for me . . . [that] may prove decisive at the opportune time." Chaim Weizmann to Berl Locker, 18 August 1942, *Letters and Papers of Chaim Weizmann*, series A, 20:346.

98. *NYT*, 19 August 1942.

99. Compton, "Rubber Survey, August 1942."

100. "Report to the Rubber Survey Committee by the 'Chemical Engineering Group' of Consultants," 5 September 1942, box 1892, E1, RG179, NA. The report favored the butylene-glycol process developed by Seagram's and did not praise Weizmann's isoprene process nor Publicker's "Polish" process.

101. *Report of the Rubber Survey Committee*; and Conant, *My Several Lives*, 319–327. Evidently, Baruch and his colleagues had tired of the political nature of the debates—privately, Baruch described Senator Norris as a "little ga-ga" and Senator Gillette as one who put Iowa's farm interests ahead of those of the nation. See BMB to Karl Compton, 18 September 1942, series 6, vol. 55, Baruch Papers.

102. Praise for the report was widespread. See "Editorials: The Baruch Report," *RA* 51 (September 1942): 498; *DMR*, 9 September 1942; Coit, *Mr. Baruch*, 518–520; Schwarz, *The Speculator*, 394–396; as well as scores of congratulatory letters in the Baruch Papers.

103. Press statement by Subcommittee of the Senate Committee of Agriculture and Forestry, 13 September 1942, series 9, box 43; and Guy M. Gillette to BMB, 17 September 1942, series 9, box 37, Baruch Papers. "I am afraid the Baruch report is disappointing," Chaim Weizmann wrote to Felix Frankfurter, 25 September 1942, in *Letters and Papers of Chaim Weizmann*, 20:349. Stone quoted in Conant, *My Several Lives*, 327.
104. Jones, *Fifty Billion Dollars*, 411–414.
105. Morris V. Rosenbloom, *Peace Through Strength: Bernard Baruch and a Blueprint for Security* (Washington, DC: American Surveys, 1953), 245–246.
106. BMB to Sheridan Downey, 22 September 1942; and BMB to Arthur Capper, 19 September 1942, series 6, vol. 55, Baruch Papers. See also BMB to W. D. Jameison, 24 September 1942, series 9, box 46, Baruch Papers.
107. Howard A. Cowden to BMB, 19 December 1942, series 6, vol. 55, Baruch Papers. See also U. S. Senate, *Utilization of Farm Crops: Industrial Alcohol and Synthetic Rubber, Hearings Before a Subcommittee of the Committee on Agriculture and Forestry*, 78th Congress, 1st sess., 1943, 1652–1654.
108. *NYT*, 18 December 1942.
109. U.S. Senate, *Utilization of Farm Crops*, 1736–42. See also "Gillette Committee Continuing Synthetic Rubber Investigation," *RA* 52 (January 1943): 322–324; "Alcohol Developments," *RA* 52 (February 1943): 420; and *NYT*, 16 January 1943, 22 January 1943, and 26 February 1943. Note too that in 1943, the Republican Party had its eye on Gillette, a lifelong Democrat, as a potential candidate to challenge Roosevelt in the 1944 election. See *Des Moines Tribune*, 13 July 1943; and similar clippings in box 1, Gillette Papers.
110. "Gillette Committee Continuing," 322–324.
111. "House Concurrent Resolution 13" [n.d.] February 1943, box 26, E1 General Correspondence, 1942–43, RG310, NA.
112. Karl Stefan to Walter Rosicky, 18 November 1943; and Stefan to Harvey S. Firestone Jr., 2 December 1943, Karl Stefan Papers, Nebraska State Historical Society, Lincoln. See also Lloyd C. Thomas to Dwight Griswold, 23 December 1942, Papers of Governor Dwight Griswold, Nebraska State Historical Society, Lincoln.
113. Minnesota state senate resolution no. 8, 17 February 1943. See also S. C. Lind to W. C. Coffey, [February 1943?]; Lind to Coffey, 11 May 1943; and L. S. Palmer to C. H. Bailey, 17 May 1943. In the end, investigators concluded that Minnesota did not have a future as a rubber-producing state. See *Report of the Senate Investigating Committee on Synthetic Rubber as an Outlet for Minnesota Farm Products*, submitted December 1944, box 41, collection 922, Institute of the Agriculture Director's Office, University of Minnesota Archives, Elmer L. Anderson Library, Minneapolis.
114. William S. B. Lacy to Isador Lubin, 21 January 1943, series 9, box 43, Baruch Papers.
115. Weizmann quoted in HAW, *The Price of Vision: The Diary of Henry A. Wallace* (Boston: Houghton Mifflin, 1973), 189.
116. Ibid.
117. Gustav Egloff, "Synthetic Rubber from Petroleum," *RA* 52 (December 1942): 221–226.
118. Advertisements for Agripol, *Chemical Industries* 52 (January 1943): 29; and *Chemical Industries*, 52 (February 1943): 165. See also Clinton A. Braidwood, "Agripol—A Chemurgic Rubber," *Chemical Industries* 52 (March 1943): 322–324; and C. B. Fritsche to E. C. Auchter, 11 January 1943, box 26, E1, General Correspondence, 1942–43, RG310, Agricultural Research Administration, NA.

Fritsche, vice president of Reichhold, had been managing director of the Farm Chemurgic Council in the mid-1930s.

119. Herbert and Bisio, *Synthetic Rubber*, 90, quotation on 129.

120. William Jeffers to BMB, 19 August 1943, series 9, box 43, Baruch Papers.

121. Warren Woomer, "Institute Goes to War: The Buna-S Synthetic Rubber Story," in *Great Kanawha Valley Chemical Heritage, Symposium Proceedings*, compiled by Lee R. Maddex (Morgantown, WV: Institute of the History of Technology and Industrial Archeology, 2003), 59. It is significant that the Institute plant first used midwestern grain as its feedstock yet later converted to molasses and petroleum by-products as the basis for its ethyl alcohol. See *Butadiene and Styrene for Buna S Synthetic Rubber from Grain Alcohol* (New York: Carbide and Carbon Chemicals Corporation, 1943).

122. Burton K. Wheeler, speech at Ontario Paper Company's alcohol plant, 18 June 1943, box 20, Burton K. Wheeler Papers, Montana Historical Society, Helena.

123. U.S. Senate, *Investigation of the Expanded Utilization of Farm Crops: Report of the Subcommittee on Agriculture and Forestry, Senate Document No. 240, 78th Congress, 2nd Sess.* (Washington, DC: Government Printing Office, 1944).

124. Walter W. Wilcox, *The Farmer in the Second World War* (Ames: Iowa State College Press, 1947), 160–181; "Feed Crisis Near," *Business Week* 735 (2 October 1943): 14; and "Farmers Face Feed Famine," *Breeder's Gazette* 109 (May 1944): 5, 22.

125. Gillette Hearings, 87.

126. Ibid., 793–795. Sweeney asserted that "it seemed very stupid indeed" for Americans to be separated from their rubber by the largest ocean in the world, and he urged development of sustainable rubber resources within the United States

127. Gillette Hearings, 87–88.

128. Ibid., 543.

129. "Statement by Dr. Earl N. Bressman," 5 June 1942, box 36, Senate 78A-F1, RG46, NA. See also Bressman Rubber Diary, 4 June and 5 June 1942. For Davies, see Marlin M. Volz and Irving M. Helman to Thomas Lynch, 8 July 1942, box 1894, E1, RG179, NA.

130. U.S. Senate, *Investigation of the Expanded Utilization of Farm Crops*, 3.

Chapter 7   Resistance to Domestic Rubber Crops and the Decline of the Emergency Rubber Project

1. See Scott, *Seeing Like a State*, 267.

2. Frank Akira Kageyama interview, 1 October 2007; and M. S. Nishimura, F. N. Hirosawa, and R. Emerson, "Rubber from Guayule," *Industrial and Engineering Chemistry* 39 (November 1947): 1477–1485.

3. Robert Emerson to Robert Millikan, 3 December 1942, Millikan Papers; "Report on Guayule Project at the Manzanar Relocation Area, for Submission to the U. S, Forest Service and the Bureau of Plant Industry, February 1943"; and Robert Emerson statement, 11 December 1944, Kageyama Family Papers; and Kageyama interview. Additional summaries of research success from Neushul, "Science, Technology, and the Arsenal of Democracy," 151–152; HHA, "Unpublished Autobiography," 156–165; and Anderson interview, 68–80.

4. R. A. Millikan to E. C. Auchter, 9 February 1943, reel 378, Berkeley Japanese American Records.

5. Robert Emerson to Sidney P. Osborn, 21 January 1943; Osborn to Emerson, 29 January 1943; and Osborn to Dillon Meyer [sic], 29 January 1943, box 23,

Governor Sidney P. Osborn Papers, RG1, SG 14, Arizona State Library, Archives and Public Records, Phoenix (hereafter cited as Osborn Papers).

6. Carl Hamilton to Hugh Fulton, 22 February 1943, box 1901, E1, RG179, NA.

7. Robert Emerson to Sidney P. Osborn, 19 February 1943, box 23, Osborn Papers; Emerson to John Collier, 29 January 1943, reel 378, Japanese American Evacuation and Resettlement Records, Bancroft Library, University of California, Berkeley; Emerson to Atherton Lee, 28 January 1943; and Millikan to E. C. Auchter, 9 February 1943, reel 27, Millikan Papers.

8. Ralph Merritt to Evan W. Kelley, 22 February 1943; and James J. Byrne to Paul Roberts, 20 August 1943, box 21, E20, RG95, NA. At least one of these interned scientists had been granted permission to travel in the "Western Defense Zone" of coastal California before the end of the war. See Curt Stern to James Walcott Wadsworth, 21 December 1945, Masuo Kodani file, Curt Stern Papers, American Philosophical Society, Philadelphia.

9. Robert A. Millikan to Ralph P. Merritt, 5 March 1943; and Robert Emerson to Merritt, 12 March 1943, box 14, UCLA Manzanar Records.

10. E. C. Auchter to Milton S. Eisenhower, 27 January 1943, box 322, E16 Subject-classified General Records, RG210, Records of the War Relocation Authority, NA.

11. Ralph P. Merritt to Dillon S. Myer, box 14, 29 August 1943, UCLA Manzanar Records.

12. Ken Emerson, "Comments," in McGinnies and Haase, *An International Conference on the Utilization of Guayule,* 157.

13. For instance, G. Ledyard Stebbins Jr. and Masuo Kodani, "Chromosonal Variation in Guayule and Mariola," *Journal of Heredity* 35 (June 1944): 163–172; M. S. Nishimura, R. Emerson, T. Hata, and A. Kageyama, "The Propagation of Guayule from Cuttings," *American Journal of Botany* 31 (October 1944): 412–418; R. C. Rollins, "The Evidence for Genetic Variation Among Apomictically Produced Plants of Several F_1 Progenies of Guayule (*Parthenium agentatum*) and Mariola (*P. incanum*)," *American Journal of Botany* 32, no. 9 (1945): 554–560; Katherine Esau, "Apomixis in Guayule," *Proceedings of the National Academy of Science* 30 (1944): 352–255; and Fredrick T. Addicott, "The Anatomy of Leaf Abscission and Experimental Defoliation in Guayule," *American Journal of Botany* 32 (May 1945): 250–256. See also *Manzanar Free Press,* 20 March 1943; and Arthur W. Galston, "An Accidental Plant Biologist," *Plant Physiologist* 128 (March 2002): 786–787.

14. Paul H. Roberts to Ralph Merritt, 24 August 1943, box 14, UCLA Manzanar Records; and Roberts to Gordon Salmond, 3 September 1943, box 16, E20, RG95, NA.

15. Gordon Salmond to Paul H. Roberts, 9 September 1943, box 14, E20, RG95, NA.

16. Ansel Adams, *Born Free and Equal: Photographs of the Loyal Japanese-Americans at Manzanar Relocation Center, Inyo County, California* (New York: U.S. Camera, 1944), 88. See also Ansel Adams with Mary Street Alinder, *Ansel Adams: An Autobiography* (Boston: Little, Brown, 1985), 257–259.

17. David Spence to Atherton Lee, 14 June 1943, Kageyama Family Papers.

18. Robert Emerson to Ralph P. Merritt, 7 January 1944, box 14, UCLA Manzanar Records.

19. Kageyama interview; Robert Emerson to HHA, 20 September 1946; and Emerson to HHA, 18 November 1946, Kageyama Family Papers; and Marjorie Noble to Nisei Soldier, 9 January 1945, box 41, Japanese American Research

Project, Department of Special Collections, Young Research Library, University of California, Los Angeles.

20. The sentences read, "Certain of the Japanese internees were interested in guayule. They did a considerable amount of research work, particularly in extraction." *Final Report*, 43.

21. *Bakersfield Californian*, 7 December and 10 December 1942.

22. *Bakersfield Californian*, 19 December and 31 December 1942.

23. "Typescript Testimony Before the Subcommittee in Bakersfield," 6 April 1944, box 11, Poage Papers. For a parallel case, see Hales, *Atomic Spaces*, 167.

24. A good summary of the farmers' objections can be found in a confidential report prepared in the aftermath: E. L. Perry, memorandum to the record, 29 March 1943, box 28, E20, RG95, NA. See also "Guayule Rubber Plantings: A Hearing on Studies Under the Auspices of Kern County Chamber of Commerce Agricultural Council, December 10, 1942," box 28, E20, RG95, NA; *Bakersfield Californian*, 9 December, 11 December 1942 and 19 December 1942, clippings, box 27, E20, RG95, NA; and *Los Angeles Times*, [12?] December 1942 and 20 February 1943, clippings, Crowl Scrapbook. On nematodes, see Gerald Thorne to G. Steiner, 10 December 1942, box 199, file 44814, E2, Correspondence of the Bureau Chief, 1940–1952, RG54, NA. In brief, AAA regulations stated that farmers could not be reimbursed for cotton production if cotton was not planted for at least one of three years. Because guayule required four to five years to reach maturity, it inherently threatened farmers' access to AAA benefits. For details on leasing arrangements, see *Final Report*, 65–75.

25. *Wall Street Journal*, 1 February 1943. See also other articles in series about guayule, *Wall Street Journal*, 28 and 29 January 1943.

26. From E. L. Perry, memorandum to the record.

27. *Los Angeles Times*, 12 March 1943, clipping, Crowl Scrapbook.

28. *Santa Maria (California) Daily Times*, 20 February 1943 and 22 February 1943, box 27, E20, RG95, NA.

29. *Woodland (California) Democrat*, 25 February 1943; and *Los Angeles Times*, 1 March 1943, box 27, E20, RG95, NA.

30. *Woodland (California) Democrat*, 4 March 1943, clipping, box 27, E20, RG95, NA; and unidentified clipping, probably April 1943, Crowl Scrapbook.

31. E. L. Perry, memorandum to the record. With this possibility in mind, Perry suggested that USDA experts continue research on guayule so that it could return as a "first-class crop" in postwar California.

32. William Jeffers to Claude Wickard, 4 March 1943, quoted in *Final Report*, 5.

33. *NYT*, 31 March 1943 and 2 April 1943.

34. Neushul, "Science, Technology, and the Arsenal of Democracy," 154.

35. *Final Report*, 194.

36. *Russian Dandelion*, 145–147, 177. By the end of the year, the ERRL sent that rubber to the major companies; B. F. Goodrich produced about fifty kok-sagyz truck tires, the quality of which were said to be "much superior" to those produced from either guayule or synthetic rubber. See "Tires from Kok-Sagyz," *Chemurgic Digest* 2 (15 June 1943): 82; and "Kok-sagyz Progress Report," *Chemurgic Digest* 2 (30 January 1943): 14–16.

37. Knudson, "Rubber Investigations at Cornell University."

38. *Summary of Kok-saghyz Investigations Spring and Summer 1942*; *Russian Dandelion*, esp. 160–181; H. H. Whetzel and J. S. Neiderhauser, "Investigations on Disease of the Russian Dandelion, Taraxacum Kok-Sagyz," unpublished manuscript (June 1943), box 16, Whetzel Papers; and C. E. Steinbauer to Robert W. Henderson,

23 April 1943; Steinbauer to Henderson, 24 April 1943; and Henderson to Steinbauer, 23 July 1943, Henderson Papers.

39. "Department of Botany Annual Report, 1943–1944," box 3, Carl E. F. Guterman Papers, Division of Rare and Manuscript Collections Cornell University Archives, Ithaca, New York.

40. Claude Wickard to Bradley Dewey, 20 September 1943, box 55, E90, RG54, NA.

41. C. E. Steinbauer to St. Paul Research Staff Members, 3 January 1944, Henderson Papers.

42. For example, Robert M. Salter to Harold W. Mowry, 14 June 1944, box 200, E2 Correspondence of Bureau Chief, 1940–1952, RG54, NA.

43. Robert W. Henderson to H. E. Warmke, 24 January 1944; and Henderson to D. E. Stephens, 17 December 1944, Henderson Papers. See also Roderick K. Eskew, "Natural Rubber from Russian Dandelion," *IRW* 113 (January 1946): 517–520. It also seems clear that enthusiasm for kok-sagyz dimmed in the Soviet Union. See Klaus Möbius, "Die Kautschukindustrie in den Ostblockstaaten," *Gummi, Asbest, und Kunststoffe* 22 (October 1969): 1021–1031.

44. TAF, "Cryptostegia Report," [June 1943?]; and J. T. Curtis, "Cryptostegia Research Program," 3 June 1943, reel 15, HAW Papers at the Franklin D. Roosevelt Presidential Library; and interview with Thomas Dudley Fennell, 7 February 2008.

45. Russ Symontowne, "Cryptostegia Research in Haiti—I," *IRW* 108 (May 1943): 148–150.

46. Loren G. Polhamus to Gordon Salmond, 27 December 1943, box 11, E90, RG54, NA. For MIT's Ernst Hauser, chronic problems with impurities seriously hindered any hopes of using cryptostegia on a large scale. Ernst A. Hauser, "Cryptostegia Rubber," unpublished report, submitted 25 March 1943, box 1899, E1, RG179, NA; "Research Pertaining to Natural and Synthetic Rubbers Carried out Under the Supervision of Dr. Ernst A. Hauser at the Massachusetts Institute of Technology, Cambridge Massachusetts, Third Report" unpublished manuscript, 12 March 1943, box 107, Compton-Killian Papers.

47. Russ Symontowne, "Cryptostegia," *New York Sunday News*, 3 October 1943, clipping, box 27, E20, RG95, NA.

48. "Tapping Latex Now at Cayes," *A Propos de SHADA* 1 (November 1943), 1, 6, 8; and Russ Symontowne, "Cryptostegia Research in Haiti—II," *IRW* 108 (June 1943): 259–261.

49. Interview with Thomas Dudley Fennell. See also "SHADA Planting Food," *A Propos de SHADA* 1 (June 1943): 1, 6.

50. TAF to John C. White, 20 February 1943, file 838.50/40, box 4708, decimal file, RG59, NA; and Thomas H. Young, military attaché report, 15 May 1943, 838.51/ Cooperative Program/69, box 4719, decimal files, RG59, NA. See also the ambassador in Haiti (Wilson) to the secretary of state, 24 July 1944, in U.S. Department of State, *Foreign Relations of the United States, Diplomatic Papers, 1944*, vol. 7, *The American Republics* (Washington, DC: Government Printing Office, 1967), 1174–1175 (hereafter cited as *FRUS*).

51. State Department telegram to U.S. embassy, Port-au-Prince, 28 October 1943, 838.51/Cooperative Program/107, box 4719, decimal files, RG59, NA.

52. Thomas H. Young, military attaché report, 1 June 1943, 838.51/Cooperative Program/74, box 4719, decimal files, RG59, NA.

53. J. K. Ball to R. E. Ball, n.d., 838.51/Cooperative Program/79, box 4719, decimal files, RG59, NA.

54. John C. White to Willard F. Barber, 23 November 1943, 838.51/Cooperative Program/113, box 4719, decimal files, RG59, NA. See also "Condensed Crptostegia Report," *A propos de SHADA* 1 (October 1943): 7; and "A Statement by Thomas A. Fennell, President of SHADA," *A propos de SHADA* 1 (November 1943): 1, 8.

55. John C. White to Willard F. Barber, 17 August 1943, 838.51/Cooperative Program/83, box 4719, decimal files, RG59, NA.

56. TAF to J. W. Bicknell, [23 October 1943?], series 9, box 43, Baruch Papers.

57. Hugh Butler, *Expenditures and Commitments by the United States Government in or for Latin America: Senate Document 132, 78th Congress, 1st Sess.* (Washington, DC: Government Printing Office, 1943). Charges widely disseminated in Hugh Butler, "Our Deep Dark Secrets in Latin America," *Reader's Digest* 43 (December 1943): 21–25.

58. Volney Jones to Dorothy Schulte, 11 November 1950, Volney H. Jones Papers, Bentley Historical Library, University of Michigan, Ann Arbor.

59. Symontowne, "Cryptostegia."

60. Leslie N. Gooding to Loren G. Polhamus, 14 February 1944, box 47, E90, RG54, NA.

61. Summary of charges from M. Chauvet, [March 1944?], 838.51/Cooperative Program/144, box 4719, decimal files, RG59, NA. For a defense of SHADA, see TAF, "Cryptostegia Rubber," 329–332.

62. The chargé in Haiti (Chapin) to the secretary of state, 14 April 1944, *FRUS*, 1169–1171.

63. "Haitian Rubber Timeline." See also Howard, "The Role of Botanists During World War II," 221; and Volney Jones to Dorothy Schulte, 20 October 1950, Volney H. Jones Papers, Bentley Historical Library, University of Michigan, Ann Arbor.

64. Jones, *Fifty Billion Dollars*, 420–426.

65. Summary of dispatch from Mr. Tuson, 11 May 1944; and A. A. L. Tuson to Anthony Eden, 11 May 1944, "Hayti: Failure of Cryptostegia Rubber Plantations, 1944," FO 371/38281, NA, Kew, Surrey, United Kingdom. See also the secretary of state to the chargé in Haiti (Chapin), 27 May 1944, *FRUS, 1944*, 7:1171–72; and Bradley Dewey to Leo Crowley, 17 May 1944; Crowley to the president, 10 May 1944; and J. W. Bicknell to Dewey, 16 June 1944, series 6, vol. 64, Baruch Papers. In August, the United States agreed to provide $175,000 to assist the thirty-five thousand to forty thousand families displaced by the project—just a few dollars per family. See "Haitian Rubber Timeline."

66. See Claus Müller-Stolberg, "The Caribbean in the Second World War," in *General History of the Caribbean*, vol. 5, *The Twentieth Century*, ed. Bridget Brereton (London: UNESCO, 2004), 116–118; and Smith, "Shades of Red in a Black Republic," 87–88.

67. Bayard Hammond to Loren G. Polhamus, 14 January 1943; and Polhamus to Gordon E. Salmond, 18 January 1943, box 10, E90, RG54, NA; "Goldenrod Rubber," *Chemurgic Digest* 2 (15 April 1943): 50; and *SMN*, 14 February 1943.

68. Melchior, "The Production of Goldenrod as a Source of Emergency Rubber."

69. "Extraction, Characterization, and Utilization of Goldenrod Rubber: Work at the Southern Regional Research Laboratory Under the Emergency Rubber Project, U.S. Forest Service, July 1942 to June 1944," box 5, E90, RG54, NA. See also Forrest L. McKennon and J. Raymond Lindquist, "A Solution for the Compounding of Goldenrod Rubber," *RA* 56 (December 1944): 289–292; Mary L. Rollins et al., "Microscopial Studies in Connection with the Extraction of Rubber from

Goldenrod," *IRW* 113 (October 1945): 75–78; and Vanderbilt, *Thomas Edison, Chemist*, 312–314.

70. M. N. Walker to Loren G. Polhamus, 7 September 1943, box 48, E90, RG54, NA; and Gordon R. Salmond to D. F. J. Lynch, 19 February 1944; John S. Bowen to G. M. Granger, 1 March 1944; Salmond to Regional Forester, 22 April 1944; and David A. Bisset to Polhamus, 14 May 1945, box 10, E90, RG54, NA.

71. Melchior, "The Production of Goldenrod," 11.

72. Ibid., 125–147. Quotation on p. 147.

73. Ibid., 121–122. Volunteer goldenrod remained a problem for the government in 1945. See D. A. Bisset to R. H. Marchman, 9 February 1945; and Bisset to J. L. Mahoney, 9 April 1945, Manly Farm file, Bamboo Farms and Coastal Gardens, Savannah, Georgia.

74. John Caswell, "Report on Guayule Plantings of the USDA (USFS), July 4–28, 1943," box 28, E20, RG95, NA.

75. E. L. Perry to the director, 31 July 1943, box 28, E20, RG95, NA.

76. Governor Sidney Osborn to Paul Burgess, 26 July 1943, box 23, Osborn Papers. For similar view in Texas, see A. B. Connor to W. R. Poage, 10 November 1943, box 9, Poage Papers; and Editorial, "Synthetic Tire Program," *DMN*, 25 October 1943.

77. Caswell, "Report on Guayule Plantings of the USDA."

78. *Final Report*, 194.

79. LeRoy Powers, "Seed Collecting of Guayule from Mexico and the Trans-Pecos Area (Big Bend Country) of Texas," unpublished manuscript, 7 December 1942, box 33, Reed Rollins Papers, Archives of the Gray Herbarium, Harvard University, Cambridge, Massachusetts (hereafter cited as Rollins Papers).

80. *El Paso Herald-Post*, 29 December 1943 and 13 January 1944, clippings; and Abner Foskette to Paul Roberts, 8 March 1944, box 28, E20, RG95, NA.

81. Hart, *Empire and Revolution*, 417–418; RFC, *The Development of Wild and Cultivated Natural Rubber in Western Hemisphere* (Washington, DC: Reconstruction Finance Corporation, 1945), 8–11; and IRC, *Annual Report*, 1943–1945.

82. Fred J. Hart to Edward Tickle, 25 March 1943; McCargar to Earl Warren, 27 March 1943; and press release, 3 April 1943, Administrative Files, Agriculture—Guayule Rubber Program, 1943, Earl Warren Papers, California State Archives, Sacramento (hereafter cited as Warren Papers 1943); and *Salinas Californian*, 23 March 1943. For background on Warren's political ambitions, see Leo Katcher, *Earl Warren: A Political Biography* (New York: McGraw-Hill, 1967).

83. Fred J. Hart to Earl Warren, 10 January 1944; and Fred S. McCargar to Alexander Heron, 18 January 1944, Administrative Files, Agriculture—Guayule Rubber Program, 1944–46, Earl Warren Papers, California State Archives, Sacramento (hereafter cited as Warren Papers 1944–1946); and "Minutes of California State Guayule Committee, Bakersfield, 15 February 1944," box 28, E20, RG95, NA.

84. Fred J. Hart to Earl Warren, 21 February 1944; and W. J. Cecil to Warren, 1 March 1944, Warren Papers, 1944–1946; *Bakersfield Californian*, 18 November 1944 and 31 January 1944, clippings, box 27, E20, RG95, NA; and *Final Report*, 201–205.

85. "Minutes of California State Guayule Committee"; and Kern County Guayule Committee to State Reconstruction and Reemployment Commission, 18 December 1944, Warren Papers, 1944–1946.

86. *Congressional Record*, House, 15 February 1944, 78th Congress, 2nd sess., 1675–1676.

87. *Salinas Californian*, 3 April and 4 April 1944.

88. "Hearings Held Before Committee on Agriculture, House of Representatives, Washington, D.C., at Salinas, Calif., Before Sub-Committee Investigating Guayule, April 3, 1944 and April, 4, 1944," box 9; typescript testimony Before the Subcommittee in Bakersfield, 6 April 1944, box 11; Carlyle Thorpe, "Draft Summary of the Testimony Before the Poage Committee on April 3, 4, 5, 6," submitted to Poage, 12 April 1944, box 11, Poage Papers. For more general complaints, see *Bakersfield Californian*, 31 January 1944; *San Francisco Chronicle*, 17 March 1944; *Bakersfield Californian*, 22 March 1944; and *Oceanside (California) Daily Blade Tribune*, 24 March 1944, clippings, box 25, E20, RG95, NA.
89. *NYT*, 22 March 1944.
90. *Agricultural Appropriation Bill for 1945: Hearings Before the Subcommittee of the Committee on Appropriations, U.S. Senate, 78th Congress, 2nd Sess., on HR 4443* (Washington, DC: Government Printing Office, 1944), 218–246. The first public acknowledgement that I have seen of Eisenhower's confidential report occurred in the *Congressional Record* in December 1943. By mid-1944, guayule supporters invoked Eisenhower's warnings consistently. Very few Americans, however, had actually seen the Eisenhower and Wilkes report. Guayule booster Fred McCargar charged that government officials routinely denied requests for copies, even those from U.S. senators. See McCargar to Poage, 14 September 1944, box 9, Poage Papers.
91. *Congressional Record*, House, 8 June 1944, 78th Congress, 2nd sess., 5570–5571; *Congressional Record*, House, 20 June 1944, 78th Congress, 2nd sess., 6284–6290; and *Congressional Record*, House, 23 June 1944, 78th Congress, 2nd sess., A3432–A3435.
92. *Agricultural Appropriation Bill for 1946: Hearings Before the Subcommittee of the Committee on Appropriations, U.S. Senate, 79th Congress, 1st Sess., on HR 2689* (Washington, DC: Government Printing Office, 1945), 122–124; *Final Report*, 95–96, 205–208; "Memorandum: Guayule Program," 24 May 1945, folder: Guayule Rubber, 1941–1978, Firestone Records; and *Bakersfield Californian*, 26 June 1946.
93. W. R. Poage to Phil Ohanneson, 8 January 1945, box 11, Poage Papers.
94. David Spence to Phil Ohanneson, 16 February 1945, box 11, Poage Papers.
95. *A Study of Rubber in United States, Mexico, and Haiti*, report no. 2098, 78th Congress, 2nd sess., 2 January 1945; and minutes of subcommittee testimony, 2 March 1945, discussion of HR 2347, box 11, Poage Papers. Poage and his allies recognized that one private company, the IRC, was in the best position to take over the ERP's guayule assets. Because the ERP had been born amid charges of the IRC's alleged attempts at war profiteering, however, it was politically out of the question to consider that possibility. See also W. R. Poage to Fred McCargar, 26 February 1945; McCargar to Poage, 13 March 1945; and Poage to McCargar, 16 March 1945, box 11, Poage Papers.
96. *Final Report*, 209–222; *Bakersfield Californian*, 12 December 1945 and 4 January 1946; and USFS, "Announcement of Offer of Sale of Guayule Rubber Processing Plant," 28 December 1945, box 16, IRC Records.
97. *Final Report*, 113, 210–223; *Salinas Californian*, 4 February 1946, clipping; and James M. Platz to HHA, 1 May 1946, box 10, Poage Papers; *Salinas Californian*, 4 September 1946; and *Bakersfield Californian*, 4 January 1946.
98. J. A. Hall to chief, Forest Service, 30 November 1945, box 29, E20, RG95, NA.
99. For accusations or hints of an alleged conspiracy, see entry of 12 February 1943, in Wallace, *The Price of Vision*, 189; Anderson was especially blunt: "The oil companies prevailed on the government to plough it all up. . . . [They] wanted to

take advantage of the synthetic rubber plants that the government had built for the war effort, and now turn over to them for a song to the oil companies. They couldn't stand the competition of high quality rubber in this country, such as guayule." See Julie L. Benton, "Interview with Hugh Harris Anderson," Pasadena Oral History Project (Pasadena, CA: Pasadena Historical Society, 1982), 74. See also Fred S. McCargar to HHA, 2 December 1945, box 10, Poage Papers.

100. *Congressional Record*, 21 December 1945, 79th Congress, 1st sess., A5780–5781.
101. Another example is John Reber, who envisioned guayule farms as the basis for vast communities for displaced "Okies" who wanted to settle in the American Southwest. See John Reber to W. R. Poage, 24 January 1945, box 10, Poage Papers; and *Wall Street Journal*, 1 February 1943, clipping, box 27, E20, RG95, NA.
102. According to McCargar, "my lettuce growers have been giving me hell" for his excessive attention to guayule. See Fred S. McCargar to Fulton Lewis, 29 March 1945, box 10, Poage Papers. See also "The Guayule Emergency," *Pacific Rural Press* (10 November 1945), clipping, carton 3, Rollins Papers.
103. Two examples include: David Sherman Beach, an eccentric Connecticut native who insisted that Osage orange yielded enough rubber to solve the nation's shortages (and also claimed to have "run for president" in every election since 1920); and Elliott Simpson, who in 1945 led teams of explorers carrying shotguns, bolo knives, and Seminole guides into the Big Cypress Swamp of southern Florida where they claimed to find wild rubber trees by the millions. Simpson's search party reportedly included the actor Chico Marx. For Beach, see W. R. Poage to David Sherman Beach, 2 May and 4 May 1944, box 11; *Milford (Connecticut) Citizen*, 5 August 1943; and an unidentified clipping, 7 May 1944, box 11, Poage Papers. For Simpson, see "Rubber in Everglades," *RA* 56 (January 1945): 427; and Elliot E. Simpson, "Conclusions up to Now of Rubber in the Everglades and its Future Possibilities," n.d.; and statement of W. D. Outman, 19 February 1945, box 10, Poage Papers.
104. "Import Rubber or Grow It?" *Wall Street Journal*, 2 April 1945, clipping, E20, RG95, NA. For a description of the toast, see diary entry of 14 October 1943, Wallace, *The Price of Vision*, 260–261.
105. See K. E. Knorr, *Rubber After the War* (Stanford, CA: Stanford University Food Research Institute, 1944), 44–45.
106. John Krige, *American Hegemony and the Postwar Reconstruction of Science in Europe* (Cambridge, MA: MIT Press, 2006); and Clark A. Miller, "'An Effective Instrument of Peace': Scientific Cooperation as an Instrument of U.S. Foreign Policy, 1938–1950," in *Global Power Knowledge: Science and Technology in International Affairs*, ed. John Krige and Kai-Henrik Barth, *Osiris* 21 (2006): 133–160. For emerging tensions over rubber, see Bradley Dewey to A. B. Petrenko, 13 April 1944, series 9, box 43, Baruch Papers; and Donald M. Nelson to Mr. Mikoyan, 3 January 1944, E1, RG179, NA.
107. Slosson, *Creative Chemistry*.
108. All quotations from Haynes, *The Chemical Front*, v, vii, 188, 264. See also Williams Haynes, *This Chemical Age: The Miracle of Man-made Materials* (New York: Knopf, 1942); and Haynes and Hauser, *Rationed Rubber*.
109. Ernst A. Hauser, "Our Rubber Problem," *Journal of Chemical Education* 20 (April 1943): 203–205.
110. P. W. Litchfield, "Guayule: Asset or Liability?" June 1945, box 10, Poage Papers; and "Opposition to Guayule," *RA* 57 (July 1945): 448.
111. Litchfield to Poage, 10 August 1945, box 10, Poage Papers.
112. *Cheyenne Tribune*, 6 May 1943, clipping, box 27, E20, RG95, NA.

113. Jeffers to Franklin D. Roosevelt, 1 September 1943, series 9, box 43, Baruch Papers.
114. Press release of the Rubber Manufacturers Association, 29 July 1945, box 10, Poage Papers. See also Brendan J. O'Callaghan, "Rubber in World War II: A History of the United States Government's Natural and Synthetic Rubber Programs in World War II," box 11, E26, Administrative Histories of the RFC's Wartime Programs, 1943–1954, RG234, Records of the Reconstruction Finance Corporation, NA.
115. [Firestone Tire and Rubber Company], "Rubber," pamphlet, 1946. The American Education Press Inc. prepared this booklet for school use.
116. John L. Collyer, "Will America Have to Jack Up Its 29,000,000 Automobiles?" B. F. Goodrich Company pamphlet, [1943?]; and similar pamphlets in box 20, John Lyon Collyer Papers, Division of Rare and Manuscript Collections, Cornell University, Ithaca, New York.
117. See Hugh Allen, *The House of Goodyear: A Story of Rubber and Modern Business* (Cleveland: Corday and Gross, 1943), 101–108. See also *Bricks Without Straw: The Story of Synthetic Rubber As Told Within the B. F. Goodrich Company* (Akron: B. F. Goodrich Company, [1944?]); and Lief, *The Firestone Story*, esp. 354.
118. Frank A. Howard, *Buna Rubber: The Birth of an Industry* (New York: Van Nostrand, 1947), vii–viii, 241. The book is dedicated "to the chemical engineer."
119. John Lyon Collyer, *The B. F. Goodrich Story of Creative Enterprise* (New York: Newcomen Society of North America, 1952), 5, 19–23.
120. In 2002, world consumption of natural rubber was about 6.8 million metric tons, compared with about 1.4 million metric tons in 1941. The U.S. share in 2000 was about 1.4 million metric tons, compared with about 700,000 metric tons in 1941. See Food and Agricultural Organization of the United Nations, *Profiles and Relevant WTO Negotiating Issues* (2002), available at http://www.fao.org; and Clay, *World Agriculture and the Environment*, 335.
121. Influential works include Meikle, *American Plastic*; Joseph J. Corn, ed., *Imagining Tomorrow: History, Technology, and the American Future* (Cambridge, MA: MIT Press, 1986); Eric Schatzberg, "Ideology and Technical Choice: The Decline of the Wooden Airplane in the United States, 1920–1945," *Technology and Culture* 35 (January 1994): 34–69; and Pap A. Ndiaye, *Nylon and Bombs: DuPont and the March of Modern America* (Baltimore: Johns Hopkins University Press, 2007).

Chapter 8 *From Domestic Rubber Crops to Biotechnology*

1. David Fairchild to Alvin H. Hansen, 25 March 1934, unidentified collection, Bertram Zuckerman Garden Archives, Fairchild Tropical Garden Research Center, Miami, Florida.
2. Davis, *One River*, 363.
3. Alan I Marcus and Amy Sue Bix, *The Future Is Now: Science and Technology Policy in America Since 1950* (Amherst, NY: Humanity, 2007), 22–25, 34–36. See also Alex Roland, "Science and War," in *Historical Writing on American Science*, ed. Sally Gregory Kohlstedt and Margaret W. Rossiter, *Osiris* 2 (1985): 247–272. For postwar agricultural issues, see John H. Perkins, *Geopolitics and the Green Revolution: Wheat, Genes, and the Cold War* (New York: Oxford University Press, 1997); and Jack Ralph Kloppenburg Jr., *First the Seed: The Political Economy of Biotechnology*, 2nd ed. (Madison: University of Wisconsin Press, 2004).
4. Drayton, *Nature's Government*; and Pauly, *Fruits and Plains*.
5. Harry Truman, "To the Congress of the United States," 7 February 1947, box 244, RG46, Records of the U.S. Senate, Committee of Judiciary, 80th Congress, 1st

sess., NA. On the slow decline of chemurgy in postwar America, see Finlay, "Old Efforts at New Uses," 33–46.

6. "A Fifty-year Stretch," 55–57; IRC, *Annual Report* 1947–1952; and *NYT*, 26 September 1953.

7. Clay, *World Agriculture and the Environment*, 335.

8. The 1950s insurgency in Malaya certainly put pressure on rubber plantations, but it did not cause a crisis in international supply. Production hit new records in 1950 and did not fall much thereafter despite the insurgence. See Anthony Short, *The Communist Insurrection in Malaya, 1948–1960* (London: Frederick Muller, 1975); and *NYT*, 9 December 1951 and 7 May 1953.

9. Statement by Lyndon Johnson, 18 January 1951, box 41, E90, RG54, NA. See also "Senate Looks to Chemurgy for Defense," *Chemurgic Digest* 10 (November 1951): 9.

10. Marion N. Walker to Loren G. Polhamus, 8 November 1948, box 40; H. M. Tysdal, "Field Trip to Texas, April 6 to 12, 1953," box 39; and "Establishment of Reserves of Guayule on Waste Lands," [1952 or 1953?], box 40, E90, RG54, NA.

11. *NYT*, 4 January 1951; clippings, *Cleveland Plain Dealer*, 1 July 1951; and *Akron Beacon Journal*, 21 March 1952, folder: "Guayule Rubber, 1941–1978," Firestone Records; and testimony of J. Wayne Whitworth, in U.S. Senate, *Guayule Rubber Development: Hearing Before the Subcommittee on Regional and Community Development of the Committee on Environment and Public Works, 95th Congress, 2nd Sess.* (Washington, DC: Government Printing Office, 1978), 70–72.

12. Text of Concurrent Resolution 582, House, 86th Congress, 2nd sess., box 44, Legislative Series, Carl Albert Papers, Carl Albert Center of Congressional Research and Studies, Norman, Oklahoma. See also Davis, *One River*, 367–368.

13. James Bonner to Ed Flynn, 19 April 1973 and 7 July 1978, James Bonner Papers, Institute Archives, California Institute of Technology, Pasadena.

14. HHA to Kitty Schirmer, 14 January 1977, box C-12, White House Central Files, Jimmy Carter Presidential Library, Atlanta; and TAF to James Schlesinger, 17 May 1977, box 8, Rexford Guy Tugwell Papers, Franklin D. Roosevelt Presidential Library, Hyde Park, New York. Within two weeks of the inauguration, Carter's team launched an assessment of the guayule possibility.

15. Gerald C. Cornforth et al., *Guayule—Economic Implications of Production in the Southwestern United States* (College Station: Texas Agricultural Experiment Station, Texas A&M University System, 1980), 1–3.

16. National Academy of Sciences, *Guayule*, vii.

17. In apomixis, embryos arise from unfertilized seed without undergoing meiosis and thus yield a seed that is a clone, genetically identical to the parent plant.

18. See also Enzo R. Grilli, Barbara Bennett Agostini, and Maria J. 't Hooft-Welvaars, *The World Rubber Economy: Structure, Changes, and Prospects* (Baltimore: Johns Hopkins University Press, 1980), 3–4, 92–99.

19. "Rubber Plant," *Newsweek*, 11 April 1977, 66; *NYT*, 30 March 1977; *Boston Sunday Globe*, 10 April 1977; *Boston Globe*, 24 November 1977; Dick Griffen, "Natural Rubber Has a Future After All," *Fortune* 97 (24 April 1978): 78–81; "Wy-oo-lee Rebound," *Science News* 119 (6 June 1981): 365–366; and Noel P. Vietmeyer, "Rediscovering America's Forgotten Crops," *National Geographic* 159 (May 1981): 702–712. On the beginning of the "peak oil" thesis, see Noel Grove, "Oil, the Dwindling Treasure," *National Geographic* (June 1974): 792–825.

20. *Guayule Rubber Development*, 34–36, 40–66, 97–99; William G. McGinnies and Edward F. Haase, eds., *An International Conference on the Utilization of Guayule, November 17–19, 1975* (Tucson: University of Arizona, 1975); *A Sociotechnical*

Survey of Guayule Rubber Commercialization (Tucson, AZ: Office of Arid Land Studies and Midwest Research Institute, 1979); *A Technology Assessment of Guayule Rubber Commercialization* (Tucson, AZ: Office of Arid Land Studies and Midwest Research Institute, 1980); Daniel M. Bragg and Charles W. Lamb Jr., *The Market for Guayule Rubber* (College Station: Texas Engineering Experiment Station, n.d.); and *Report of the Feasibility of Commercial Development of Guayule in California* (Sacramento: California Department of Food and Agriculture, 1982).

21. Emphasis in original. *Guayule Rubber Development*, 67.

21. Emphasis in original. *Guayule Rubber Development*, 67.

22. [Quentin Jones and H. T. Huang], "Summary of Current and Visualized Future Effort Toward Commercialization of Guayule," [1979?], carton 4, Rollins Papers.

23. "Guayule Projects Sponsored by the National Science Foundation," [1979?], carton 4, Rollins Papers. See also Noel D. Vietmeyer, "Guayule: Domestic Natural Rubber Rediscovered," in *New Agricultural Crops*, ed. Gary A. Ritchie (Boulder, CO: Westview, 1979), 167–176.

24. See "California Forges Ahead with Guayule Project," *Rubber and Plastic News* (15 October 1979), clipping, carton 3, Rollins Papers; and unidentified notes from speech, [1981 or 1982?]; and "Briefing on Guayule," August 1982, Guayule file, Jim Hightower Papers, Center for American History, University of Texas, Austin.

25. Clipping from *Akron Beacon Journal*, 12 March 1978, folder: Guayule Rubber, 1941–1978, Firestone Records; and *NYT*, 4 November 1980.

26. U.S. Congress, House, *Guayule Research: Joint Hearing Before the Subcommittee on Science, Research and Technology of the Committee of Science and Technology and the Subcommittee on Departmental Operations, Research, and Foreign Agriculture of the Committee on Agriculture 98th Congress, 1st Sess.* (Washington, DC: Government Printing Office, 1984). According to one scholar, the collapse of the Native Latex Commercialization Act stands out a classic illustration of how government bureaucrats can hinder and stall technological innovation. See Francis W. Wolek, "Guayule: A Case Study in Civilian Technology," *Technology in Society* 7 (1985): 11–23. See also *NYT*, 2 February 1982.

27. For the extensive literature that emerged from this era, see Deidre Campbell, compiler, *Guayule Bibliography, 1980–1988* (Tucson: Office of Arid Land Studies, University of Arizona, 1988); and Kenneth E. Foster, N. Gene Wright, and Susan Fitzgerald Fansler, *Guayule Natural Rubber Commercialization: A Scale-up Feasibility Study* (Tucson: Office of Arid Lands Studies, University of Arizona, 1991). In 1979, a small group of crop researchers calling themselves "Guayuleros" formed the Guayule Rubber Society and began to publish a rather informal research journal. Members voted in 1988 to broaden their mission and reincorporate as the Association for the Advancement of Industrial Crops. That group is small but still active, with annual meetings at which a handful of scholars present the latest developments in rubber crop research. See Himayat H. Naqvi, "A Society for All New Industrial Crops," *El Guayulero* 9 (Summer–Fall 1987): 1–2; F. S. Nakayama, "New Crops, New Changes, New Challenges," *El Guayulero* 11 (Spring–Summer 1989): 1; and *AAIC Newsletter* 23 (October 2002) and 24 (April 2003). See also http://www.aaic.org.

28. http://www.pewtrusts.org/our_work_detail.aspx?id=442, accessed 11 July 2006.

29. Katrina Cornish, "Guayule Latex Process Is Patented," *Agricultural Research* 45 (May 1997): 23; Katrina Cornish, "Latex from Desert Shrub Blocks Viruses, Bacteria," *Agricultural Research* 48 (March 2000): 23; Alex Halperin, "High Hopes of a Would-be Rubber Baron," *BusinessWeek Online*, 27 December 2005; John Rosenthal, "Another Rubber Tree Plant," *Fast Company Magazine*, 115 (May 2007); Dennis Mitchell, "Center in Maricopa Leads to Development of Desert

Crops," Cronkite News Service, accessed at http:///www.yulex.com, 6 July 2007; "Southwestern U.S. Perennial Offers New Biofuel Potential," *Agripulse* 2 (10 October 2006): 7–8; and Yulex Corporation, "Guayule Tech Report" (2007), accessed http://www.guayuletech.com, 15 November 2007; Richard Banks, "Future Bumper Crop," *Farm Life* (Spring 2011): 14-16; and Pure Latexbliss and Biorubber Producer Yulex Corporation Announce Exclusive Partnership," accessed at http:///www.yulex.com, 8 Decemeber 2012.

30. "Guayule Is For Real," Interview from *Polymer Tyres-Asia* posted at http://www.panaridus.com/news/mike-fraley-interview-with-polymers-tyre-asia/; and "All in the Genes," http://www.tyre-asia.com/index.php?option=com_content&view=article&id=271:rubber&catid=1:sections, both accessed 8 December 2012.

31. Many aspects of this research are described at http://www.eu-pearls.eu/UK/, accessed 8 December 2012. See also "Researchers in Ohio Working to See If Dandelions Can Be Future of the Tire Industry," Cleveland.com, 20 September 2012.

32. Edward Noga, "Stories We Beat to Death," *Rubber and Plastics News*, 3 May 2004.

33. See Michael T. Klare, *Resource Wars: The New Landscape of Global Conflict* (New York: Henry Holt, 2001); and Paul Roberts, *The End of Oil: On the Edge of a Perilous New World* (Boston: Houghton Mifflin, 2004).

34. Don M. Huber, "Anticrop Bioterrorism," in *The Science of Homeland Security*, ed. Sandra F. Amass et al. (West Lafayette, IN: Purdue University Press, 2006), 151–191; and Committee on Advances in Technology and the Prevention of Their Application to Next Generation Biowarfare Threats, *Globalization, Biosecurity, and the Future of the Life Sciences* (Washington, DC: National Academies Press, 2006), 90–98.

35. Animesh Roul, "Biological Weapon, Infectious Disease, and India's Security Imperatives," Society for the Study of Peace and Conflict, paper no. 7 (September 2006), http://www.sspconline.org, accessed 21 August 2007.

36. *NYT*, 22 February and 26 February 2007.

37. Davis, *One River*, 339, 371.

38. For instance, Jeremy Rifkin, *The Biotech Century: Harnessing the Gene and Remaking the World* (New York: Penguin, 1998); Walter Isaacson, "The Biotech Century," *Time*, 11 January 1999, 42–43; and Stan Davis and Christopher Meyer, "What Will Replace the Tech Economy?" *Time*, 22 May 2000, 76–77.

39. Quoted in Marvin Duncan, "U.S. Federal Initiatives to Support Biomass Research and Development," *Journal of Industrial Ecology* 7, nos. 3–4 (2004): 194.

40. Robert C. Brown, *Biorenewable Resources: Engineering New Products from Agriculture* (Ames: Iowa State University Press, 2003); Robert Anex, "Something New Under the Sun? The Industrial Ecology of Biobased Products," *Journal of Industrial Ecology* 7, nos. 3–4 (2004): 1–4; and Samuel K. Moore, "Natural Synthetics," *Scientific American* 276 (February 1997): 36–37. These trends may bring changes to the rubber industry, for the heavy metals used in vulcanization make rubber production one of the most toxic industrial processes in the world. These heavy metals remain unsafe long after production as well. See Terry B. Councell et al., "Tire-wear Particles As a Source of Zinc to the Environment," *Environmental Science and Technology* 38 (2004): 4206–4214.

41. See Christine Meisner Rosen, "Industrial Ecology and the Transformation of Corporate Environmental Management," in *Inventing for the Environment*, ed. Arthur Moella and Joyce Bedi (Cambridge, MA: MIT Press, 2003), 319–338.

42. Dave Zielasko, "Warning! Warning! Gingo Tells IISRP Shortage of Natural Rubber Possible," *Rubber and Plastics News*, 30 May 2005; and Miles Moore, "Soaring NR Prices Create Quandary for Tire Makers," *Tire Business*, 5 June 2007.

Index

Page numbers in italics indicate illustrations.

About the Author

MARK R. FINLAY is a professor of history at Armstrong Atlantic State University. He lives in Savannah, Georgia.